Beyond Bioethics

Beyond Bioethics

Toward a New Biopolitics

Edited by
OSAGIE K. OBASOGIE AND
MARCY DARNOVSKY

Foreword by Troy Duster
Afterword by Patricia J. Williams

University of California Press

University of California Press, one of the most distinguished university presses in the United States, enriches lives around the world by advancing scholarship in the humanities, social sciences, and natural sciences. Its activities are supported by the UC Press Foundation and by philanthropic contributions from individuals and institutions. For more information, visit www.ucpress.edu.

University of California Press
Oakland, California

Library of Congress Cataloging-in-Publication Data

Names: Obasogie, Osagie K., editor. | Darnovsky, Marcy, editor.
Title: Beyond bioethics : toward a new biopolitics / edited by Osagie K. Obasogie and Marcy Darnovsky ; foreword by Troy Duster ; afterword by Patricia J. Williams.
Description: Oakland, California : University of California Press, [2018] | Includes bibliographical references and index.
 Identifiers: LCCN 2017030709 (print) | LCCN 2017035245 (ebook) | ISBN 9780520961944 (ebook) | ISBN 9780520277823 (cloth : alk. paper) | ISBN 9780520277847 (pbk. : alk. paper)
Subjects: LCSH: Biopolitics—United States. | Biopolitics. | Bioethics.
Classification: LCC JA80 (ebook) | LCC JA80 .B47 2018 (print) | DDC 172—dc23
LC record available at https://lccn.loc.gov/2017030709

Manufactured in the United States of America
27 26 25 24 23 22 21 20 19 18
10 9 8 7 6 5 4 3 2 1

Contents

Illustrations

Foreword

Closing the Gap between Bioethics and Biopolitics

Troy Duster

At the end of the Second World War, startling revelations of the horrors of the Holocaust and the medical experiments on prisoners in German concentration camps produced a revulsion that in turn generated international agreements on the need for governments to a) protect groups from genocide, and b) protect individuals from involuntary medical experiments. Indeed, the field of bioethics was born out of this very recent history—ultimately riveting its attention on the right of the individual to consent to medical procedures or experimental drug trials.

From a platform grounded in this perspective, "informed consent" catapulted to the heart of this framework and became fundamental bedrock. Matters of privacy, choice, and the right to decline participation followed course. But this often admirable concern for individual autonomy and personal sovereignty can cloud, even obscure and divert our attention away from group harms and the vulnerabilities of collectives and communities. There is a reason for the gap between these sometimes divergent but sometimes interrelated concerns.

Genocide is the extreme form of group harm, and the world has seen several monstrous mass genocidal insurgencies in the post-Holocaust era. Only a few have been actively countered by internationally coordinated interventions, and Rwanda stands out as a dramatic failure of the United Nations to address the problem. But far less dramatic group harms are routinely ignored in this traditional frame of bioethics' concern for the individual versus the group. What is to be done, for example, when researchers from technologically advanced societies wish to do genetic research on populations where "individual consent" is an alien idea—where, for example, the elder of the group typically makes such decisions?

THE EVOLUTION AND ASCENDANCY OF INDIVIDUAL CHOICE

In a treatise on reproductive choice titled *Children of Choice*, John A. Robertson acknowledges that social and economic constraints, such as access to employment, housing, and childcare, often play a significant role in the decision to have a child. However, the overarching theme to which Robertson returns again and again is that reproduction "is first and foremost an individual interest."[1] Framed as individual choice, debate about reproductive decisions takes place in the decontextualized vacuum of individual rights: to have a child or not, to have a male or a female child, to have a child with Down syndrome or cleft palate, to produce a clone. But rather than an either-or formulation, a continuum is a better analytic device to deploy for arraying an understanding of strategies and options—from individual choice through embedded but powerful social pressures (stigma and ridicule), through economic pressures, and even through the coercive power of the state to penalize, as in China's former one-child policy and state-sponsored sterilization programs in the West. Yet discussions of individual rights are routinely lifted from such contexts.

Many circumstances exist, however, in which latent and responsive state-level interventions reveal that the individual is not the appropriate unit of analysis. Amniocentesis is a relatively expensive procedure for the poor, for example, and some states provide assistance to women seeking amniocentesis. In the 1980s, however, the California Department of Maternal and Child Health noted that the state was providing financial support for amniocentesis primarily to wealthier women past their midthirties. Mindful of the eugenic history in the state, officials initiated a program to persuade poorer women to accept the service. But because the poor tend to have their children at an early age, this policy was not consequential.

LOCATING THE APPROPRIATE UNIT OF ANALYSIS: INDIVIDUAL VERSUS SOCIAL

As will be revealed in several contributions to this volume, refusal to address the matter of informed consent at any level other than that of the autonomous individual can twist and distort the determined focus on the individual into unlikely contortions when dealing with non-Western cultures. But there are also subtle and unexamined ethical issues inside Western societies when we insist on ignoring the social reality of group interests and possibly the need for some element of group "consent." Here is a case in point.

Huntington's disease is a late-onset neurological disorder that usually strikes after the age of 35–40. The race to locate the Huntington's gene(s) resulted in a triumphant discovery in the early 1990s. There is now a genetic test that can be performed to determine whether the person at risk for the disease actually carries the gene. Within a few short years of the discovery, neuroscientists in Denmark published a study in which they concluded that males with Huntington's are twice as likely to commit crimes as those who do not have the disease. The authors reported that when they applied for permission to pursue this research, they made it clear to the human subjects review panel that *no individuals* would be harmed by participating in the study. They noted that when analyzing the data, only serial numbers were used and all identifiable personal information was removed.

Yet this research report can implicate all those males in a group category—that is, all those diagnosed as having Huntington's. This number includes far more people than those individuals who "participated" in the statistical manipulations that were the fundamental methodological techniques used in the study. The deeply embedded assumption of the ethics committee was that if no harm were done to the individuals participating in the study, then no other ethical questions would deserve scrutiny or consideration. Nevertheless, the results of the study have implications for and potentially stigmatize all those with Huntington's. To the extent that the researchers found evidence that there is a general association between crime and Huntington's, all persons with the disease are vulnerable to being stigmatized by this association. I am not suggesting that this study actually established a strong link between criminal activity and Huntington's disease. That is a topic for a different analysis. Rather, what is most important for this line of argument is that the human subjects protection committee did not even have on its agenda a radar screen for picking up the matter of "group interests" (all those males with Huntington's) in its review of the research protocol. Understandably, the Institutional Review Board would have a difficult time determining and establishing "who speaks for the group" (all males with Huntington's) in such a situation.[2]

This example highlights one of the central reasons why there is a need to expand our notion of bioethics to biopolitics: namely, to address the effects of policies (and patterned practices) that have varying impacts on groups and communities and on classes and cohorts and aggregates, not just on individuals. On the surface, direct-to-consumer genetic testing seems all and only about the individual and the assessment of health risk generated by test results. Indeed, the Presidential Commission for the Study of Bioethics published a report at the end of 2013, in which all seventeen recommendations address individual level rights and privileges and the corresponding responsibilities of genetic-testing companies, clinicians, and researchers to individual consumers.[3]

However, on closer inspection, the kinds of statistical probabilities generated depend on aggregated data that reveal the centrality of categories of race, ethnicity, class, and gender to gene-testing companies. The major player in this drama is the genetic-testing company 23andMe. In the first five years of its existence, the company collected DNA samples through a self-selection process of "individuals" who paid for kits and then sent in their genetic information. It turned out that for a combination of cultural, social, and economic reasons, the database was made up overwhelmingly of samples from middle-class whites. When the company noted that only a small fraction of its databank included African Americans, it set about to "correct" the problem by actively recruiting both African Americans and sub-Saharan Africans.

The reasons are complex, but one simple example might serve to illuminate why the company had such a determination to achieve a better balance. African American males are approximately twice as likely to suffer from prostate cancer than white males. Whether the explanation lies at the level of genetics is not the issue, and in any case, the question cannot be answered definitively by the identification of possible DNA markers. However, since we know that the phenotypical expression is markedly different by self-identified race, a general statement of probability could raise concerns and suspicions about appropriate reference populations. Several of the bioethics commission's recommendations address the matter of just how detailed data-collecting companies need to be in reporting back to consumers about sampling procedures and data collection methods.[4]

PERSONALIZED MEDICINE VERSUS GROUP-BASED THERAPEUTIC TREATMENT REGIMENS

The development of genetic tests has ushered in the promise of "personalized medicine." The idea is simple enough—once a clinician has access to an individual's genetic make-up, he or she will be able to fine-tune medications to that specific individual. One of the most popular candidates for this is warfarin, commonly used as a blood thinner for heart patients. Indeed, in the United States alone, approximately two million people take warfarin to prevent excessive clotting or coagulation, and there is high variability among different patients. Calibrating the correct amount of warfarin to prescribe for an individual patient is difficult, because taking too much of the drug will produce bleeding problems, while taking too little will not stop the serious health risks of clots forming.

To address this matter, researchers at the Perelman School of Medicine at the University of Pennsylvania assessed a gene-based method for selecting patient's dosage levels. They published their results in late 2013 in the *New England Journal of Medicine,* and here is how the lead author characterized the significance of the study: "Warfarin therapy has served as a model for the promise of a gene-based approach to patient care, but we needed a large, prospective clinical trial to determine if a patient's genetic information provides the added benefit above and beyond what can be obtained simply with clinical information."[5]

At the individual level, the results of the four-year study (at eighteen clinical sites involving more than a thousand patients) were disappointing. Patients were divided into two groups—for the first group, medication was prescribed on the basis only of traditional clinical information, whereas for the second group, genetic information was also taken into account. The disappointment came from the fact that there was no difference among the two groups in the key measure for efficacy. That might have been the takeaway, but the reason why this study is of particular relevance to a discussion of bioethics versus biopolitics lies in the following finding: "There was, however, a statistically significant difference by race. Among African Americans, the same PTTR (percentage of time in therapeutic range) for the pharmacogenetic-guided dosing group was less than that for the clinical-guided dosing group—35.2 versus 43.5 percent respectively. Pharmacogenetic-based dosing also led to more over-coagulation and a longer time to first therapeutic levels of the warfarin among African Americans."[6]

Here we see the potential inversion of personalized medicine to racialized medicine. In other words, while an individual's genetic profile does not provide evidence relevant for calibrating warfarin dosage, the designated race of the individual could well influence a clinician's decision to adjust dosage levels. If this sounds hyperbolic, here is a quotation from the *Annual Review of Pharmacology and Toxicology:* "Ethnicity affects the average warfarin dose required to maintain therapeutic anticoagulation. . . . White patients require higher warfarin doses than Asians to attain a comparable anticoagulant effect. Chinese patients required a ~50% lower average maintenance dose of warfarin than white patients to obtain comparable anticoagulation."[7]

In sum, physicians are prescribing warfarin doses on the basis of the race and ethnicity of the patient, not on the patient's individual genome. Here is where a biopolitical analysis of group issues can and should trump a riveting focus on the bioethics of individual interests—because it is the group-designated-individual at greater risk for misdiagnosis.

THE PUBLIC GOOD VERSUS THE PRIVATE CHOICE

In *From Chance to Choice*, Allen Buchanan and his associates argue that assessment of the consequences for the general public good is vital to a discussion of the distinction between treatment and enhancement.[8] Before 1960, it was possible to achieve consensus that the public good was well served by elimination or mitigation of diseases such as smallpox, cholera, tuberculosis, yellow fever, and typhoid, as well as sexually transmitted diseases. However, the discovery of many genetic disorders did not place the general population at risk but rather only groups determined to be at highest risk for conditions such as Tay-Sachs, beta-thalassemia, cystic fibrosis, and muscular dystrophy. For this reason, a new set of issues and new kinds of concerns have arisen regarding who has control over genetic screening and testing. This is true not just for categories of race and ethnicity but also for sex, and thus for the sex ratio, in any society.

Whereas it is true that individuals make choices, they do so in a social and economic context that can be coercive. This may be relatively obvious to someone looking at other communities, yet may nonetheless be less well understood by people examining their own societies. The contextualized nature of choice is substantially obscured because individual choice is deeply embedded in taken-for-granted assumptions about decision making. For example, long before the advent of prenatal detection technologies, preference for a male child in India and China was so great that a significant fraction of the population practiced infanticide of newborn girls. The widespread practice of sex selection in India and China dramatically illustrates how and why a focus on individual choice can obscure the demonstrably collective aspect of socially patterned individual choices. Today, China is contending with a large gender gap among those of childbearing age—a direct consequence of the legacy of a one-child policy and of a preference for males.[9] But China is simply the largest case of a practice that is common across much of the globe.[10]

Once technologies for prenatal determination of sex became available, the quest for disclosure of the sex of a fetus took a momentous turn for public policy in India. In 1971, India passed the Medical Termination of Pregnancy Act, which stipulates that a woman may undergo an abortion only if her situation is life threatening or if grave injury to her physical or mental health would result from the pregnancy. Use of amniocentesis began in India in 1974, but there were early reports that the test was being used less to detect birth defects than to determine the sex of a fetus. In August 1994, the Parliament of India passed a law that stiffened the penalties for screening a fetus to determine the sex. As an example of the imbalance the law attempted to address, the populous northern state Haryana had, by this time, an astonishingly low sex ratio of 874 females to every 1,000 males.[11] However, a large loophole in the law made it practically unenforceable, and the practice of screening for sex has continued at a high rate.

INDIVIDUAL DECISIONS AND UNEXAMINED GROUP PATTERNS

The example of sex selection preferences in India suggests that what appear to be individual familial choices may often be better understood as predictable reflections of social and cultural norms. For example, in early 1994 *Nature* published "China's Misconception of Eugenics," an article portraying certain state policies—such as the Chinese government policy of prohibiting couples with certain diseases from procreating—as having a distinctively distasteful eugenic quality.[12] Although the article was forthright in denouncing the use of state power as a vehicle for discouraging procreation, it implied that the decision to interrupt a pregnancy is entirely personalistic and individualistic. The health minister for China, Chen Minzhang, announced the plan to enforce a new law that would not only prohibit screening of a fetus for sex determination but also ban marriages of people with diagnoses of diseases that "may totally or partially deprive the victim of the ability to live independently, that are highly possible to recur in generations to come and that are medically considered inappropriate for reproduction."[13]

Once it is possible to determine soon enough for termination of a pregnancy whether a fetus has a condition that is regarded as a defect, who is entitled to make the decision about carrying to term or aborting? This decision is not a simple binary matter of voluntarism versus state power. There is considerable evidence to support the observation that what European and North American Western societies characterize as personal or individual decisions are on closer inspection (as with sex selection in India) actually socially patterned.

During a time of rapid social change in which there are disruptions of the established order and attendant challenges to authority and tradition, genetic explanations and eugenic solutions have a special appeal to the most privileged strata of society. The power of the state to control its population can be awesome, and thus when the state puts forward eugenic programs in a post-Holocaust world, critics are well prepared to react with revulsion. The government of Singapore came under fire during the 1990s for its program of rewarding middle-class and wealthy families for having more children while actively discouraging the poor from having large families. Singapore later generated a more sophisticated series of state-sponsored "baby bonus" packages that were less obviously eugenic but had consequences that continued to favor reproductive choices for the wealthy.[14] Critics appear to have given less attention to the fact that, in many countries, people now choose what they consider better human stock through sperm and egg banks, many of which provide detailed information about the health, the appearance, the accomplishments, and other characteristics of the men and women who provide gametes.

The Industrial Revolution and rapid urbanization wreaked havoc on traditional life and traditional social roles in nineteenth-century Europe and the United States. Extended kinship systems that had been valued as giving farmlands an economic advantage were often inverted and became economic liabilities when those families were forced off the land and moved to the teeming cities. Unemployment, homelessness, mental illness, and a host of other social problems seemed to especially victimize the poor, whose visibility if not sheer numbers dominated the public sphere of urban life.

Cholera, yellow fever, typhoid, and tuberculosis were the scourge of city dwellers, and the poor were the most likely victims. But as Sylvia Noble Tesh noted, some considered the poor, typically characterized as living in unclean conditions, not as the victims but the cause of the problems.[15] Hygiene came first as both an explanation for the better fortunes of the privileged and middle classes and later as a challenge to the poor.[16]

As wealthier families began to have fewer children and as their resources for hiring the poor as servants to help them clean decreased, some observers began to notice what they thought was a disturbing pattern. The more well-to-do members of society were procreating less while the poor continued to have large families. The Malthusian prediction about population explosion had taken an elitist turn. If people are to learn anything from the past, it is imperative that they understand the appeal and popularity of eugenics and why it was compelling to a wide range of thinkers of all political persuasions at the beginning of the twentieth century. Much like the sister concept of hygiene and the strong association between cleanliness and order, eugenics became associated with progress.

Just as hygiene was seen as the normal value of cleanliness to which all should aspire, eugenics was widely accepted and actively promoted by the major public figures of the period. University presidents, medical doctors, judges, academic scholars, writers, intellectuals, and political figures on both the left and right of the political spectrum espoused the idea that the betterment of humankind would result from practices and techniques that would prevent the procreation of criminals, prostitutes, homosexuals, alcoholics, gamblers, and people who were called at the time imbeciles and mental retards.[17]

THE OFTEN SUBTLE BIOPOLITICS OF GENETIC EXPLANATIONS OF HUMAN ACHIEVEMENTS AND FAILURES

In early 2013, *Science* published a study with the title, which was also its conclusion, "GWAS [Genome Wide Association Study] of 126,559 individuals identifies genetic variants associated with educational attainment."[18] But when reading the study more carefully, it becomes clear that the three genetic variants "associated with" educational attainment account for only 2 percent of the explanation. In other words, the remaining 98 percent of the explanation are attributable to other, mainly environmental factors. Yet as Jonathan Latham points out, the press release says nothing about the 98 per cent, and instead emphasizes the 2 percent.[19] It has been two decades since Dorothy Nelkin and M. Susan Lindee published *The DNA Mystique*, as prescient a work as one can find in the literature, which both predicted and explained this otherwise very strange, even inexplicable development.[20] Well into the first two decades of the twenty-first century, our cultural proclivity to be mesmerized by DNA "explanations" of everything from violent crime to IQ, from religious preference to voting behavior, is on display weekly in media accounts of ongoing research.

Just a few months later, in early fall 2013, the *Sunday Times* of London published an article announcing the forthcoming book by Kathryn Asbury and Robert Plomin, which was bound to generate controversy.[21] The main thesis of the book *G Is for Genes* is that data from a study of more than eleven thousands twins suggest that genetics accounts for 60 percent of *individual* performance in schools. This echoes the same claim—60 percent of IQ is explained by genetics—that appears in *The Bell Curve*, published almost twenty years earlier.[22] And very much as Piaget proposed in 1905, Asbury and Plomin argue that genetic tests should be developed and used to assign students to the appropriate learning levels.

A bioethics of genetic explanations that directs our attention to the measurement of intelligence using an IQ test can obscure the sociopolitical uses of such explanations to assign aggregates to their purportedly deserved and earned places at the bottom rung of school performance tests and, ultimately, to our prisons. Only when we are able to step back and analyze the striking parallels between the dramatic population dislocations of the late nineteenth century (rural to urban, sharp shift to industrial economies, and massive immigration) and the late twentieth century (urban to exurban, postindustrial tertiary economies, and massive cross-national migrations) can we establish an analytic purchase to explain the seductive appeal of genetic explanations of group achievements and failures. The latter requires a biopolitics of group interests to supplement, complement, and sometimes even displace the bioethics of informed consent.

NOTES

1. John A. Robertson, *Children of Choice: Freedom and the New Reproductive Technologies* (Princeton, NJ: Princeton University Press, 1994), 22.

2. Per Jensen, Kirsten Fenger, Tom G. Bolwig, and Sven Asger Sørensen, "Crime in Huntington's Disease: A Study of Registered Offences among Patients, Relatives, and Controls," *Journal of Neurology, Neurosurgery, and Psychiatry* 65, no. 4 (1998): 467–71.

3. Presidential Commission for the Study of Bioethical Issues, *Anticipate and Communicate: Ethical Management of Incidental and Secondary Findings in the Clinical, Research, and Direct-to-Consumer Contexts* (Washington, DC, 2013).

4. Ibid.

5. Stephen E. Kimmel, Benjamin French, Scott E. Kasner, Julie A. Johnson, Jeffrey L. Anderson, Brian F. Gage, Yves D. Rosenberg, Charles S. Eby, Rosemary A. Madigan, Robert B. McBane, Sherif Z. Abdel-Rahman, Scott M. Stevens, Steven Yale, Emile R. Mohler III, Margaret C. Fang, Vinay Shah, Richard B. Horenstein, Nita A. Limdi, James A.S. Muldowney III, Jaspal Gujral, Patrice Delafontaine, Robert J. Desnick, Thomas L. Ortel, Henny H. Billett, Robert C. Pendleton, Nancy L. Geller, Jonathan L. Halperin, Samuel Z. Goldhaber, Michael D. Caldwell, Robert M. Califf, and Jonas H. Ellenberg, "A Pharmacogenetic versus a Clinical Algorithm for Warfarin Dosing," *New England Journal of Medicine* 369, no. 24 (2013): 2283–93.

6. University of Pennsylvania School of Medicine, "No Benefit Found to Selecting Dose of Blood Thinner Based on Patients' Genetic Makeup," *ScienceDaily*, November 19, 2013, accessed November 18, 2016, www.sciencedaily.com/releases/2013/11/131119153021.htm.

7. Hong-Guang Xie, R.B. Kim, A.J. Wood, and C.M. Stein, "Molecular Basis of Ethnic Differences in Drug Disposition and Response," *Annual Review of Pharmacology and Toxicology* 41 (2001): 818.

8. Allen Buchanan, Dan W. Brock, Norman Daniels, and Daniel Wikler, *From Chance to Choice: Genetics and Justice.* (Cambridge: Cambridge University Press, 2000).

9. Valerie M. Hudson and Andrea M. den Boer, *Bare Branches: Security Implications of Asia's Surplus Male Population* (Cambridge, MA: MIT Press, 2004).

10. Mara Hvistendahl, *Unnatural Selection: Choosing Boys over Girls, and the Consequences of a World Full of Men* (Philadelphia: Perseus Books, 2011).

11. John F. Burns, "India Fights Abortion of Female Fetuses," *New York Times,* August 27, 1994, accessed November 18, 2016, www.nytimes.com/1994/08/27/world/india-fights-abortion-of-female-fetuses.html.

12. "China's Misconception of Eugenics," *Nature* 367 (January 6, 1994), 1–2.

13. "New Chinese Law Prohibits Sex Screening of Fetuses," *New York Times,* November 15, 1994, accessed November 18, 2016, www.nytimes.com/1994/11/15/world/new-chinese-law-prohibits-sex-screening-of-fetuses.html.

14. Shirley Hsiao-Li Sun, *Population Policy and Reproduction in Singapore: Making Future Citizens* (London: Routledge, 2012).

15. Sylvia Noble Tesh, *Hidden Arguments: Political Ideology and Disease Prevention Policy* (New Brunswick, NJ: Rutgers University Press, 1988).

16. Victoria F. Nourse, *In Reckless Hands: Skinner v. Oklahoma and the Near-Triumph of Eugenics* (New York: W. W. Norton, 2008).

17. Daniel J. Kevles, *In the Name of Eugenics: Genetics and the Uses of Human Heredity* (New York: Alfred A. Knopf, 1985); Gregory Michael Dorr, *Segregation's Science: Eugenics and Society in Virginia* (Charlottesville: University of Virginia Press, 2008).

18. Cornelius A. Rietveld et al., "Genome Wide Association Study of 126,559 Individuals Has Identified Genetic Variants Associated with Educational Attainment," *Science* 340 (6139) (June 21, 2013), accessed November 18, 2016, http://science.sciencemag.org/content/340/6139/1467.

19. Jonathan Latham, "Science as Social Control: Political Paralysis and the Genetics Agenda," *Independent Science News,* July 31, 2013, accessed November 18, 2016, www.independentsciencenews.org/science-media/science-and-social-control-political-paralysis-and-the-genetics-agenda/.

20. Dorothy Nelkin and M. Susan Lindee, *The DNA Mystique: The Gene as a Cultural Icon* (New York: W. H. Freeman, 1995).

21. Kathryn Asbury and Robert Plomin, *G Is for Genes: The Impact of Genetics on Education and Achievement* (Chichester, UK: John Wiley and Sons, 2014).

22. Richard Herrnstein and Charles A. Murray, *The Bell Curve: Intelligence and Class Structure in American Life* (New York: Simon and Schuster, 1994).

Acknowledgments

Our first thanks go to the contributors to this volume. *Beyond Bioethics: Toward a New Biopolitics* is inspired by a rich and diverse network of scholars and public interest advocates, all of them working in various ways to understand and shape how human biotechnologies intersect with societies and cultures, politics and policies, families and other forms of kinship—and what it means to be human.

The idea of collecting some of the key writings produced by these scholars and advocates germinated during the Tarrytown Meetings, a three-year initiative of annual convenings organized by the Center for Genetics and Society (CGS). Several participants from the Tarrytown Meetings are contributors to *Beyond Bioethics*. While some of them, and some other contributors, may not see themselves as part of an explicit effort to build a "new biopolitics," all have importantly shaped our thinking about that project, which we see as an urgent and consequential undertaking for the future of social justice and human rights.

We are especially grateful to our colleagues at the Center for Genetics and Society, with whom both of us have longstanding and deep connections. Richard Hayes, CGS's executive director from its founding in 2001 until the end of 2012, remains a special motivating influence. Our own intellectual and public interest efforts have been significantly stimulated and supported during the years we have been contemplating and compiling this collection by CGS staff members, advisory board members, fellows, and consultants. They include Diane Beeson, Emily Beitiks, Francine Coeytaux, Jessica Cussins, Alexander Gaguine, Emily Galpern, Charles Garzón, Katie Hasson, Elliot Hosman, Lisa Ikemoto, Sujatha Jesudason, Gina Maranto, Brendan Parent, Douglas Pet, Jesse Reynolds, Dorothy Roberts, Pete Shanks, Alexandra Minna Stern, Diane Tober, and David Winickoff.

xxvi / *Acknowledgments*

We could not have completed this anthology without the skills and dedication of CGS staff members who provided editing, formatting, and technical assistance and took on the huge task of seeking copyright permissions for the book's many contributions. Many thanks to Jonathan Chernoguz, Kate Darling, Leah Lowthorp, Victoria Massie, Natalie Oveyssi, Emily Stehr, and Kayla Tolentino for their incredible efforts.

We also want to extend our appreciation to those who have formally or informally advised CGS on a range of projects during this time period. Among the many whose advice and support have sustained us are: George Annas, Françoise Baylis, Ruha Benjamin, Deborah Bolnick, Mark Brown, Katayoun Chamany, Nathaniel Comfort, Marsha Darling, Daisy Deomampo, Donna Dickenson, Troy Duster, George Estreich, Susan Berke Fogel, Duana Fullwiley, Rosemarie Garland-Thomson, Michele Goodwin, Debra Greenfield, Jeremy Gruber, Hille Haker, Charles Halpern, Jaydee Hanson, J.P. Harpignies, Eric Hoffman, Ben Hurlbut, Rosario Isasi, Sheila Jasanoff, Jonathan Kahn, Andrew Kimbrell, David King, Paul Knoepfler, Susan Lindee, Abby Lippman, Laura Mamo, Jonathan Marks, Bill McKibben, Shree Mulay, Ari Ne'eman, Stuart Newman, Judy Norsigian, Dana Perls, Radhika Rao, Milton Reynolds, Marsha Saxton, Tania Simoncelli, Tina Stevens, Charis Thompson, Sally Whelan, Patricia Williams, and Silvia Yee. And we are deeply thankful to the Appleton Foundation for their commitment to this work.

A collection such as this one inevitably omits as much as it includes. We could have filled two or even more volumes with insightful, beautifully written essays that make important contributions to the project of building a new biopolitics. Our hope is that this anthology helps inspire other steps toward that end.

Our editor at UC Press, Naomi Schneider, has been unfailingly gracious and supportive as we have worked toward the completion of *Beyond Bioethics*. For that, and for her editorial wisdom and commitments to social justice and human rights, we thank her sincerely.

Note to Readers

This anthology is a collection of influential articles that together build an understanding of the "new biopolitics," an emerging way of understanding, evaluating, and addressing the social and policy implications of human genetic and assisted reproductive technologies. The foreword by Troy Duster and afterword by Patricia J. Williams, along with the volume and section introductions, appear here for the first time. All other contributions have been previously published; original publication information appears in a footnote on the first page of each article.

Please note that some of the contributions included here have been edited for length. We have used ellipses to indicate that text ranging from a few words to several pages has been omitted. In addition, we have not included supplementary materials such as abstracts and have edited the footnotes and endnotes that accompanied some of the articles in their originally published form. Contributions that have been edited are indicated in a footnote on the first page of each article.

Introduction

Osagie K. Obasogie and Marcy Darnovsky

We begin with three vignettes.

The 1997 announcement that researchers in Scotland had produced the first cloned mammal, a sheep named Dolly, gave rise to rampant discussions about the prospects of cloned human beings. Many scientists, bioethicists, and others weighed in to explore how we should respond—morally, politically, socially—to this new possibility. Public opinion surveys found that about 90 percent of respondents opposed the prospect of cloning as a form of human reproduction. But among those who supported it were some reproductive rights advocates, who were reluctant to endorse anything that might be construed as a limit on reproductive freedom.

In that context, Planned Parenthood Federation of America embarked on a process aimed at adopting an organizational position on human reproductive cloning, part of which involved bringing the issue to a 2001 meeting of board members and senior staff members. To gauge participants' sentiments, the senior Planned Parenthood staffer charged with facilitating the discussion, herself troubled by some of her colleagues' hesitation to oppose human reproductive cloning, prepared several hypothetical scenarios and presented them to breakout groups at the meeting. In one of these scenarios, titled "Happy Workers: Creation of a Free Market Empire,"

> a company [is] making a large profit with sales of a gadget for which consumer demand is inexhaustible. The owners face a challenge, however, because the final step in the manufacture of the gadget must be done manually, and "small, dexterous human hands are needed."
>
> They identify a worker who is particularly skilled at this task, and propose to produce 1,000 clones of her. They approach her with an offer of $1 million now, and another $5 million after the clones are able to perform the job once they reach 16 years of age. The company will

assume complete responsibility for rearing and caring for the clones. And "since the sole purpose of the clones will be to do the work in the last step of manufacture, they will be engineered to require little sleep and to have few ambitions other than performance of their work."[1]

Each discussion group was asked to decide whether it should be "permissible to produce clones for such a purpose." The scenarios' author later reported that although most participants found the situation disturbing, none of the breakout groups were willing to support legislation that would prohibit human reproductive cloning. All were persuaded by the counterargument that "the decision should be a matter of individual choice." Fortunately, Planned Parenthood did not adopt this sentiment as policy.

*

In 2006, the Massachusetts Institute of Technology hosted a conference on a new drug called BiDil, which had recently been approved by the US Food and Drug Administration to treat heart failure. As Jonathan Kahn recounts in "Race in a Bottle" (part 9 of this volume), the FDA's action had triggered quite an uproar, because BiDil was the first drug ever to be approved for a particular racial group. The agency had given its blessing to BiDil with the stipulation that it would be indicated for African Americans—that is, sold with a label on the bottle that essentially said the drug was for black people only. To many observers, this governmental support for race-specific pharmaceuticals ratified the false idea that human races are biologically distinct and encouraged the conclusion that health disparities are caused by biological differences that can be resolved through race-specific medicines.

The MIT conference was convened to explore this new twist on the old tension between social and biological explanations for racial health disparities. Meeting organizers pointed out that the questions raised by BiDil's approval were especially pressing in light of new claims emerging from the field of human genetics about the ability to detect and speak to racial differences at a molecular level.

Legal scholar Dorothy Roberts, who explores these issues further in "The Problem with Race-Based Medicine" (part 9), presented a paper at the conference on the diversity of opinions within the black community about race-based medicines like BiDil. She explained that some black people are skeptical about BiDil because of the long history of medical exploitation of the community, while others see race-specific medicines as a form of atonement for years of neglect by medical researchers. Roberts's main point was that there was no unified perspective about BiDil in the black community. During the question-and-answer portion of the talk, Juan Cofield, president of the NAACP's New England Area Conference, stood up and vociferously endorsed BiDil, declaring that "There is a consensus in the Black community that this drug is good for Black people." To which Roberts coolly responded, "There isn't a consensus among the black people *in this room*."[2]

*

In 2015, the acronym "CRISPR" hit mainstream discussions. CRISPR is the latest of several recently developed techniques that allow scientists to "edit" segments of DNA and produce changes to the genetic code of any organism. In scientific and news reports, many researchers and others expressed particular excitement about the prospect of using CRISPR for *human* gene editing. Some focused on the potential of genetic modifications to treat diseases, including both inherited conditions such as Huntington's disease and beta thalassemia and pervasive illnesses like cancer.

Others spoke of a very different vision for CRISPR: using it to edit the genes of human embryos in order to control the traits of future children and generations. While most of these commentators focused on preventing the transmission of serious inherited diseases, they often ignored or downplayed the existence of safer alternatives for meeting the same objective, including embryo screening. And some jumped easily to enthusiasm for reproductive gene editing to create offspring with preferential traits—height, hair color, perhaps even intelligence—that would be permanent and heritable for all future generations.

This vision represents the most recent in a long line of proposals to alter the human germ line in perpetuity—that is, using science to engineer future generations according to current social preferences. Human germline engineering raises a host of thorny ethical questions. Who would decide which social or medical traits to insert into or weed out of the gene pool? Would it be solely up to parents? What could happen when individual preferences—for a particular kind of musculature, a certain eye color, or even skin color—align with historically fraught concepts about which children are more socially valuable and which groups should be marginalized?

These questions have been addressed in policy conversations, intellectual debates, and public deliberations around the world for the past several decades. Many have concluded that if human biotechnology can deliver the power and incentive to engineer enhancements and inequality into the germline, the potential for new forms of discrimination and social conflict is profound. As a result, dozens of countries have adopted legal prohibitions on human germline modification.

It is in this context that Nobel Prize–winning molecular biologist Craig Mello explored the implications of human gene editing during a radio interview on WBUR, Boston's public radio station. When asked about the ethics of altering the human germline and the potential for new forms of inequality, Mello replied, "If [gene editing technologies] were safe, and if we have the knowledge to make improvements in the human germline, then it might be unethical *not* to do so."[3]

. . .

Medical ethics, a collection of professional norms designed to give guidance to doctors on how to manage health care dilemmas in a thoughtful and humane manner, dates back many centuries. Its most notable antecedent is the Hippocratic Oath, the modern version of which implores physicians to, in essence, "Do no harm." In the broadest sense, modern medical ethics has provided professional standards meant to guide physicians through the significant moral challenges that confront those engaged in the work of healing. Over the course of the nineteenth century and early to mid-twentieth century, medical ethics became, as Albert Jonsen states in his definitive work, *The Birth of Bioethics*, "almost synonymous with rules for professional cohesion and respectability."[4]

In other words, medical ethics organizes the private practice of medicine in a manner that aligns with the social virtues and preferences of Western thought as expressed through a mixture of religious, philosophical, and secular traditions. Key to this ideal is the notion that physicians and other medical practitioners, guided by these ethical norms, will typically make the right choices without any meaningful oversight beyond their peers or professional organizations.

To the extent that early twentieth-century medical ethics situated itself as a field that could rely largely on physician and researcher discretion to promote health and well-being, its limitations were grossly exposed in the wake of World War II and the Holocaust. The Nuremberg trials drew worldwide attention to the central role that physicians and researchers played in facilitating unconscionable human suffering in a twisted pursuit of advancing medical science. The role of medical doctors and researchers in concentration camps was nothing short of ghastly; people were subjected to inhumane experiments and purposefully given certain diseases and poisons just so researchers could study the precise mechanisms through which they suffered and died. However, while Nazi researchers received the bulk of the global criticism for these cruel practices, questionable research and exploitation of vulnerable populations were not unknown in other parts of the world, particularly the United States.

The Nuremberg trials began the process of raising questions about the wisdom of relying on professional norms to protect patients and research subjects from harmful practices. During this same postwar period, other scandals such as the Tuskegee syphilis experiments came to light, creating public pressure for physicians and researchers to be more thoughtful and accountable.[5] Amid these developments and questions came a unique set of deliberations that allowed a new scholarly field—bioethics—to emerge, and a new professional figure—the bioethicist—to claim expertise.

Thus, in many ways, "bioethics was born out of the ashes of the Holocaust."[6] The field formalized in the latter half of the twentieth century with new technological developments, such as recombinant DNA, assisted reproduction, organ transplants, and life support mechanisms that blurred the line between life and death. As public discussion of these developments increased, the field of bioethics evolved. It staked a claim to providing a detached, secular approach to these profound questions of life and death, one that could be endorsed by governments, private organizations (hospitals, professional associations, etc.), and practitioners regardless of religious affiliation or philosophical belief.

Notably, the principles set forth in the Belmont Report in the late 1970s and most famously explored in Tom Beauchamp and James Childress's 1979 book *Principles of Biomedical Ethics* (now in its 7th edition) have come to dominate bioethical thinking. Known as principlism, this modern approach, which is closely associated with Beauchamp and Childress, provides a common framework by embracing four principles as the foundation of bioethical deliberation: respect for autonomy (antipaternalism, or people should have agency in their decision making), nonmaleficence (do no harm), beneficence (help people), and justice (being thoughtful about the distribution of benefits and burdens).

In the context of the technological questions and demands for greater transparency that were rapidly arising at the time, the professional field of bioethics that congealed was certainly an important step forward. Bioethics helped transition the manner in which medicine and scientific research handled ethical issues from a mostly private and individually contemplative endeavor largely hidden from public view to a more transparent, outward-facing set of principles that, in theory, could be applied across fields and topics to produce desirable and predictable outcomes for patients and research subjects. Moreover, as public awareness grew about startling new developments such as in vitro fertilization and artificial organs, bioethicists played an important role not only in helping practitioners think through their professional duties but also in assuring the public that an independent party would be an important part of conversations and decisions about moving these new developments forward in an appropriate manner.

But bioethics as a field has at least three significant limitations. First, as John Evans discusses in "A Sociological Account of the Growth of Principlism" (part 2), the principlism that drives much of the conversation has an inherently individualist orientation. Principlism largely conceives of bioethical dilemmas as occurring in the relationships between doctors and patients, or researchers and human subjects. As an example, principlism offers a coherent way to determine the types of information that should be communicated in an informed consent procedure for a research study. But it provides a thin vision of how group and social dynamics might be crucial for a realistic understanding of "consent." For instance, it leaves much to be desired in terms of how and whether "informed consent" can be meaningful when recruiting subjects from vulnerable populations without access to basic medical care. While principlism's justice component offers a foothold for thinking through such issues, this approach has nevertheless proved inadequate for incorporating ethical values outside of individual relationships or transactions.

Second, as Carl Elliott describes in "The Ethicists" (part 2), the notion that bioethics provides a detached and independent assessment of the ethical issues surrounding science and medicine does not conform to on-the-ground bioethical practice. Bioethicists are often embedded in the same professional institutions and contexts as the very technologies and industries they are ostensibly overseeing. This can lead to situations in which financial or other professional incentives can obscure bioethicists' willingness or ability to consistently comport themselves in a manner that aligns with the public interest.

And lastly, while bioethics can provide a set of guidelines or professional norms, these principles or ideals are typically not translated into enforceable policy mechanisms. Bioethics can inform decision making by institutional review boards, help develop professional standards, and shape a handful of specific regulations such as those pertaining to human research protection. But without any broader regulatory mechanisms to implement these norms, bioethics is largely without teeth in dealing with the deeply profound and transformative social power of many new developments.

The departure point for this volume is to highlight how these limitations become fully exposed when bioethics' principlist approach is put in conversation with the social and policy challenges raised by new human genetic and reproductive technologies. This suite of powerful tools and practices carries social and political implications that go far beyond the limited parameters that can be addressed by autonomy, nonmaleficence, beneficence, and justice. The three vignettes at the beginning of this introduction highlight these limitations and tensions. None of the scenarios can be meaningfully assessed without guidance beyond principlism.

The "Happy Workers" scenario, for example, draws attention to the particular political context that underlay early responses to the prospect of human reproductive cloning. In an era of relentless attacks on abortion rights, people otherwise sympathetic to concerns about exploitation were perhaps understandably tempted to overlook it and double down on their commitment to "choice." Planned Parenthood's initial unwillingness to foreclose any avenue of reproduction—even one that involved producing "cloned insomniac slaves"[7]—fortunately did not persist. But this episode dramatizes the inadequacy of a principlist approach for a complex issue playing out on difficult social and political terrain.

Similarly, the dynamics surrounding race-based medicine cannot be meaningfully understood without considering the remarkably twisted (and at times brutal) history of race and medicine, or taking into account the commercial imperatives leading to BiDil's development. Questions about, for example, autonomous decision making, or about "doing harm" versus "doing good" to individual patients, are tangential to the deeper social and political implications of the US government giving its stamp of approval to medicine premised on the dubious notion of biological race.

The gene-editing technologies at issue in the third vignette represent perhaps the most consequential of these new technologies—the ability to engineer and redesign human beings, and even humanity. Yet when faced with the prospect of the significant social and political challenges created by such developments, a leading scientist's ethical sensibilities leads him to ponder issues of individual safety to the exclusion of potentially great social harm.

In the mid-twentieth century, new technological advances and changing social contexts created the conditions for the private, inward-looking nature of medical ethics to be supplemented by a more public, outward-facing bioethical discourse. Similarly, in the late twentieth and now early twenty-first centuries, new human genetic and assisted reproductive technologies, coupled with the increasingly commercialized conditions of their development and deployment in addition to other social changes, are prompting new and far more public conversations about the proper relationships and negotiations between science and society. This approach places social justice, human rights, and the public interest at the center of its analysis. We call this emerging field, which will inevitably remain in conversation with medical ethics and bioethics, the *new biopolitics.*

An entry for the term *biopolitics* in the *Encyclopedia of Bioethics* (co-authored by Darnovsky) asserts the following:

> The social and ethical challenges posed by human biotechnologies in the early twenty-first century encompass much broader issues, and capture much greater public attention, than was the case in the early days of the field of bioethics. Biopolitical controversies play out in social realms, including academia, news and online media, and popular culture, and on political stages including the courts, legislatures, and even national elections. And biopolitics is increasingly a focus for civil-society constituencies and public-interest organizations.
>
> In contrast to bioethics, then, biopolitics focuses on broad social and political dynamics more than on encounters in institutional settings between doctors and patients or between researchers and human subjects. It emphasizes social values and policy proposals more than procedural recommendations and professional guidelines. Though bioethics and biopolitics are in constant conversation with each other, biopolitics is situated largely outside the organizational structures (such as academic departments, hospitals and clinics, institutional review boards, and corporate advisory boards) that are most closely associated with bioethics.[8]

The term *biopolitics* is used in a number of academic disciplines, including bioethics, sociology, anthropology, philosophy, and science and technology studies. It is most widely associated with the work of French philosopher Michel Foucault.[9] What we are calling the "new biopolitics" is distinct from these prior articulations yet remains a project in formation, with many of its contours and characteristics still blurry. But it is useful to sketch at least five particular concerns that further distinguish it from mainstream bioethical approaches:

- *Reckoning with the role of commerce and markets in biomedicine and biotechnology.* The new biopolitics is sensitive to the ways in which commercial pressures and market incentives can warp deliberations on the potential social impacts of new human genetic and reproductive technologies. It also pays close attention to how science and medicine are not only healing endeavors but also profit-seeking enterprises that, like all market-oriented ventures, need regulation and oversight.

- *Understanding the human genome as part of the common heritage of humanity.* The new biopolitics takes seriously that if gene editing or cloning were used for human reproductive purposes, either could alter basic understandings of what it means to be human in ways likely to have cascading effects for all subsequent generations. This approach also suggests greater humility and a precautionary approach in questioning whether we have the wisdom to understand the full range of social and biological consequences associated with, for example, inserting genetic enhancements or deleting genetic challenges from our gene pool. It appreciates statements by international bodies, including the United Nations Educational, Scientific and Cultural Organization; the World Medical Association; and the Human Genome Organization, Ethical, Legal, and Social Issues Committee (the international scientific coordinating body for the Human Genome Project), asserting that the human genome should be considered a symbol and a part of the common heritage of humanity.[10]

- *Avoiding technical developments and genetic narratives that embed social and political preferences at the molecular level.* The new biopolitics is sensitive to the discriminatory attitudes that may allow social preferences to guide the way human biotechnologies are implemented and to the deleterious impacts this can have on the preconditions for, and basic notions of, social justice. Human biotechnologies have the ability to reaffirm social preferences and lines of difference at a molecular level. As one example, race-specific medicines treat health disparities as a function of natural differences rather than as products of social inequality. As another example, using assisted reproductive technologies to identify and eliminate embryos that contain conditions many consider fully compatible with health and happiness effectively reinforces the view that bodies rather than social and political context are always the disabling factor. (See "Disability Equality and Prenatal Testing: Contradictory or Compatible?" by Adrienne Asch in part 8.)

- *Ensuring democratic oversight of powerful human biotechnologies.* The new biopolitics expresses skepticism about the ability of private companies or associations, professional bioethicists, or self-enforced rules to ensure that human biotechnologies truly serve the public interest. The choices being put in front of us by new human biotechnologies are profoundly political; they create winners and losers, shape our societies (and in some cases our very selves), and implicate society's deepest values about who we are and what we want to be. It would be remarkably unwise to hand such decision making to any one group of experts or professionals and extremely reckless to leave it to "the market." Therefore, a new biopolitical approach not only promotes greater transparency but also much greater civic participation and inclusive democratic engagement in decisions about whether and how new human genetic and assisted reproductive technologies should be developed, deployed, and governed. It calls for "public engagement" and insists that this involve robust support for thoughtful and extended deliberations, so as to enable participation by civil society, public interest and faith-based organizations, community groups, labor unions, and others.

- *Steering clear of a new market-driven eugenics.* A central concern of the new biopolitics is that particular human genetic and assisted reproductive technologies may lead to a new form of eugenics. In contemplating this prospect, some observers argue that we no longer need to fear state involvement in citizens' reproductive choices or government efforts to weed out undesirable populations—that the sterilization laws, immigration restrictions, and outright genocides justified by eugenics are earmarks of a bygone (late nineteenth- and early twentieth-century) era. But a "new" eugenics, driven by market forces as opposed to state discrimination, could look eerily similar. Collective efforts to enhance future generations along the lines of socially acceptable aesthetic norms, or market dynamics that encourage using IVF-based screening technologies to avoid certain traits, may allow racist, sexist, and ableist norms to dictate who is an accepted member of society.

· · ·

Our ambition for this anthology is that it will crystallize and promote the growth of the new biopolitics as a field of public, policy, and scholarly concern. Though they focus on developments in the United States, the volume's essays and articles articulate a range of concerns and perspectives prompted by new human biotechnologies worldwide. They explore the troubling directions in which these technologies and associated ideologies can lead without proper care, and offer an alternative vision for negotiating the power-laden fault lines among the life sciences, the biotechnologies they spawn, and society. The volume's specific focus—forgoing traditional bioethical topics[11] such as end-of-life care and organ transplants—stems from the particular dilemmas created for bioethics by contemporary financial and political conditions and by new and emerging technologies. Many of the contributions also explore how these new dilemmas speak to the field's original core themes, like eugenics, the relationship of science and medicine to vulnerable populations, and the need for greater oversight of researchers and medical professionals.

The new biopolitical perspective this volume presents may provide useful insights for study and critique in other fields (such as environmental studies or urban planning) that involve social justice, human rights, and public interest values. We leave such extensions to future scholars, and focus here on the ways in which the profound challenges created by powerful new human biotechnologies call for a fresh framework of analysis and assessment. The new biopolitics doesn't provide definitive answers but guides us toward questions that prioritize social justice and human rights and suggests fruitful ways to explore them.

The articles in parts 1 through 10 are all reprinted or excerpted from previous publications. This heterogeneous collection exemplifies what has until now been a loosely aligned group of writings that focus on the shortcomings and limitations of mainstream bioethical approaches, in search of deeper inquiry into the intensely social and political nature of human biotechnologies. Not all contributors to this volume explicitly identify with this new biopolitical vision. As editors, our goal is to highlight the connections that some authors themselves may not immediately see but that, taken as a web of concerns, give birth to an alternative understanding of the appropriate relationship between science, medicine, and society. Inasmuch as multiple excerpts touch on a similar theme or topic, this book draws strength from demonstrating the many standpoints from which particular events can be understood, challenged, and critiqued. The original foreword by Troy Duster and original afterword by Patricia J. Williams are extraordinarily rich additions to this field-in-formation.

Part 1, "The Biopolitical Critique of Bioethics: Historical Context," and part 2, "Bioethics and Its Discontents," situate the new biopolitics in the contexts from which it emerges. The contributions in part 1 consider historical examples of techno-scientific abuses and begin to demonstrate the many linkages between past, present, and future. Part 2 collects articles that call attention to some of the limitations of mainstream bioethics.

The articles in part 3, "Emerging Biotechnologies, Extreme Ideologies: The Recent Past and Near Future," focus on the technological enthusiasts in the United States and elsewhere, who advocate the unfettered use of biotechnologies for extreme scenarios that include "designer babies" and "posthumans." Several contributors consider how such visions and ideologies threaten social justice, human equality, and the common good.

The next three sections examine the commercial context of biotechnological research and development. Contributors to part 4, "Markets, Property, and the Body," consider how powerful market forces affect researchers, doctors, patients, clinical trial participants, universities, institutional review boards, and the field of professional bioethics. Part 5, "Patients as Consumers in the Gene Age," takes a critical look at "precision medicine," "big genomics," and similar declarations of revolutions in health care, asking how we should weigh their promises against their perils and understand other costs and opportunities. In part 6, "Seeking Humanity in Human Subjects Research," contributors focus on problems in biomedical research involving human subjects, examining the economic, career, and other pressures and incentives that underlie them.

Part 7, "Baby-Making in the Biotech Age," and part 8, "Selecting Traits, Selecting Children," turn to assisted reproductive technologies. Contributors to part 7 look through a social justice lens at troubling aspects of the fertility industry, focusing on the United States and on cross border fertility arrangements. Part 8 explores the selection technologies currently in use in the context of assisted reproduction, including how they are changing the experience of pregnancy and the ethical and social challenges they pose.

Part 9, "Reinventing Race in the Gene Age," examines the scientific fallacies behind the resurgence of race as a biological concept and the dangerous social consequences this entails. In part 10, "Biopolitics and the Future," we look forward, exploring opportunities and mechanisms for incorporating new biopolitical ways of thinking into scientific discourses, policy debates, and public understandings of human biotechnologies to foster a new biopolitical imagination that is unafraid to confront the social and moral challenges they raise.

NOTES

1. Marcy Darnovsky, "Political Science," *Democracy* 13 (2009): 46.

2. Anne Pollock, "Medicating Race: Heart Disease and Durable Preoccupations with Difference" (PhD diss., MIT, 2007), 243.

3. "Re-engineering Human Embryos," *On Point*, WBUR 90.9, April 28, 2015, accessed November 17, 2016, www.wbur.org/onpoint/2015/04/28/human-embryo-genetic-engineering-china.

4. Albert R. Jonsen, *The Birth of Bioethics* (New York: Oxford University Press, 2003), 8.

5. Susan Reverby, *Examining Tuskegee: The Infamous Syphilis Study and Its Legacy* (Chapel Hill: University of North Carolina Press, 2009).

6. Arthur Caplan, quoted in George Annas, *American Bioethics: Crossing Human Rights and Health Law Boundaries* (Oxford: Oxford University Press, 2004), 161.

7. Darnovsky, "Political Science."

8. Marcy Darnovsky and Emily Beitiks, "Biopolitics," in *Bioethics*, 4th ed., ed. Bruce Jennings (Farmington Hills, MI: Gale Cengage Learning, 2014).

9. Michel Foucault, *The Birth of Biopolitics: Lectures at the Collège de France, 1978–1979*, ed. Michel Senellart, trans. Graham Burchell (Basingstoke, UK: Palgrave Macmillan, 2008).

10. United Nations Educational, Scientific and Cultural Organization Declaration on the Human Genome and Human Rights, Article 1 (Paris: UNESCO, 1997); World Medical Association, "Declaration on the Human Genome Project (September 1992)," *Bulletin of Medical Ethics* 87 (1993): 9–10; Human Genome Organization, Ethical, Legal, and Social Issues Committee, "Statement on the Principled Conduct of Genetics Research" (1996), accessed on November 17, 2016, www.eubios.info/HUGO.htm.

11. Other anthologies on bioethics explore these traditional areas in more depth. See, for example, Vardit Ravitsky, Autumn Fiester, and Arthur L. Caplan, eds., *The Penn Center Guide to Bioethics* (New York: Springer, 2009); and Jessica Pierce and George Randels, *Contemporary Bioethics: A Reader with Cases* (Oxford: Oxford University Press, 2009).

PART I

The Biopolitical Critique of Bioethics

Historical Context

A full appreciation for what is at stake with new human biotechnologies requires an examination of past social and political efforts to use biological explanations to understand human differences and group disparities. From skull measurements attempting to compare the intelligence of different racial groups to hereditarian theories of criminality, the scientific method has at times been used to make disparate social and health outcomes appear to be a function of who people are rather than the different treatments afforded to them or the biases embedded in the methods themselves. Some of these efforts took place under the banner of eugenics, the science of improving a human population through controlled breeding.

Many people associate eugenics solely with the Nazi regime and the Holocaust and assume that eugenic beliefs and practices were limited to that nightmarish chapter of European history. In fact, eugenics has a longer history, a broader reach, and greater persistence.

In the early twentieth century, eugenic ideologies and practices drew on genetic theories that provided scientific cover for policy decisions about who should and shouldn't reproduce—decisions largely informed by discriminatory attitudes toward marginalized groups. In the United States, a widespread eugenics movement led to the forced sterilization of tens of thousands of people considered "unfit" to reproduce, stringent immigration restrictions on undesired populations, and public policies that encouraged "fitter families" to produce more children.

Eugenic ideas and rhetoric pioneered in the United States were taken up by the Third Reich, where they were used to justify the extermination of Jews, people with disabilities, and others. Revelations of these horrors led eugenic policies and practices to largely recede from public life. But as later sections of this anthology explain, these beliefs—along with more subtle but still pernicious assumptions about the relationship between biology and inequality—persisted throughout the twentieth century and are still with us.

In "The Biological Inferiority of the Undeserving Poor," Michael B. Katz argues that in the late eighteenth and early nineteenth centuries, "a harsh new idea of poverty and poor people as different and inferior" began to replace the ancient and biblical view that poverty reflected God's will. Part of this shift, the idea that poverty is caused by faulty heredity, fed into twentieth-century eugenic theories that justified racism and social conservatism. Katz considers this account to be "a cautionary note from history about the uses of science and a warning to be vigilant and prepared."

Alexandra Minna Stern's "Making Better Babies: Public Health and Race Betterment in Indiana, 1920–1935" paints a fascinating picture of the "better babies contests" that were often held at state fairs under the sponsorship of American eugenicists. "More than just a lively spectacle for fairgoers," she argues, "these contests brought public health, 'race betterment,' and animal breeding together in a unique manner." Stern suggests that the "better babies contests" demonstrate that public health and eugenics were not, as is often assumed, "antithetical movements separated by a conceptual gulf between environment and heredity."

In "Eugenics and the Nazis: The California Connection," Edwin Black discusses the considerable influence of the American eugenics movement on Adolf Hitler and the Nazi genocide. Black briefly sketches the prewar intellectual and political exchanges between German and American eugenicists and notes that Germans who were tried for genocide after the war cited California's eugenic sterilization law in their defense. Black suggests that today, as the capabilities and reach of human genetic technologies expand, it is dangerous to ignore the histories and legacies of German and American eugenic crimes.

"Why the Nazis Studied American Race Laws for Inspiration" is a provocative piece by James Q. Whitman that expands our understanding of the intellectual and ideological synergies between Nazi Germany and the United States during this period. It explores the extent to which Jim Crow and other forms of institutional racism served as models for Nazi lawmakers as they remade German society. The admiration of Nazis for the United States' race laws is chilling and suggests, as Whitman notes, that Americans may need to engage in deeper reflection to understand our nation's role in producing the atrocities committed during the Holocaust.

Lennard J. Davis examines the concepts of "the norm" and "the normal body" in "Constructing Normalcy: The Bell Curve, the Novel, and the Invention of the Disabled Body in the Nineteenth Century." He notes that eugenicists, who aim to normalize those who are nonstandard, tend to group together all traits that they consider undesirable. They may mention, in the same breath, "criminals, the poor, and people with disabilities." Davis argues that "revers[ing] the hegemony of the normal" and "institut[ing] alternative ways of thinking about the abnormal" are important for developing consciousness of disability issues.

In the final contribution of this section, Osagie K. Obasogie considers the legacy of Robert Edwards, who was honored with a 2010 Nobel Prize for his work developing in vitro fertilization. Unnoted in the coverage of the award was Edwards's long and active membership in Britain's Eugenics Society and his understanding of IVF and other reproductive and genetic technologies in a clearly eugenic framework. "Although there is nothing inherently eugenic about IVF," Obasogie writes, "being able to manipulate human conception outside of the womb is an essential platform technology for any modern eugenic goal."

1. The Biological Inferiority of the Undeserving Poor

Michael B. Katz

> ... If the misery of our poor be caused not by the laws of nature, but by our institutions, great is our sin. ...
>
> CHARLES DARWIN (1839)

For most of recorded history, poverty reflected God's will. The poor were always with us. They were not inherently immoral, dangerous, or different. They were not to be shunned, feared, or avoided. In the late eighteenth and early nineteenth centuries, a harsh new idea of poverty and poor people as different and inferior began to replace this ancient biblical view. In what ways, exactly, are poor people different from the rest of us became—and remains—a burning question answered with moral philosophy, political economy, social science, and, eventually, biology. Why did biological conceptions of poverty wax and wane over the last century and a half? What forms have they taken? What have been their consequences?

The biological definition of poverty reinforces the idea of the undeserving poor, which is the oldest theme in post-Enlightenment poverty discourse. Its history stretches from the late eighteenth century through to the present. Poverty, in this view, results from personal failure and inferiority. Moral weaknesses—drunkenness, laziness, sexual promiscuity—constitute the most consistent markers of the undeserving poor. The idea that a culture of poverty works its insidious influence on individuals, endowing them with traits that trap them in lives of destitution, entered both scholarly and popular discourse somewhat later and endures to this day. Faulty heredity composes the third strand in the identification of the undeserving poor; backed by scientific advances in molecular biology and neuroscience, it is enjoying a revival. The historical record shows this idea in the past to

The Undeserving Poor: America's Enduring Confrontation with Poverty 2e by Michael B. Katz (2013): pp. 29–49. By permission of Oxford University Press USA. Due to space limitations, portions of this chapter have been deleted or edited from the original.

have been scientifically dubious, ethically suspect, politically harmful, and, at its worst, lethal. That is why we should pay close attention to its current resurgence.

This article excavates the definition of poor people as biologically inferior. It not only documents its persistence over time but emphasizes three themes. First, the concept rises and falls in prominence in response to institutional and programmatic failure. It offers a convenient explanation for why the optimism of reformers proved illusory or why social problems remained refractory despite efforts to eliminate them. Second, its initial formulation and reformulation rely on bridging concepts that try to parse the distance between heredity and environment through a kind of neo-Lamarkianism. These early bridges invariably crumble. Third, hereditarian ideas always have been supported by the best science of the day. This was the case with the ideas that ranked "races," underpinned immigration restrictions, and encouraged compulsory sterilization—as well as those that have written off the intellectual potential of poor children.

In its review of the biological strand in American ideas about poverty, this article begins in the 1860s with the first instance of the application of hereditarian thought I have discovered and moves forward to social Darwinism and eugenics, immigration restriction, and early IQ testing. It then picks up the story with Arthur Jensen's famous 1969 article in the *Harvard Educational Review*, follows it to the *Bell Curve*, and ends with the astonishing rise of neuroscience and the field of epigenetics. It concludes by arguing that despite the intelligence, skill, and good intentions of contemporary scientists, the history of biological definitions of poor persons calls for approaching the findings of neuroscience with great caution.

In 1866 the Massachusetts Board of State Charities, which had oversight of the state's public institutions, wrote, "The causes of the evil [the existence of such a large proportion of dependent and destructive members of our community] are manifold, but among the immediate ones, the chief cause is inherited organic imperfection, vitiated constitution or *poor stock*" ([MA] State Board of Charities 1866). This early proclamation of the biological inferiority of the undeserving poor arose as a response to institutional failure. Recurrent institutional and programmatic failure has kept it alive in writing about poverty ever since, supported always by scientific authority.

Beginning in the early nineteenth century, reformers sponsored an array of new institutions designed to reform delinquents, rehabilitate criminals, cure the mentally ill, and educate children. Crime, poverty, and ignorance, in their view, were not distinct problems. The "criminal," "pauper," and "depraved" represented potentialities inherent in all people and triggered by faulty environments. Poverty and crime, for instance, appeared to cause each other and to occur primarily in cities, most often among immigrants. This stress on the environmental causes of deviance and dependence, prominent in the 1840s, underpinned the first reform schools, penitentiaries, mental hospitals, and, even, public schools (Katz [1968] 2001).

By the mid-1860s it had become clear that none of the new institutions built with such optimism had reached their goals. They manifestly failed to rehabilitate criminals, cure the mentally ill, reeducate delinquents, or reduce poverty and other forms of dependence. The question was, why? Answers did not look hard at the failures in institutional design and implementation or at the contexts of inmates', prisoners', and patients' lives. Rather, they settled on individual-based explanations: inherited deficiencies. The Massachusetts Board of State Charities supported its belief that the inheritance of acquired characteristics (later known as Lamarkianism) reproduced the undeserving poor as well as criminals, the mentally ill, and other depraved and dependent individuals with scientific evidence from physiologists which emphasized the toxic impact of large amounts of alcohol on the stimulation of the "animal passions" and the repression of *"will"* ([MA] State Board of Charities 1866). . . .

By the 1920s, two initially separate streams—social Darwinism and eugenics—converged in the hard-core eugenic theory that justified racism and social conservatism. Social Darwinism attempted to apply the theory of Darwinian evolution to human behavior and society. Social Darwinists—whose leading spokesperson, Herbert Spencer, enjoyed a triumphant tour of the United States in 1882—insisted on the heritability of socially harmful traits, including pauperism, mental illness, and criminality, and on the harmful effects of public and private charities that interfered with the survival of the fittest. They viewed the "unfit" not only as unworthy losers but as savage throwbacks to a primitive life. Hereditarian beliefs thus fed widespread fears of "race suicide" giving an urgency to the problem of population control. The "ignorant, the improvident, the feeble-minded, are contributing far more than their quota to the next generation," warned Frank Fetter of Cornell University (Bender 2009, 202).

The English scientist Francis Galton originally coined the term eugenics in 1883 to denote the improvement of human stock by giving "the more suitable races or strains of blood a better chance of prevailing speedily over the less suitable." In the United States, eugenic "science" owed more to the genetic discoveries of Gregor Mendel, first published in 1866 but unrecognized until the end of the century, than to mathematical genetics as practiced by Galton and his leading successor Karl Pearson. In 1904 Charles Davenport, the leading US eugenics promoter, used funds from the newly established Carnegie Corporation to set up a laboratory at Cold Springs Harbor on Long Island. Davenport looked forward to the "new era" of cooperation between the sociologist, legislator, and biologist, who together would "purify our body politics of the feeble-minded, and the criminalistic and the wayward by using the knowledge of heredity." Eugenics entered public policy through its influence on immigration restriction and social reform as well as through state sterilization laws. Indiana passed the first of these in 1907. By the end of the 1920s, twenty-four states passed laws permitting the involuntary sterilization of the mentally unfit, a practice upheld by the US Supreme Court in 1920 in *Buck v. Bell* (Kevles 1985).

In the United States, the application of evolutionary and genetic ideas to social issues gained traction in the late nineteenth century as a tool for explaining and dealing with the vast changes accompanying industrialization, urbanization, and immigration. Eugenics drew support from both conservatives and progressives and underlay the emerging consensus on the need for immigration restriction that resulted in the nationality-based immigration quotas legislated by Congress in 1924. "In the early twentieth century," point out Hilary Rose and Steven Rose in *Genes, Cells and Brains*, "barring Catholics, eugenics commanded the support of most EuroAmerican intellectuals—not just racists and reactionaries but feminists, reformers, and Marxists" (Rose and Rose 2012, 129). Conservatives found in eugenics and social Darwinism justification for opposing public and private charities that would contribute to the reproduction of the unfit. But eugenics found enthusiasts as well in birth control advocate Margaret Sanger and in settlement house workers preoccupied with the alleged degeneracy of an immigrant working class. Like their predecessors on the Massachusetts Board of State Charities decades earlier, they turned to the heritability of acquired characteristics and the plasticity of human nature to reconcile their belief in the biological foundation of physical and moral degeneration with their commitment to the power of social reform to build character and instill habits.

Nonetheless, by the 1920s, cracks appeared in the bridge that linked the environmentalists and hereditarians. Hereditarians took an increasingly hard line, manifest in the new science of intelligence tests as well as in their continued advocacy of sterilization. Developed by the French psychologist Alfred Binet, intelligence tests were brought to the United States in 1880 by American psychologist Henry H. Goddard who first applied them at the Vineland, New Jersey, Training School for Feeble-Minded Boys and Girls—

he directed its new laboratory for the study of mental deficiency. Other psychologists picked up Goddard's work on intelligence testing, extended it to other populations, and experimented with different methods. Lewis Terman at Stanford, one of the most prominent ... proponent[s] of the hereditarian view of intelligence, introduced the term "IQ," which stood for "intelligence quotient," a concept developed in 1912 by William Stern, a German psychologist. Intelligence testing, which at first aroused skepticism and hostility, received a tremendous boost during World War I, when a trial of the tests on more than 1.7 million people during the war dramatically brought them to public attention. The tests purported to show that nearly one-fourth of the draft army could not read a newspaper or write a letter home and, by implication, that the mental ages of the average white and black Americans were, respectively, thirteen and ten (Gould 1980).

Davenport, Goddard, and others blamed the results for whites on the immigration of inferior races and used them as ammunition in their advocacy of immigration restriction. The tests, they argued, demonstrated the genetic heritability of mental deficiency. These ideas worked their way into public education in the 1920s, underpinning the educational psychology taught in teacher preparation courses and the massive upsurge in testing used to classify students, predict their futures, and justify unequal educational outcomes. "Terman and other psychologists," points out historian Paul Fass, "were quick to point out that opening up avenues of opportunity to the children of the lower socioeconomic groups probably made no sense; they did not have the I.Q. points to compete." In the minds of its prominent advocates, intelligence testing was linked with beliefs that science had demonstrated the primacy of heredity over environment and that the immigration of inferior races was driving America toward a dysgenic future (Fass 1989).

Even before the 1920s, strains between eugenicists and reformers had opened fissures in the consensus around the heritability of mental and character defects. Eugenicists' commitment to "germ plasm" pulled them away from the environmental and neo-Lamarkian theories underpinning Progressive reform. Then, after the 1920s, biochemistry and the rise of the Nazis combined to drive eugenics into eclipse and disrepute. The more research revealed about the complexity of human genetics, the less defensible even reform genetics appeared. The American Eugenics Society praised Hitler's 1933 sterilization law while German eugenicists flattered their American counterparts by pointing out the debt they owed them, and the Nazi regime welcomed and honored prominent American eugenicists (Bender 2009).

The fall of eugenics left the field open to environmental explanations. Nurture rather than nature became the preferred explanation for crime, poverty, delinquency, and low educational achievement. The emphasis on environment fit with the emergent civil rights movement, which rejected racial, or biological, explanations for differences between blacks and whites—explanations that had been used to justify slavery, lynching, segregation, and every other form of violent and discriminatory activity. Hereditarian explanations fit badly, too, with the optimism underlying the War on Poverty and Great Society that assumed the capacity of intelligent government action to ameliorate poverty, ill health, unemployment, and crime.

Nonetheless, by the late 1960s a new eugenics began to challenge the environmental consensus. Its appearance coincided with the white backlash against government-sponsored programs favoring African Americans and the disenchantment following on what appeared to be the failure of programs of compensatory education designed to make up for the culturally deficient homelife of poor, especially poor black, children. Psychologist Arthur R. Jensen's 1969 article in the *Harvard Educational Review*, "How Much Can We Boost IQ and Scholastic Achievement?" led the revival of hereditarianism. "Compensatory education," Jensen argued, "has been tried and it apparently has failed." The reason was that compensatory education programs ran up against a genetic wall. Poor, minority children lacked the intelligence to profit from them (Jensen 1969).

Jensen's article provoked a furious counter-attack. Nonetheless, the controversy breathed new life into research and writing on the influence of heredity on intelligence and seeped into the rationales for failure offered by educators. (I recall sitting in a meeting in the early 1970s with a high-level Toronto school administrator who, in a discussion of the low achievement of poor students, said, in effect, "well, Jensen has told us why.")The new field of sociobiology, founded by Harvard zoologist E.O. Wilson, a leading authority on insect societies, reinforced the renewed emphasis on heritability. Sociobiology, Wilson wrote, focused on "the study of the biological basis of social behavior in every kind of organism, including man" (1975, 39). This new emphasis on heritability, however, met strong scientific as well as political criticism and failed to clear away the taint that still clung to eugenics and genetically based theories of race, intelligence, and behavior. The idea that the undeserving poor were genetically inferior had not been wiped from the map by any means, but it remained muted, unacceptable in most academic circles.

In 1994, in their widely publicized and discussed *The Bell Curve*, Richard Herrnstein and Charles Murray—whose notorious *Losing Ground* had served as a bible for anti–welfare state politicians—challenged the reigning environmentalist view of intelligence. Success in American society, they argued, was increasingly a matter of the genes people inherit. Intelligence, in fact, had a lot to do with the nation's "most pressing social problems," such as poverty, crime, out-of-wedlock births, and low educational achievement. They wrote that *"low intelligence is a stronger precursor of poverty than low socioeconomic background."* Poverty, they argued, "is concentrated among those with low cognitive ability," which, itself, was largely inherited. It also was racially tinged because blacks, they found, revealed lower cognitive ability at every socioeconomic level. Evidence points "toward a genetic factor in cognitive ethnic differences" because "blacks and whites differ most on tests" measuring "g, or general intelligence," which is a fixed, inherited index of mental capacity (Herrnstein and Murray 1994, 117, 127, 371, 270).

. . . Despite assaults in the public media and by scholars, hundreds of thousands of copies of the 800-plus page hardcover edition of the book were sold. *The Bell Curve* is best understood not as a popularization of science but as an episode in the sociology of knowledge. Clearly, even if it often did not dare speak its name, the suspicion remained alive that heredity underlay the growth and persistence of the "underclass" and the black-white gap in educational achievement, which seemed to many impervious to increased public spending or reform. . . .

From the 1990s onward, a profusion of new scientific technologies has provided the tools with which to explore mechanisms underlying the linkages between biology and society and fostered the astounding growth of the bioscience industry in genetics (the Human Genome Project); stem cell research; and, most recently, neuroscience. Teachers, point out Hilary and Steven Rose, "report receiving up to seventy mailshots a year promoting a variety of neurononsense. . . . The snake-oil entrepreneurs are in there selling hard to teachers who are without the protection provided by clinical trials" and other tools available to physicians (2012, 275).

With astonishing acceleration, neuroscience, evolutionary psychology, genomics, and epigenetics emerged as important scientific fields—in practice, often combined in the same programs. Neuroscience and other biological advances promised new ways of explaining social phenomena, like crime, and medical issues, such as the black-white gap in cardiovascular diseases, the increase in diabetes, the rise of obesity, and the origins and treatment of cancer-related disease. They promised, as well, the possibility of understanding how the brain ages and how Alzheimer's disease and dementia might be mitigated or delayed. Research focuses, too, on how the environmental stresses associated with poverty in childhood could damage aspects of mental functioning and learning capacity with lasting impact throughout individuals' lives, and, some scientists believe, beyond through the inheritance of acquired deficiencies (Stricker 2009).

In its January 18, 2010, cover story, *Time* announced, "The new field of epigenetics is showing how your environment and your choices can influence your genetic code—and that of your kids." Epigenetics, the article explained, "is the study of changes in gene activity that do not involve alterations to the genetic code but still get passed down to at least one generation. These patterns of gene expression are governed by the cellular material—the epigenome—that sits on top of the genome, just outside it. . . . It is these 'epigenetic' marks that tell your genes to switch on or off, to speak loudly or whisper. It is through [epigenetic] marks that environmental factors like diet, stress and prenatal natal nutrition," which "can make an imprint on genes," are transmitted across generations. More soberly, the eminent child psychiatrist Sir Michal Rutter offered this definition: "The term 'epigenetics' is applied to mechanisms that change genetic effects (through influences on gene expression) without altering gene sequence." Hilary and Steven Rose report [that] . . . "genes are no longer thought of as acting independently but rather in constant interaction with each other and with the multiple levels of the environment in which they are embedded." . . .

Epigenetics found such a receptive audience, in part, because once again scientific advance coincided with a major conundrum—the persistent "achievement gap" between blacks and whites which bedeviled educators. A large literature suggested a variety of sources, most of which focused in one way or another on the handicaps associated with growing up in poverty while the proponents of hereditary explanations lurked in the background. What the environmentalists lacked was a mechanism that explained exactly how the environment of poverty was translated into low school achievement. This is what epigenetics offered. It promised as well to parse the acrimonious differences between environmentalists and hereditarians in explaining the sources of criminality and virtually all other behavior (Noguera 2013).

The breathless embrace of epigenetics ran ahead of the evidence about the heritability of acquired characteristics and limits of existing epigenetic knowledge. . . . Epigenetics has facilitated and revived the reconciliation of hereditarianism and reform that flourished before social Darwinism in the late 1860s and then again in the Progressive Era, before splitting apart in the 1920s. Epigenetics promises to move beyond the long-standing war between explanations for the achievement gap, persistent poverty, crime, and other social problems based on inheritance and those that stress environment. It gives scientific sanction for early childhood education and other interventions in the lives of poor children. As with earlier invocations of science, popular understanding fed by media accounts threatens to run ahead of the qualifications offered by scientists and the limits of evidence.

Herein lies the danger. In the past, the link between hereditarianism and reform proved unstable, and when it broke apart the consequences were ugly. Even when in place the link supported racially tinged immigration reform and compulsory sterilization—all in the name of the best "science." Indeed, every regime of racial, gender, and nationality-based discrimination and violence has been based on the best "science" of the day. "It is when scientists and doctors insist that their use of race is purely biological," cautions legal scholar and sociologist Dorothy Roberts, "that we should be most wary." Philosopher Jesse J. Prinz warns that "when we assume that human nature is biologically fixed, we tend to regard people with different attitudes and capacities as inalterably different. We also tend to treat differences as pathologies" (2012, 4).

It is not a stretch to imagine epigenetics and other biologically based theories of human behavior used by conservative popularizers to underwrite a harsh new view of the undeserving poor and the futility of policies intended to help them. This is not the aim, or underlying agenda, of scientists in the field, or a reason to try to limit research. It is, rather, a cautionary note from history about the uses of science and a warning to be vigilant and prepared.

REFERENCES

"Arthur R. Jensen Dies at 89: Set Off Debate around IQ." 2012. *New York Times*, November 2.

[Massachusetts] State Board of Charities. 1866. Second Annual Report of the [Massachusetts] State Board of Charities to Which Are Added the Reports of the Secretary and the General Agent of the Board. Public Document 19.

Bender, Daniel. 2009. *American Abyss: Savagery and Civilization in the Age of Industry*. Ithaca, NY: Cornell University Press.

Boyce, Thomas, Maria B. Sokolwski, and Gene E. Robinson. 2012. "Toward a New Biology of Social Adversity." *PNAS* 109 (supplement 2): 17143.

Carey, Nessa. 2012. *The Epigenetics Revolution: How Modern Biology Is Rewriting Our Understanding of Genetics, Disease, and Inheritance*. New York: Columbia University Press.

Dain, Norman. 1964. *Concepts of Insanity in the United States, 1987–1865*. New Brunswick, NJ: Rutgers University Press.

Davis, David Brion. 1957. *Homicide in American Fiction, 1978–1860: A Study in Social Values.* Ithaca, NY: Cornell University Press.

Fass, Paula. 1989. *Outside In: Minorities and the Transformation of American Education.* New York: Oxford University Press.

Fischer, Claude S., Michael Hout, Martín Sánchez Jankowski, Samuel R. Lucas, Ann Swidler, and Kim Voss. 1996. *Inequality by Design: Cracking the Bell Curve Myth.* Princeton, NJ: Princeton University Press.

Gould, Stephen J. 1980. *The Mismeasure of Man.* New York: W. W. Norton and Company.

Hanson, Jamie, Nicole Hair, Amitabh Chandra, Ed Moss, Jay Bhattacharya, Seth D. Polk, and Barbara Wolfe. 2012. "Brain Development and Poverty: A First Look." In *The Biological Consequences of Socioeconomic Inequalities,* edited by Barbara Wolfe, William N. Evans, and Teresa E. Seeman. New York: Russell Sage Foundation.

Heckman, James. 2008. "Schools, Skills, and Synapses." *Economic Inquiry* 46 (3): 289.

———. 2012. "Promoting Social Mobility." *Boston Review,* September /October.

Herrnstein, Richard J. 1971. "I.Q." *Atlantic,* September, 63.

Herrnstein, Richard J., and Charles Murray. 1994. *The Bell Curve: Intelligence and Class Structure in American Life.* New York: Free Press.

Jensen, Arthur R. 1969. "How Much Can We Boost IQ and Scholastic Achievement?" *Harvard Educational Review* 39 (1).

Katz, Michael B. (1968) 2001. *The Irony of Early School Reform: Educational Innovation in Mid-nineteenth Century Massachusetts.* Cambridge, MA: Harvard University Press, Teachers College Press.

Kevles, Daniel J. 1985. *In the Name of Genetics: Genetics and the Uses of Human Heredity.* New York: Knopf.

Lewis, W. David. 1965. *From Newgate to Dannemora: The Rise of the Penitentiary in New York, 1796–1848.* Ithaca, NY: Cornell University Press.

Noguera, Pedro. 2013. "The Achievement Gap and the Schools We Need: Creating the Conditions Where Race and Class No Longer Predict Student Achievement." In *Public Education under Siege,* edited by Michael B. Katz and Mike Rose, 180. Philadelphia: University of Pennsylvania Press.

Roberts, Dorothy. 2012. *Fatal Invention: How Science, Politics, and Big Business Re-create Race in the Twenty-First Century.* New York: New Press.

Rose, Hilary, and Steven Rose. 2012. *Genes, Cells and Brains: The Promethean Promises of the New Biology.* London: Verso Books.

Rutter, Michael. 2012. "Achievements and Challenges in the Biology of Environmental Effects." *PNAS* 109 (supplement 2): 17151.

Schwartz, Harold. 1956. *Samuel Gridley Howe, Social Reformer, 1801–1876.* Cambridge, MA: Harvard University Press.

Shonkoff, Jack. 2012. "Leveraging the Biology of Adversity to Address the Roots of Disparities in Health and Development." *PNAS* 109 (supplement 2): 17302.

Stricker, Edward M. 2009. "2009 Survey of Neuroscience Graduate, Postdoctoral, and Undergraduate Programs." April 7, 2013, http://docslide.us/education /2009-survey-report-of-neuroscience-departments-and-programs-2009- .html.

Thompson, Heather Ann. 2013. "Criminalizing the Kids: The Overlooked Reason for Failing Schools." In *Public Education under Siege*, edited by Michael B. Katz and Mike Rose, 131. Philadelphia: University of Pennsylvania Press.

Wilson, Edward O. 1975. *Sociobiology: The New Synthesis*. Cambridge, MA: The Belknap Press of Harvard University Press.

Wolfe, Barbara, William Evans, and Teresa E. Seeman, eds. 2012. *The Biological Consequences of Socioeconomic Inequalities*. New York: Russell Sage Foundation.

2. Making Better Babies

Public Health and Race Betterment in Indiana, 1920–1935

Alexandra Minna Stern

By 8 a.m. on the morning of September 3, 1929, dozens of mothers were lined up in front of the Better Babies Building at the Indiana State Fair, eagerly waiting for the doors to open.[1] Since 1920, and in increasing numbers, babies from nearly every Indiana county had been weighed, measured, and tested at the state fair by physicians and psychologists affiliated with the State Board of Health's Division of Infant and Child Hygiene. During the 1920s, this division launched a multifaceted program of "child saving" and maternal education, which included radio talks, mother's classes, the screening of hygiene films, statistical reports, and consultation clinics. The Better Babies Contest, however, was by far the division's most spectacular and beloved event, drawing hundreds of young entrants and thousands of curious onlookers to the state fairgrounds during the week of Labor Day.

Each year, more and more Hoosiers—as Indianans like to refer to themselves—crowded into the Better Baby facilities. They watched nurses demonstrate infant feeding techniques, collected free pamphlets such as the *Indiana Mother's Baby Book,* or perused displays about nutrition and the virtues of sterilized and sparkling bathrooms and kitchens. While individual girls and boys, twins, and triplets competed for blue ribbons and cash prizes, tired mothers could find refuge at the rest tent, and noncontestant children could romp in the playground or nap peacefully in the nursery. According to many physicians, the *Indianapolis News,* and the promotional

"Making Better Babies: Public Health and Race Betterment in Indiana, 1920–1935." Alexandra Minna Stern. *American Journal of Public Health* 92, no. 5 (May 2002): 742–752. doi: 10.2105/AJPH.92.5.742. Reprinted by permission of Sheridan Press. Due to space limitations, endnotes of this chapter have been deleted or edited from the original.

newsletter the *Hub of the Universe,* the Better Babies Contest was one of Indiana's most anticipated yearly events.[2]

At the helm of the better babies program was Dr. Ada E. Schweitzer. Over the course of little more than a decade, Schweitzer, appointed director of the newly created Division of Infant and Child Hygiene in 1919, assembled one of the most vibrant public health agencies in the nation. Immediately before she was ousted and the division was disbanded in 1933, Schweitzer counted four physicians, four nurses, and five assistants on her core staff.[3]

During her fourteen-year reign, Schweitzer worked sedulously to lower infant and maternal death rates and convince Indianans of the importance of scientific motherhood and child rearing. She lectured to hundreds of neighborhood and civic associations, penned voluminous articles and poems, assessed the physical condition of babies in every one of the state's ninety-two counties, and fastidiously managed the affairs of her industrious division. Seemingly unfazed by a taxing travel schedule, Schweitzer could frequently be found adding miles to the division's child hygiene mobile, which had been equipped with a generator to project movies and lantern slides in remote towns and villages. She was even known to take to the air in a two-seater airplane to arrive punctually for speaking engagements.[4] In part owing to Schweitzer's efforts, Indiana's infant mortality dropped by one-third, from 8.2 percent in 1920 to 5.7 percent in 1930.[5]

In this article, I explore not only Schweitzer's better baby crusade but also the particular circumstances that gave rise to such a dynamic child welfare project in Indiana from 1919 to 1933. This work flourished because of the state's concern with public health and eugenics. By 1907, for example, Indiana had a pure food statute and a vital statistics act on the books and, furthermore, had passed the country's first eugenic sterilization law. In 1915, the Indiana State Board of Health was ranked sixth nationwide, in terms of effectiveness, by the American Medical Association.[6]

During this period many Indiana health reformers, including Schweitzer, frightened by what they perceived to be an escalating menace of the feebleminded, joined the Indiana State Mental Hygiene Association. Through legal and educational means, Indiana Progressives sought to control procreation and endorsed only the birth of the "best" and healthiest babies. For many Hoosiers, born and raised as farmers, breeding superior children was just a step away from producing heartier corn, pigs, and cattle.[7]

The activities of the Division of Infant and Child Hygiene multiplied markedly in the 1920s owing to the resources made available by the federal Sheppard–Towner Act, passed in 1921.[8] Administered by the US Children's

Bureau, this act provided matching funds to states that approved "enabling legislation" and established agencies devoted to infant and maternal welfare. Schweitzer, who had preexisting ties to the Children's Bureau, astutely took advantage of the support granted by Sheppard–Towner. With Schweitzer serving as the intermediary, the convergence of state and federal infant and maternal hygiene programs proved exceptionally efficacious in Indiana.

The success of the Division of Infant and Child Hygiene was also facilitated by Indiana's demography and topography. In 1920, the state's population hovered at close to three million residents, 95 percent of whom were native-born and 97 percent of whom were white. The bulk of Indiana's African American and immigrant communities lived a marginalized and segregated existence in the cities of Indianapolis, Gary, and East Chicago.[9] Unlike Progressive reformers in diverse, multilingual states such as New York, Illinois, and California, Schweitzer and other Indiana child savers did not need to translate their message into foreign languages or tailor their "Americanization" campaigns for Polish, Italian, or Mexican newcomers.[10] Instead, the primary targets of public health and race betterment efforts in Indiana were poor and working-class whites, especially impoverished farm dwellers living beyond the orbit of urbanization and industrialization.

Despite Indiana's unusual makeup, it has often been characterized as the quintessentially American state, a reputation most decidedly earned by the 1929 publication of Robert and Helen Lynd's *Middletown: A Study in Modern American Culture,* which examined the city of Muncie.[11] If Middletown encapsulated the values of America as idealized in the 1920s, then tracing the emergence of its better babies movement should reveal a great deal about the largely understudied interplay between public health and race betterment in the country as a whole during the first half of the 20th century.[12]

THE INDIANA CHILD CREED

> Every child has the inalienable right to be born free from disease, free from deformity and with pure blood in its veins and arteries.
>
> Every child has the inalienable right to be loved; to have its individuality respected; to be trained wisely in mind, body, and soul; to be protected from disease, from evil influences and evil persons; and to have a fair chance. In a word, to be brought up in the fear and admonition of the Lord.
>
> That state is delinquent which does not ceaselessly strive to secure these inalienable rights to its children. (Indiana Child Creed)

The Indiana Child Creed entered the Hoosier vernacular in 1915 when it debuted as the epigraph of the *Indiana Mothers' Baby Book*.[13] Published by the State Board of Health, this advice manual was distributed free of charge, along with a letter of introduction, to every mother who registered her newborn with the state.[14] Over the subsequent two decades, this creed, an awkward patchwork of eugenic, public health, Protestant, and Progressive ideas, would be printed in hundreds of articles and tracts and recited resolutely by the state's health advocates. Although it is impossible to gauge how many Hoosiers absorbed or heeded the child creed, its appearance signified the inception of better baby work in Indiana. Promoted on three interconnected levels—local, state, and federal—child welfare programs took shape and began to coalesce in the Hoosier heartland in the late 1910s and early 1920s.

Indiana's burgeoning interest in infant hygiene reflected broader trends at the turn of the century, as reformers from coast to coast began to embrace the doctrine of "progressive maternalism."[15] According to Molly Ladd-Taylor, progressive maternalists occupied the middle ground between feminists and proponents of sentimental motherhood. Whereas the former waged a fierce battle for sex equality on the streets and in the halls of Congress, the latter saw no place for women outside the home. Progressive maternalists combined and tempered these two perspectives, asserting that the biological and social experience of motherhood endowed women with a heightened sense of moral duty that was beneficial to both family and nation. They politicized maternity by arguing that female citizens, who carried the well-being of future generations in their wombs, were entitled to suffrage as well as leadership roles in society and government.

This logic was employed by suffragettes, who asserted that New Zealand's low infant mortality rate was a direct result of more than twenty years of female enfranchisement. One leaflet issued by the National Woman Suffrage Publishing Company, for instance, portrayed a toddler looking warily at a door that was barely ajar and swarming with deadly microbes. The accompanying caption read, "I wish my mother had a vote—to keep the germs away."[16] Linking the language of bacteriology and biology, progressive maternalists charted an agenda for the country in which national strength, better babies, and the political visibility of women went hand in hand.

On a national scale, the ethos of progressive maternalism was best exemplified by the US Children's Bureau. Established in 1912 by Progressives long committed to immigrant and infant welfare, the Children's Bureau was the first government agency directed by a woman, Julia Lathrop, who in

1921 was succeeded by Grace Abbott. Both were veterans of Chicago's Hull House, one of the first immigrant settlement homes in the United States.[17]

Immediately after taking charge of the Children's Bureau, Lathrop settled on the reduction of infant mortality as the agency's cardinal objective. In 1910, between one and two hundred of every thousand infants born in the United States perished, a figure that had been lowered discernibly since 1900 but still exceeded rates in New Zealand and a handful of European countries.[18]

Although the bureau interpreted the manifold problems of children through the prism of public health and medicine, few members of its staff were physicians. The preliminary composition and outlook of the bureau were altered substantially over time, however, as doctors—many of whom were male—began to claim jurisdiction over most arenas of children's health. One of the historical ironies is that by professionalizing infant and maternal welfare and urging mothers to consult private pediatricians, the Children's Bureau enhanced the authority of doctors and bolstered the notion that private primary care was the most creditable mode of child health.[19] The development of better baby work in Indiana emulated this pattern, arising in large part from a groundswell of women's volunteerism in the first decade of the twentieth century but thoroughly controlled by male pediatricians by the mid-1930s. Much to her dismay, Schweitzer, a physician *and* a progressive maternalist, paved the way for this gendered transfer of power in the Hoosier heartland.

Just two years after its founding, the Children's Bureau initiated what would blossom into a fruitful relationship with the Indiana State Board of Health when it dispatched its designated exhibit expert, Dr. Anne Louise Strong, to Indianapolis to preside over an upcoming child welfare display.[20] Two years later, in January 1916, the Children's Bureau returned to Indiana. This time, however, the agency came for four months, to lay the foundations for a comprehensive infant and maternal hygiene program. During the second decade of the twentieth century, the Children's Bureau rotated its field agents throughout the country to galvanize child welfare initiatives.[21] To rural Indiana, it sent Florence Brown Sherbon, a Kansas physician, who in the 1920s became a member of the American Eugenics Society and one of the most vocal exponents of "fitter families" contests at agricultural fairs. She was later joined by Mary Mills West, author of the acclaimed Children's Bureau tracts *Infant Care* and *Prenatal Care*, and Elizabeth Moore, who had helped Strong with the child welfare exhibit less than two years earlier.[22]

During the winter and spring of 1916, Sherbon, Moore, and West preached the gospel of child saving throughout Indiana. In places often inaccessible by rail or asphalt, they showed movies; handed out pamphlets; examined the eyes, ears, and mouths of children; illustrated infant feeding techniques; and dispensed scales of normal development and nutrition charts, all the while compiling birth and death statistics.[23] Following on the heels of American Farm Bureau agents, who had begun canvassing the countryside at the turn of the century to further modern agriculture, Sherbon and her team used the language of crops and breeding to persuade Hoosiers to apply scientific knowledge to the procreation and bringing up of children.[24] They implored farmers to shed superstition for science, vowing that if they heeded their instructions about milk sterilization, nourishment, and parturition, their sons and daughters would "grow up strong and well."[25]

The crux of the Children's Bureau mission in Indiana, however, was the "babies' health conferences" Sherbon and her colleagues orchestrated in towns and cities such as Lagrange, Butler, Kendalville, Petersburg, and Washington. According to the informational brochure, the aim of these gatherings was to "show the physical condition of the children examined and indicate the points at which their health and vigor may be improved by the efforts of the parents."[26] Spending four days in each locale, bureau representatives inspected children younger than six years and exposed Hoosiers to films, written materials, and visual aids.[27]

Aside from ushering in innovative notions about child rearing, one of the topmost goals of these conferences was to expedite the continuation of better baby work in Indiana. In every town on their itinerary, Sherbon and her team scrutinized the feasibility of follow-up efforts among local women's clubs and medical groups, many of which had independently begun grassroots campaigns.

ESTABLISHING THE DIVISION OF INFANT AND CHILD HYGIENE

These activities sparked moderate interest in child hygiene but certainly did not bring about the sweeping program envisioned by the Children's Bureau. The seeds for a more far-reaching plan were planted at the babies' health conference, held in conjunction with Indiana University's Extension University in Bloomington in March 1916.[28] Sherbon and her colleagues were convinced that the movement would be guaranteed "a much better chance for permanent survival" if it were housed in a state division, not scattered among local groups. Hence, at that meeting they broached the possibilities for such a public health unit with John N. Hurty, director of the Indiana State Board of Health from 1896 to 1922 and an ardent eugenicist and outspoken supporter of the sterilization and marriage laws.[29]

Not surprisingly, given his concerns with race betterment and child saving, Hurty responded enthusiastically and, in spite of an immediate lack of funds, conveyed his willingness to submit a proposal in the next legislative cycle.[30] Three years after these conversations and the Children's Bureau babies' health conferences, the Indiana State Board of Health's Division of Child and Infant Hygiene was created and Schweitzer was appointed its director.[31]

Schweitzer's relationship with Hurty and the Board of Health began in 1906, when she was hired as an assistant bacteriologist at the state laboratory. Born in 1873 in the northern town of Lagrange, Schweitzer grew up on a farm where her Scottish-Irish mother and German father raised mint and purebred poultry. After attending Lima High School, she left Indiana to obtain her baccalaureate at Michigan State Normal College, returning permanently in 1902 to pursue her medical degree at Indiana Medical College.

While in school, Schweitzer conducted bacteriological studies at the state laboratory, concentrating on the prevalence and morbidity of children's epidemics such as measles, diphtheria, and typhoid. Inspired by ideas of progressive maternalism and Hurty's principles of racial uplift, Schweitzer soon became Indiana's leading champion of infant hygiene, and by the second decade of the twentieth century she was spearheading child welfare projects. In 1916, for example, she represented the State Board of Health at the Children's Bureau conference in her hometown of Lagrange, delivering two talks titled "Personal Hygiene" and "Sanitation in the Home."[32]

In 1918 she was elected chairwoman of the Indiana branch of the American Association for the Study and Prevention of Infant Mortality.[33] That same year she authored a survey of infant mortality in Gary on special assignment for the Children's Bureau. In the fall of 1918, she traveled to the South to realize a similar investigation but soon was attending to the crisis engendered by the influenza outbreak. As she was vacating her temporary post in Georgia, Schweitzer received word of Hurty's invitation to become director of the Division of Infant and Child Hygiene.[34]

When the division's starting appropriation of ten thousand dollars became available in October 1919, Schweitzer and her staff of three—a nurse, a chauffeur, and a stenographer—swung into action.[35] In a detailed letter, Hurty boasted to Lathrop about the extensive and trailblazing endeavors of Schweitzer and her underlings. Hurty explained that Schweitzer was "carrying the news into the rural regions beyond the railways" and venturing "deep into the country" to find areas that had scarcely been reached by health officials or the *Indiana Mothers' Baby Book*. Above all, Hurty was proud of the division's child mobile, a Dodge truck outfitted with "a Delco electric apparatus" that lit up "country school houses or churches" and activated the "stereopticon and moving picture machine." In most towns, Schweitzer's entrance behind the wheel of the mobile—customarily adorned with flags—was a festive event always "announced by a bugle." In town after town the routine was repeated: "Mothers are invited to bring their babies for physical examination. Advice and circulars are given to them, and then the Division moves on to the next stand, which as said, is always advertised beforehand."[36]

As she launched Indiana's better baby movement, Schweitzer reenergized the mission of the Children's Bureau. Crisscrossing Indiana from county to county, Schweitzer and her corps handed out pamphlets, mounted exhibits, delivered lectures, screened films, and demonstrated techniques for nursing and preparing formula. Each month, Schweitzer personally scrutinized the health of hundreds of children—assessing their teeth, height, weight, vision, hearing, tonsils, adenoids, possible infections, defects, eating habits, hours of sleep, access to fresh air, and home surroundings.[37]

Schweitzer's 1920 annual report revealed that the division had convened conferences in twenty-seven counties, examined eight thousand children, and presented lectures or films in 290 towns. Continually striving for lay involvement, the division had collaborated with 476 local, 53 state, and 63 national organizations on joint projects.[38] Furthermore, by 1921, many Indiana mothers, worried about the health of their babies, had sent "Dear Dr. Schweitzer" letters to the division's headquarters at the State House.

If the division grew steadily in the early 1920s, it expanded exponentially after the passage of the federal Maternity and Infancy, or Sheppard–Towner Act. Although the Indiana State Medical Association—like its parent the American Medical Association—loudly opposed Sheppard–Towner and labeled it intrusive state medicine, Hurty's national prestige and Indiana's entrenched eugenic and public health programs guaranteed endorsement of the necessary "enabling legislation" by the state assembly.[39] With a budget three times her original one, in 1923 Schweitzer substantially broadened and reconfigured the division. She hired additional nurses and assistants, amplified the radius of the child health conferences, founded maternal and infancy centers, augmented public nursing efforts, and realized increasingly ambitious statistical and clinical studies.[40] As in similar agencies across the country, the bulk of her staff were female non-physicians, an arrangement that provoked the ire of a vocal segment of Indiana's predominantly male medical establishment.

During this period Schweitzer initiated mothers' classes to teach pregnant women the fundamentals of prenatal and baby care.[41] In addition to inculcating scientific motherhood and basic precepts of public hygiene, these courses also furnished a venue for Schweitzer to expound on the virtues of Indiana's eugenic marriage and sterilization laws, which she believed ensured the robustness of Hoosier babies.[42] In 1925, 16,649 women—more than 50 percent of all attendees nationwide—took mothers' classes in Indiana under the aegis of the Division of Infant and Child Hygiene.[43]

That same year, Schweitzer wrote to the Children's Bureau, "Our work is growing so fast that it is difficult to plan so far ahead."[44] By 1926, the division's operating funds had climbed to $60,000, and it counted more than twenty full-time and temporary employees.[45] Moreover, according to the census, Indiana's infant mortality rates had fallen to the fourth lowest in the country, a decrease due to several intertwined factors, including the division's campaigns.[46] After one decade, the division had examined the health of 77,584 children, enrolled 55,171 mothers in instructional classes, shown health films to 606,364 viewers, and reached almost half of the state's population of 3 million through the distribution of 1,216,577 pamphlets.[47]

Schweitzer's crusade indubitably altered attitudes about health, maternity, and childhood in Indiana. In Muncie, for example, the Lynds found that mothers were voracious readers of pamphlets and installment books on prenatal and infant care, always on the lookout for "every available resource for help in training their children."[48] Moreover, according to the Lynds, most Muncie parents readily embraced the latest pediatric advice.

Some, however, were bewildered by this avalanche of new instructional materials and were averse to renouncing tried and true practices that had been handed down from generation to generation through female relatives. Like the Children's Bureau during its 1916 tour of Indiana, since 1919 the Division of Infant and Child Hygiene had sought to persuade inhabitants of Muncie and the rest of Indiana of the imperative of the rules of scientific motherhood and child rearing. From the perspective of Schweitzer and other reformers, the integrity of Hoosier health and citizenship depended on the mass adaptation of infant and maternal hygiene. As Schweitzer was fond of saying, only this would enable Indiana to become a good parent.[49]

THE BETTER BABIES CONTESTS

The centerpiece of Schweitzer's quest to groom Indiana into an enlightened guardian of Hoosier children were the better babies contests, inaugurated in 1920 and, until their discontinuation in 1932, one of the most popular events at the state fair. The significance of the contests was layered and complex. As manifestations of the state fair in miniature, each year the better babies contest served as a venue for Hoosiers to negotiate past and present, nostalgia and modernity. They acquainted Indianans with the most up-to-date opinions of child specialists, thereby reinforcing emergent pediatric norms and imbuing university-trained experts with ultimate authority over matters pertaining to the biology, physiology, and psychology of children.

The contests also commercialized this process, through advertising in and sponsorship by the *Indianapolis News,* by soliciting patronage from businesses such as the Hoosier Fence Company and the Weber Milk Company, and by fostering a competitive climate in which the winner received cash prizes and a trophy.[50] Finally, while the contests bolstered professionalized child medicine and brand name consumerism, they simultaneously depicted babyhood as a time of innocence and purity that was under assault by twentieth-century urbanization and industrialization.

Moreover, by excluding African American children, the contests reinforced patterns of segregation in Indiana and promoted the idea that only white babies could achieve perfection and symbolize the Hoosier state.[51] Schweitzer reportedly ordained the contest "a school of education in eugenics" and countenanced the use of categories that made "some allowance for familial and racial types."[52] More implicitly than overtly, she furthered Indiana's racial divisions as she strove to improve the overall health of Hoosier children and modernize rural mothers through science.

Despite their immense popularity at the state fair, better babies contests did not originate in Indiana. Adumbrated by nineteenth-century beauty pageants, the contests began at the Iowa State Fair in 1911 when clubwoman Mary T. Watts asked, "You are raising better cattle, better horses, and better hogs, why don't you raise better babies?"[53] To judge infants like livestock, Watts and another rural reformer, Margaret Clark, devised scorecards that tallied level of physical health, anthropometric traits, and mental development. Soon thereafter, the widely read magazine *Woman's Home Companion* embarked on its Better Baby Campaign by sending one of its editors, Anna Steese Richardson, to Colorado to advance the contests.[54] Soon they were all the rage, and by 1914, *Woman's Home Companion* claimed "that contests had been held in every state except West Virginia, New Hampshire, and Utah, and that more than 100,000 children had been examined."[55]

With its fieldworkers already dispersed around the country taking part in local infant hygiene efforts, the Children's Bureau became involved in the contests as well. Lathrop, however, while supporting the educational aspect of the contests, was disturbed by the competitiveness they fostered, the commercialism they endorsed, and their glaring lack of a standardized scoring system. Thus, she arranged for the bureau to join forces with the American Medical Association to develop a scorecard acceptable to the pediatric establishment and also began to sponsor an alternative, the children's health conferences, which contained most of the elements of the better babies contests without numerical rankings. Indeed, during Sherbon's reconnoitering of Indiana in 1916, she and her colleagues complained in several towns about crowded, confusing, and ill-managed contests that had been inspired by *Woman's Home Companion*.[56] They hoped that the children's health conference would "successfully demonstrate a different method."[57]

Given her close ties to the Children's Bureau, Schweitzer was initially reluctant to incorporate better babies contests into her division's activities. Hence in 1920, when Charles F. Kennedy, then secretary of the State Board of Agriculture, proposed that she oversee a contest at the state fair that year, she evinced skepticism. Kennedy, who had 'conducted a similar contest in Grand Rapids, Mich., was convinced it would be a wonderful addition to the fair.[58] Schweitzer was soon swayed by Kennedy's petition and in 1920 presided over one of the fair's most crowd-pleasing features.[59] Within no time, she was a fervent defender of the contests, justifying them as completely professional endeavors, guided by the firm principles of pediatrics and child psychology. Explaining her decision to avidly back the contests at the state fair, Schweitzer wrote to Dr. Talafierro Clark of the US Public Health Service, "I had numerous consultations with men skilled in pediatrics and specialists" as "we needed to place the contest on as high a plane as possible, in order to free it as near as could be from objectionable features."[60]

Schweitzer consistently distinguished the state fair contest from its makeshift and unregulated imitations in small rural towns and villages. Schweitzer wanted all of Indiana's baby contests to be directed solely by the division and regularly pleaded with rural reformers to erect an alternative, the baby rest tent, where toddlers would be shielded from dust, crowds, possible exploitation, and the disappointment of losing in an amateur competition.[61] To meet the benchmark of professionalism, she used a scorecard based directly on the template formulated by the American Medical Association and the Children's Bureau.[62]

By the mid-1920s, Schweitzer was showcasing the better babies contests on the radio and in articles in the *Monthly Bulletin of the ISBH* and the *Hub of the Universe.* She contended that the value of the contests resided in the fact that they "set the best standards of health before the parents that they may compare these with the actual condition of their child."[63] For the most part, Schweitzer believed that better babies contests provided a level playing field on which infants could be judged according to their own merits.

While Schweitzer certainly viewed the contests as a facet of a more extensive race betterment project, she alleged that the "gates of heredity" were closed after the baby left the womb. It was essential to first restrict birth to only the most fit, through marriage and sterilization laws, and then create only the most desirable children through scientific child rearing and motherhood. Reflecting her particular blend of eugenics and public health, Schweitzer told one Muncie reformer, "You can not make a silk purse out of a sow's ear, neither can we make a citizen out of an idiot or any person who is not well born."[64]

MAKING INDIANA A GOOD PARENT

Schweitzer frequently extolled the benefits of the contests, claiming, for example, that the lessons taught by the contests had helped to lower the percentage of underweight contestant babies from 10 percent in 1920 to 2 percent in 1929.[65] To publicize this annual September event, she wrote announcements, published fact sheets, and explained scoring procedures in laborious detail.[66] With each passing year, the contests became more popular among Indianans. In 1920, for example, 78 babies were examined; by 1925 this number had risen more than tenfold, to 885, and in 1930 when 1,301 young entrants were counted, enthusiasm was so overwhelming that Schweitzer opted to cap the number of entrants at 1,200 the following year.[67]

In 1923 the *Indianapolis News* began to sponsor the contest, giving it a big boost; not only did the paper devote more space to articles, but it also began to print full-length pages with individual photos of hundreds of contestant babies whose mothers had sent in their registration forms by the deadline.[68] In 1924 the contest grounds were enlarged when a window-paneled Better Babies Building was erected, thanks to a ten thousand dollar donation from J. E. Oliver of Oliver Chilled Plow Works. This new edifice housed exhibits, examinations of noncontestant babies, and demonstrations.[69] In 1927, the contests themselves were moved from a partition of the Woman's Building, where they had been held faithfully since 1920, to a brand-new Better Babies Contest Building. Constructed as part of the state fair's Diamond Jubilee, this building was financed by a special five thousand dollar appropriation from the State Board of Agriculture and the legislature.[70]

As the contest grew under Schweitzer's commanding presence, it also became more streamlined and efficient. For the weeklong event, she contracted a general pediatrician—almost always Dr. James C. Carter—as well as an optometrist, an otorhinolaryngologist, and several extra nurses and orderlies. In addition, assistance was provided by the Girl and Boy Scouts, who escorted mothers from station to station.

The contest procedure was well honed. Before the event began, the registered infants were divided into groups based on age (12–24 months or 24–36 months), sex, and place of residence. Those categorized as city babies lived in places with ten thousand inhabitants or more; the others were considered rural. With their children classified, parents—usually mothers—arrived at the better babies complex at the state fair at a designated time. As the mothers entered the building, they submitted their enrollment forms to a female attendant, who recorded the names of the mother and child. Then the baby was whisked away to the next booth, where its overall health history was taken by a nurse.

Mental tests—distinct for each age group—followed. Psychologists observed whether the children could stand, walk, and speak; how they manipulated blocks and balls; and how they responded to questions such as "How does the doggie do?" and "Who is the baby in the mirror?"[71] Mental tests completed, the babies were undressed and their clothes placed in a paper bag with an identifying number. Identically robed in flannel togas, the babies were weighed and measured. From here each baby passed from the optometrist to the general pediatrician and finally to the otorhinolaryngologist. After being weighed and measured a second time, each baby was dressed in his or her own clothing and bedecked with a bronze medal on a blue ribbon, courtesy of the *Indianapolis News*.

Scores were calculated along the way. From a starting score of 1000, deductions were calculated for a wide host of physical defects including unevenness of the head, scaly skin, ill deportment, delayed teething, abnormal ear size or shape, and enlarged glands. Slow reactions to the mental tests or perceived lack of muscular coordination lowered a child's score, as did deviations from the national standards for height and weight (based on age) and weight-to-height ratio. Tabulated results from the contests indicate that Schweitzer instructed her team to subtract the most infinitesimal of figures for each defect—most likely to maintain high results for every baby, thus diluting the rivalrous nature of the contests. The victor generally scored above 990; Alma Louise Strohmeyer, a one-year-old Indianapolis girl, triumphed with 999.92813 points in 1923.[72]

In spite of their professed formality and orderliness, the contests were both crowded and noisy. In 1928, for example, sixty-seven thousand people streamed through the better babies complex, watching the psychological tests, spending time at the rest tent and nursery, and taking in the infant and maternal care dioramas.[73] As thousands made their way through the contest rooms, babies howled as they were unclothed, squealed when prodded by the stethoscope, or cooed delightedly when given their blue ribbon.[74] In 1925, the *News* described the chaos as a packed room filled to overflowing with the "noisy accompaniment of more than 200 child voices. The perfection of the lungs of the babies examined Monday could scarcely be doubted, and if there are not future opera singers and booming voiced orators of the group, many of the attendants and onlookers are exceedingly poor prophets."[75]

While the impact of the contests is difficult to gauge, it is not unreasonable to accept Schweitzer's contention that they played a part in effectively reducing infant mortality rates and prompting mothers to safeguard against the bacterial infection of milk and food. The contests also provided a platform for the commercialization of public health as well as the incorporation of the "better baby" into advertising—a newborn icon that figured regularly in the 1920s, selling products such as condensed milk and infant formula. Schweitzer's correspondence, furthermore, reveals that not only did many Hoosiers regularly send her general inquiries about child rearing, but many mothers—of varying degrees of literacy—were concerned enough about their children's contest scores to contact the division. In 1922 one parent wrote to Schweitzer anxious to know her daughter's "failing points" and to find out "in what way she failed a perfect score."[76] That same year, Schweitzer received a letter from another set of fretful parents: "We are so anxious to know her defects and in what way she was lacking."[77]

These and numerous additional letters demonstrate the extent to which Indianans from every inch of the state viewed Schweitzer as a trusted expert who could direct, or at least make recommendations about, local child-saving events. They also suggest that for those who attended the

FIGURE 2.1.   Contestants (probably winners) from the 1927 Better Babies Contest, accompanied by Division of Infant and Child Hygiene nurses. (Photo courtesy of the Indiana State Archives, Indiana Commission on Public Records.)

FIGURE 2.2.   Baby Contest Building and spectators, Indiana State Fair, 1929. (Photo courtesy of the Indiana State Archives, Indiana Commission on Public Records.)

FIGURE 2.3. Spectators watching the various testing and measurement tables at the 1930 contest. (Photo courtesy of the Indiana State Archives, Indiana Commission on Public Records.)

division's many conferences, workshops, and classes, the line between public health and eugenics was nebulous or nonexistent. On the one hand, Schweitzer implored Hoosiers to adhere to the state's marriage laws and spoke out consistently in favor of the state's sterilization restrictions, which were based on a Mendelian understanding of hereditary transmission. On the other, she was just as beholden to the gospels of private hygiene, pure milk, vaccination programs, and clean air and sunshine. For Schweitzer and hundreds of other reformers, particularly the progressive maternalists, these multiple and seemingly paradoxical aspects of infant welfare and scientific motherhood coexisted quite comfortably on a wide continuum of race betterment.

EPILOGUE: A NEW DEAL FOR HOOSIER BABIES

In 1932 the last better babies contest was held, attracting thousands of spectators. In the early 1930s, the Depression, the resentment of Indiana's male pediatricians, and the election of a new Democratic administration converged, setting the stage for the end of the Division of Infant and Child Hygiene. Intent on centralizing his New Deal plan and distancing himself from his Republican predecessors, in 1933 Governor Paul V. McNutt abruptly dismissed Schweitzer and transferred the newly named

Department of Child Health and Maternal Welfare to the Indiana University School of Medicine.

Of the division's more than twenty employees, only Dr. James C. Carter—the pediatrician Schweitzer hired each year to examine better babies—was retained to serve on a committee charged with designing a new blueprint for child welfare in Indiana.[78] None of the female physicians or nurses who had so faithfully staffed Schweitzer's division for more than a decade were asked to join McNutt's revamped department, which emphasized clinical pediatric teaching instead of hands-on infant and maternal hygiene projects. Nonetheless, Schweitzer's legacy was felt in 1936 when McNutt oversaw the passage of legislation to receive Title V funds through the Social Security Act and partially revived the division. A Bureau of Maternal and Child Health, both federally and state financed in a manner akin to Sheppard–Towner, was founded, and a male physician closely affiliated with the Indiana State Medical Association was named its director.[79]

Many historians of twentieth-century America conceptualize public health and eugenics as antithetical movements separated by the conceptual gulf between environment and heredity.[80] Schweitzer's work in Indiana, and better babies contests across the country more generally, illustrate that race betterment was an expansive rubric with a great deal of space for overlapping ideas and practices. The Division of Infant and Child Hygiene's experiment demonstrates the active leadership of female reformers in infant and maternal welfare in the 1920s as well as the problematic racial and class implications of making babies better in Indiana.

NOTES

1. "Hopeful Mothers and Fathers Bring Children to Baby Contest," *Indianapolis News,* September 4, 1929, 1, 14.

2. See *Hub of the Universe, Monthly Bulletin of the Indiana State Board of Health (ISBH),* and the *Indianapolis News,* especially editorials from 1920 through 1932.

3. See "Division of Infant and Child Hygiene, Monthly Report, September 1932," *Monthly Bulletin of the ISBH* 35 (1932): 154–55.

4. Thurman B. Rice, *The Hoosier Health Officer: A Biography of Dr. John N. Hurty and the History of the Indiana State Board of Health to 1925* (Indianapolis, 1946), 316.

5. James H. Madison, *Indiana through Tradition and Change: A History of the Hoosier State and Its People, 1920–1945* (Indianapolis: Indiana Historical Society, 1982), 322. Rates from 1910 to 1925 listed in "State Fair Better Babies Demonstrations," 11–16–1, Central File (CF) 1925–28, Record Group (RG) 102, United States Children's Bureau (CB), National Archives at College Park (NACP)....

6. Madison, *Indiana through Tradition,* 309.

7. See Marilyn Irvin Holt, *Linoleum, Better Babies, and the Modern Farm Woman, 1890–1930* (Albuquerque: University of New Mexico Press, 1995), chap. 4; and Lynne Curry, *Modern Mothers in the Heartland: Gender, Health, and Progress in Illinois, 1900–1930* (Columbus: Ohio State University Press, 1999): 101–7.

8. See Richard A. Meckel, *Save the Babies: American Public Health Reform and the Prevention of Infant Mortality, 1850–1929* (Ann Arbor: University of Michigan Press, 1998).

9. Madison, *Indiana through Tradition*, chap. 1. Also see John Bartlow Martin, *Indiana: An Interpretation* (Bloomington: Indiana University Press, [1947] 1992).

10. See Howard Markel, "For the Welfare of Children: The Origins of the Relationship between US Public Health Workers and Pediatricians," *American Journal of Public Health* 90 (June 2000): 893–99.

11. Robert S. Lynd and Helen Merrell Lynd, *Middletown: A Study in Modern American Culture* (New York: Harcourt, Brace, [1929] 1957).

12. This article was inspired by the work of Martin S. Pernick, one of the few historians of medicine to trace the overlaps between public health and eugenics. See Pernick, "Eugenics and Public Health in American History," *American Journal of Public Health* 87 (1997): 1767–72.

13. *Indiana Mothers' Baby Book*, 2nd ed. (Indianapolis: Indiana State Board of Health, 1920).

14. See *Indiana Mothers' Baby Book* and John N. Hurty to Children's Bureau, June 18, 1920, 4–15–2-16, CF 1914–1920, RG 102, CB, NACP.

15. See Molly Ladd-Taylor, *Mother-Work: Women, Child Welfare, and the State, 1890–1930* (Urbana: University of Illinois Press, 1994); and Molly Ladd-Taylor, ed., *Raising a Baby the Government Way: Mothers' Letters to the Children's Bureau, 1915–1932* (New Brunswick, NJ: Rutgers University Press, 1986). . . .

16. "Better Babies" (New York: National Woman Suffrage, 1916).

17. See Kriste Lindenmeyer, *"Right to Childhood": The U.S. Children's Bureau and Child Welfare, 1912–46* (Urbana: University of Illinois Press, 1997); Ladd-Taylor, *Mother-Work* and *Raising a Baby*.

18. Lindenmeyer, *"Right to Childhood,"* 43–45.

19. See Meckel, *Save the Babies*; Lindenmeyer, *"Right to Childhood"*; and Jeffrey P. Baker, "Women and the Invention of Well Child Care," *Pediatrics* 94 (1994): 527–31.

20. See King to Lathrop, October 15, 1914; Lathrop to King, October 17, 1914; King to Lathrop, November 25, 1914; King to Lathrop, January 15, 1915; 8–1-4–2-2, CF 1914–20, RG 102, CB, NACP.

21. See Lindenmeyer, *"Right to Childhood,"* and Ladd-Taylor, *Mother-Work*.

22. See West to Sherbon, February 26, 1916, and Lathrop to West, February 15, 1916, 4–11–1-5, CF 1914–20, RG 102, CB, NACP.

23. See letters and reports from January to April 1916 in file 4–11–1-5, CF 1914–20, RG 102, CB, NACP.

24. See Holt, *Linoleum*, chaps. 1 and 4.

25. "Indiana II," January 5, 1916, 4–11–1-4, CF 1914–20, RG 102, CB, NACP.

26. "Babies' Health Conferences," 4–11–1-5, CF 1914–20, RG 102, CB, NACP.

27. Ibid. . . .

28. "Community Institute and Babies' Health Conference," February 15–18, 1916, 4–11–1-5, CF 1914–20, RG 102, CB, NACP.

29. Grace L. Meigs, director of the Child Hygiene Division, Children's Bureau, to Sherbon, March 10, 1916, 4–11–1-5, CF 1914–20, RG 102, CB, NACP.

30. Sherbon to Meigs, March 19, 1916, and Meigs to Sherbon, March 22, 1916, 4–11–1-5, CF 1914–20, RG 102, CB, NACP.

31. Hurty to Lathrop, May 8, 1919, 4–15–2-16, CF 1914–20, RG 102, NACP. . . .

32. "Community Institute and Babies' Health Conference," February 15–18, 1916, 4–11–1-5, CF 1914–20, RG 102, CB, NACP.

33. See Schweitzer to Lathrop, September 5, 1918, 9–1-2–3, CF 1914–20, RG 102, CB, NACP.

34. Rude to Schweitzer, December 12, 1918, 4–15–2-16, CF 1914–20, RG 102, CB, NACP.

35. Hurty to Lathrop, May 8, 1919, 4–15–2-16, CF 1914–20, RG 102, CB, NACP.

36. Hurty to Children's Bureau, June 18, 1920. Also see "Indiana Progress," n.d., 4–15–2-16, CF 1914–20, RG 102, CB, NACP.

37. "Report of the Division of Infant and Child Hygiene, Indiana State Board of Health, for the Year Ending September 30, 1920," 4–11–1-3 (16), CF 1921–24, RG 102, CB, NACP. . . .

38. "Annual Report of the Division of Infant and Child Hygiene, Indiana State Board of Health for the Year Ending September 30, 1921," 4–11–1-3, CF 1921–24, RG 102, CB, NACP.

39. See Rice, *Hoosier Health Officer.*

40. See "Report of the Division of Infant and Child Hygiene," *Monthly Bulletin of the ISBH* 26 (1923): 39–40; "Indiana State Board of Health, Division, Infant and Child Hygiene, Ordinance, Graf, Rules," 11–16–1, CF 1921–24, RG 102, CB, NACP.

41. "Supplementary Report for Information Concerning Plans for the Promotion of Maternal and Infant Welfare," 11–16–1, CF 1921–24, NACP.

42. "Abstract of Lectures for Mothers' Classes," 11–16–1, CF 1925–28, RG 102, CB, NACP; "Narrative Report of Maternity and Infancy Staff No. 2 for Month Ending March 31, 1924 by Dr. Wilhelmina Jongewaard, Director," *Monthly Bulletin of the ISBH* 27 (1924): 57–59. See Rima D. Apple, "Constructing Mothers: Scientific Motherhood in the Nineteenth and Twentieth Centuries," in *Mothers and Motherhood: Readings in American History,* ed. Rima D. Apple and Janet Golden, 90–110 (Columbus: Ohio State University Press, 1997).

43. See "Indiana's Work under the Maternity and Infancy Law during 1925," *Monthly Bulletin of the Indiana State Board of Health* 29 (1926): 136–38.

44. Schweitzer to Florence E. Kraker, associate director, Maternal and Infant Hygiene, CB, 21 January 1925, 11–16–1, CF 1925–28, RG 102, CB, NACP.

45. "Child Hygiene Division, Estimate Budget for 1925, 1926, and 1927," 11–16–1, CF 1925–28, RG 102, CB, NACP.

46. Schweitzer to Haines, September 28, 1926, 11–16–1, CF 1925–28, RG 102, CB, NACP.

47. "A Survey of Ten Years' Child Hygiene Work in Indiana," *Monthly Bulletin of the ISBH* 32 (1929): 173–74.

48. Lynd and Lynd, *Middletown,* 150.

49. Schweitzer, "Is Indiana a Good Parent," *Hub of the Universe* 6, no. 5 (August 1928): 1.

50. See "The Better Babies," *Monthly Bulletin of the ISBH* 31 (1928): 144–45.

51. The archival materials I consulted indicate that the contests were segregated in practice but not on paper. . . .

52. Paul Miner, *Indiana's Best! An Illustrated Celebration of the Indiana State Fairgrounds, 1852–1992* (Indianapolis: Indiana State Fair Commission, 1992), 128–30.

53. Watts to Schweitzer, June 17, 1925, DICH, ISA. See Annette K. Vance Dorey, *Better Baby Contests: The Scientific Quest for Perfect Childhood Health in the Early Twentieth Century* (Jefferson, NC: McFarland, 1999). . . .

54. See Alisa Klaus, *"Every Child a Lion:" The Origins of Maternal and Infant Health Policy in the United States and France, 1890–1920* (Ithaca, NY: Cornell University Press, 1993): 138–57. . . .

55. Ibid., 144.

56. Meigs to Sherbon, February 3, 1916; Sherbon to Meigs, January 27, 1916; 4–11–1-5, CF 1914–20, RG 102, CB, NACP.

57. Ibid.

58. Kennedy to Schweitzer, April 5, 1920, DICH, ISA; "Indiana State Fair Better Babies Activities," *Monthly Bulletin of the ISBH* 30 (1927): 136–40.

59. Miner, *Indiana's Best!*, 129–30.

60. Schweitzer to Clark, October 22, 1922, DICH, ISA.

61. Schweitzer to Clark, October 22, 1922, DICH, ISA.; Schweitzer to Mrs J. E. Pepple, June 26, 1923; Schweitzer to Miss E. Melville, July 5, 1923, DICH, ISA.

62. See Schweitzer to Mrs A. F. Bentley, June 18 and 20, 1923, DICH, ISA.

63. Schweitzer, "Why Have a Baby Contest?" *Monthly Bulletin of the ISBH* 31 (1928): 125.

64. Schweitzer to Mr George B. Lockwood, March 20, 1916, DICH, ISA.

65. "Better and Better Babies" (radio script), DICH, ISA. . . .

66. See "Better Babies at the State Fair" (several versions); "Better Babies Contest History and Rules" and "Method of Counting Scores in Baby Contests," DICH, ISA; "Meditation of a Second-Summer Baby," *Hub of the Universe* 6, no. 4 (July 1923): 1, 4.

67. "Growth of the State Fair Better Baby Work," DICH, ISA.

68. "Indiana State Fair Better Baby Activities," *Monthly Bulletin of the ISBH* 30 (1927): 136–40. . . .

69. "Indiana State Fair Better Baby Activities," *Monthly Bulletin of the ISBH* 30 (1927): 136–40; "The Better Babies Building at the Indiana State Fair," *Indianapolis News* 27, 1924, 138.

70. Schweitzer to Miss Ora Marshino, September 21, 1927, 11–16–1, CF 1925–28, RG 102, CB, NACP; "The State Fair Better Babies, 1927," *Monthly Bulletin of the ISBH* 30 (1927): 110–11.

71. "Proud Relatives Watch Better Babies Examined at Fair," *Indianapolis News,* September 5, 1927, 17; "Mothers and Babies on Hand Early at State Fair Contest," *Indianapolis News,* September 3, 1923, 1.

72. "Alma Louise Strohmeyer Best Baby Entered in State Contest," *Indianapolis News,* September 10, 1923, 1.

73. "The Better Babies," *Monthly Bulletin of the ISBH* 31 (1928): 144–45.

74. "Willy, Nilly, Every Contest Baby Undergoes Same Test," *Indianapolis News,* September 8, 1931, 1; "Babies and Their 'Trainers' Enter Ring at the State Fair," *Indianapolis News,* September 1, 1924, 1.

75. "Lung Power of the Better Babies Contest Entrants 100 Percent, Despite All Else," *Indianapolis News,* September 8, 1925, 10.

76. Mrs Cecil Rawlings to Schweitzer, September 24, 1922, DICH, ISA.

77. Mrs Stephen Sprong to Schweitzer, October 1922, DICH, ISA.

78. "The Indiana Plan for Child Health and Maternal Welfare," *Monthly Bulletin of the ISBH* 36 (1933): 86–87.

79. Madison, *Indiana through Tradition*, 325.

80. Exceptions include Pernick, "Eugenics and Public Health"; and Kathy J. Cooke, "The Limits of Heredity: Nature and Nurture in American Eugenics before 1915," *Journal of the History of Biology* 31 (1998): 263–78.

3. Eugenics and the Nazis

The California Connection

Edwin Black

Hitler and his henchmen victimized an entire continent and exterminated millions in his quest for a so-called Master Race.

But the concept of a white, blond-haired, blue-eyed master Nordic race didn't originate with Hitler. The idea was created in the United States, and cultivated in California, decades before Hitler came to power. California eugenicists played an important, although little-known, role in the American eugenics movement's campaign for ethnic cleansing.

Eugenics was the pseudoscience aimed at "improving" the human race. In its extreme, racist form, this meant wiping away all human beings deemed "unfit," preserving only those who conformed to a Nordic stereotype. Elements of the philosophy were enshrined as national policy by forced sterilization and segregation laws, as well as marriage restrictions, enacted in twenty-seven states. In 1909, California became the third state to adopt such laws. Ultimately, eugenics practitioners coercively sterilized some sixty thousand Americans, barred the marriage of thousands, forcibly segregated thousands in "colonies," and persecuted untold numbers in ways we are just learning. Before World War II, nearly half of coercive sterilizations were done in California, and even after the war, the state accounted for a third of all such surgeries.

From a syndicated column drawn from *War Against the Weak: Eugenics and America's Campaign to Create a Master Race*—Expanded Edition by Edwin Black. Copyright 2003 and 2012 Edwin Black. Original publication titled "Eugenics and the Nazis—The California Connection." *San Francisco Chronicle.* November 9, 2013. http://www.sfgate.com/opinion/article/Eugenics-and-the-Nazis-the-California-2549771.php. Portions of this chapter have been edited from the original with permission from the author.

California was considered an epicenter of the American eugenics movement. During the twentieth century's first decades, California's eugenicists included potent but little-known race scientists, such as Army venereal disease specialist Dr. Paul Popenoe, citrus magnate Paul Gosney, Sacramento banker Charles Goethe, as well as members of the California State Board of Charities and Corrections and the University of California Board of Regents.

Eugenics would have been so much bizarre parlor talk had it not been for extensive financing by corporate philanthropies, specifically the Carnegie Institution, the Rockefeller Foundation and the Harriman railroad fortune. They were all in league with some of America's most respected scientists from such prestigious universities as Stanford, Yale, Harvard and Princeton. These academicians espoused race theory and race science, and then faked and twisted data to serve eugenics' racist aims.

Stanford president David Starr Jordan originated the notion of "race and blood" in his 1902 racial epistle "Blood of a Nation," in which the university scholar declared that human qualities and conditions such as talent and poverty were passed through the blood.

In 1904, the Carnegie Institution established a laboratory complex at Cold Spring Harbor on Long Island that stockpiled millions of index cards on ordinary Americans, as researchers carefully plotted the removal of families, bloodlines, and whole peoples. From Cold Spring Harbor, eugenics advocates agitated in the legislatures of America, as well as in the nation's social service agencies and associations.

The Harriman railroad fortune paid local charities, such as the New York Bureau of Industries and Immigration, to seek out Jewish, Italian, and other immigrants in New York and other crowded cities and subject them to deportation, confinement, or forced sterilization.

The Rockefeller Foundation helped found the German eugenics program and even funded the program that Josef Mengele worked in before he went to Auschwitz.

Much of the spiritual guidance and political agitation for the American eugenics movement came from California's quasi-autonomous eugenic societies, such as Pasadena's Human Betterment Foundation and the California branch of the American Eugenics Society, which coordinated much of their activity with the Eugenics Research Society in Long Island. These organizations—which functioned as part of a closely knit network—published racist eugenic newsletters and pseudoscientific journals, such as *Eugenical News* and *Eugenics*, and propagandized for the Nazis.

Eugenics was born as a scientific curiosity in the Victorian age. In 1863, Sir Francis Galton, a cousin of Charles Darwin, theorized that if talented people married only other talented people, the result would be measurably better offspring. At the turn of the last century, Galton's ideas were imported to the United States just as Gregor Mendel's principles of heredity were rediscovered. American eugenics advocates believed with religious fervor that the same Mendelian concepts determining the color and size of peas, corn, and cattle also governed the social and intellectual character of man.

In a United States demographically reeling from immigration upheaval and torn by post-Reconstruction chaos, race conflict was everywhere in the early twentieth century. Elitists, utopians, and so-called progressives fused their smoldering race fears and class bias with their desire to make a better world. They reinvented Galton's eugenics into a repressive and racist ideology. The intent: Populate the Earth with vastly more of their own socioeconomic and biological kind—and less or none of everyone else.

The superior species the eugenics movement sought was populated not merely by tall, strong, talented people. Eugenicists craved blond, blue-eyed Nordic types. This group alone, they believed, was fit to inherit the Earth. In the process, the movement intended to subtract emancipated Negroes, immigrant Asian laborers, Indians, Hispanics, East Europeans, Jews, dark-haired hill folk, poor people, the infirm, and anyone classified outside the gentrified genetic lines drawn up by American raceologists.

How? By identifying so-called defective family trees and subjecting them to lifelong segregation and sterilization programs to kill their bloodlines. The grand plan was to literally wipe away the reproductive capability of those deemed weak and inferior—the so-called unfit. The eugenicists hoped to neutralize the viability of 10 percent of the population at a sweep, until none were left except themselves.

Eighteen solutions were explored in a Carnegie-supported 1911 "Preliminary Report of the Committee of the Eugenic Section of the American Breeder's Association to Study and to Report on the Best Practical Means for Cutting Off the Defective Germ-Plasm in the Human Population." Point No. 8 was euthanasia.

The most commonly suggested method of eugenicide in the United States was a "lethal chamber" or public, locally operated gas chambers. In 1918, Popenoe, the Army venereal disease specialist during World War I, cowrote the widely used textbook *Applied Eugenics*, which argued, "From an historical point of view, the first method which presents itself is execution. . . . Its value in keeping up the standard of the race should not be underestimated." *Applied Eugenics* also devoted a chapter to "Lethal Selection," which operated "through the destruction of the individual by some adverse feature of the environment, such as excessive cold, or bacteria, or by bodily deficiency."

Eugenic breeders believed American society was not ready to implement an organized lethal solution. But many mental institutions and doctors practiced improvised medical lethality and passive euthanasia on their own. One institution in Lincoln, Ill., fed its incoming patients milk from tubercular cows believing a eugenically strong individual would be immune. Thirty to 40 percent annual death rates resulted at Lincoln. Some doctors practiced passive eugenicide one newborn infant at a time. Others doctors at mental institutions engaged in lethal neglect.

Nonetheless, with eugenicide marginalized, the main solution for eugenicists was the rapid expansion of forced segregation and sterilization, as well as more marriage restrictions. California led the nation, performing nearly all sterilization procedures with little or no due process. In its first twenty-five years of eugenics legislation, California sterilized 9,782 individuals, mostly women. Many were classified as "bad girls," diagnosed as "passionate," "oversexed," or "sexually wayward." At the Sonoma State Home, some women were sterilized because of what was deemed an abnormally large clitoris or labia.

In 1933 alone, at least 1,278 coercive sterilizations were performed, 700 on women. The state's two leading sterilization mills in 1933 were Sonoma State Home with 388 operations and Patton State Hospital with 363 operations. Other sterilization centers included Agnews, Mendocino, Napa, Norwalk, Stockton, and Pacific Colony state hospitals.

Even the US Supreme Court endorsed aspects of eugenics. In its infamous 1927 decision, Supreme Court Justice Oliver Wendell Holmes wrote, "It is better for all the world, if instead of waiting to execute degenerate offspring for crime, or to let them starve for their imbecility, society can prevent those who are manifestly unfit from continuing their kind. . . . Three generations of imbeciles are enough." This decision opened the floodgates for thousands to be coercively sterilized or otherwise persecuted as subhuman. Years later, the Nazis at the Nuremberg trials quoted Holmes's words in their own defense.

Only after eugenics became entrenched in the United States was the campaign transplanted into Germany, in no small measure through the efforts of California eugenicists, who published booklets idealizing sterilization and circulated them to German officials and scientists.

Hitler studied American eugenics laws. He tried to legitimize his anti-Semitism by medicalizing it, and wrapping it in the more palatable pseudoscientific facade of eugenics. Hitler was able to recruit more followers among reasonable Germans by claiming that science was on his side. Hitler's race hatred sprung from his own mind, but the intellectual outlines of the eugenics Hitler adopted in 1924 were made in America.

During the 1920s, Carnegie Institution eugenic scientists cultivated deep personal and professional relationships with Germany's fascist eugenicists. In *Mein Kampf*, published in 1924, Hitler quoted American eugenic ideology and openly displayed a thorough knowledge of American eugenics. "There is today one state," wrote Hitler, "in which at least weak beginnings toward a better conception (of immigration) are noticeable. Of course, it is not our model German Republic, but the United States." Hitler proudly told his comrades just how closely he followed the progress of the American eugenics movement. "I have studied with great interest," he told a fellow Nazi, "the laws of several American states concerning prevention of reproduction by people whose progeny would, in all probability, be of no value or be injurious to the racial stock."

Hitler even wrote a fan letter to American eugenics leader Madison Grant, calling his race-based eugenics book, *The Passing of the Great Race*, his "bible."

Now, the American term "Nordic" was freely exchanged with "Germanic" or "Aryan." Race science, racial purity, and racial dominance became the driving force behind Hitler's Nazism. Nazi eugenics would ultimately dictate who would be persecuted in a Reich-dominated Europe, how people would live, and how they would die. Nazi doctors would become the unseen generals in Hitler's war against the Jews and other Europeans deemed inferior. Doctors would create the science, devise the eugenic formulas, and hand select the victims for sterilization, euthanasia, and mass extermination.

During the Reich's early years, eugenicists across America welcomed Hitler's plans as the logical fulfillment of their own decades of research and effort. California eugenicists republished Nazi propaganda for American consumption. They also arranged for Nazi scientific exhibits, such as an August 1934 display at the L.A. County Museum, for the annual meeting of the American Public Health Association.

In 1934, as Germany's sterilizations were accelerating beyond five thousand per month, the California eugenics leader C.M. Goethe, upon returning from Germany, ebulliently bragged to a colleague, "You will be interested to know that your work has played a powerful part in shaping the opinions of the group of intellectuals who are behind Hitler in this epochmaking program. Everywhere I sensed that their opinions have been tremendously stimulated by American thought.... I want you, my dear friend, to carry this thought with you for the rest of your life, that you have really jolted into action a great government of 60 million people."

That same year, ten years after Virginia passed its sterilization act, Joseph DeJarnette, superintendent of Virginia's Western State Hospital, observed in the *Richmond Times-Dispatch*, "The Germans are beating us at our own game."

More than just providing the scientific roadmap, America funded Germany's eugenic institutions.

By 1926, Rockefeller had donated some $410,000—almost $4 million in today's money—to hundreds of German researchers. In May 1926, Rockefeller awarded $250,000 toward creation of the Kaiser Wilhelm Institute for Psychiatry. Among the leading psychiatrists at the German Psychiatric Institute was Ernst Rüdin, who became director and eventually an architect of Hitler's systematic medical repression.

Another in the Kaiser Wilhelm Institute's complex of eugenics institutions was the Institute for Brain Research. Since 1915, it had operated out of a single room. Everything changed when Rockefeller money arrived in 1929. A grant of $317,000 allowed the institute to construct a major building and take center stage in German race biology. The institute received additional grants from the Rockefeller Foundation during the next several years. Leading the institute, once again, was Hitler's medical henchman Ernst Rüdin. Rüdin's organization became a prime director and recipient of the murderous experimentation and research conducted on Jews, Gypsies, and others.

Beginning in 1940, thousands of Germans taken from old age homes, mental institutions, and other custodial facilities were systematically gassed. Between fifty and a hundred thousand were eventually killed.

Leon Whitney, executive secretary of the American Eugenics Society, declared of Nazism, "While we were pussy-footing around . . . the Germans were calling a spade a spade."

A special recipient of Rockefeller funding was the Kaiser Wilhelm Institute for Anthropology, Human Heredity and Eugenics in Berlin. For decades, American eugenicists had craved twins to advance their research into heredity. The institute was now prepared to undertake such research on an unprecedented level. On May 13, 1932, the Rockefeller Foundation in New York dispatched a radiogram to its Paris office:

JUNE MEETING EXECUTIVE COMMITTEE NINE THOUSAND
DOLLARS OVER THREE YEAR PERIOD TO KWG INSTITUTE
ANTHROPOLOGY FOR RESEARCH ON TWINS AND EFFECTS
ON LATER GENERATIONS OF SUBSTANCES TOXIC FOR
GERM PLASM.

At the time of Rockefeller's endowment, Otmar Freiherr von Verschuer, a hero in American eugenics circles, functioned as a head of the Institute for Anthropology, Human Heredity and Eugenics. Rockefeller funding of that institute continued both directly and through other research conduits during Verschuer's early tenure. In 1935, Verschuer left the institute to form a rival eugenics facility in Frankfurt that was much heralded in the American eugenics press. Research on twins in the Third Reich exploded, backed by government decrees. Verschuer wrote in *Der Erbarzt*, a eugenics doctor's journal he edited, that Germany's war would yield a "total solution to the Jewish problem."

Verschuer had a longtime assistant. His name was Josef Mengele.

On May 30, 1943, Mengele arrived at Auschwitz. Verschuer notified the German Research Society, "My assistant, Dr. Josef Mengele (M.D., Ph.D.) joined me in this branch of research. He is presently employed as Hauptsturmführer (captain) and camp physician in the Auschwitz concentration camp. Anthropological testing of the most diverse racial groups in this concentration camp is being carried out with permission of the SS Reichsführer (Himmler)."

Mengele began searching the boxcar arrivals for twins. When he found them, he performed beastly experiments, scrupulously wrote up the reports, and sent the paperwork back to Verschuer's institute for evaluation. Often, cadavers, eyes, and other body parts were also dispatched to Berlin's eugenic institutes.

Rockefeller executives never knew of Mengele. With few exceptions, the foundation had ceased all eugenics studies in Nazi-occupied Europe before the war erupted in 1939. But by that time the die had been cast. The talented men Rockefeller and Carnegie financed, the great institutions they helped found, and the science they helped create took on a scientific momentum of their own.

After the war, eugenics was declared a crime against humanity—an act of genocide. Germans were tried and they cited the California statutes in their defense—to no avail. They were found guilty.

However, Mengele's boss Verschuer escaped prosecution. Verschuer reestablished his connections with California eugenicists, who had gone underground and renamed their crusade "human genetics." Typical was an exchange on July 25, 1946, when Popenoe wrote Verschuer, "It was indeed a pleasure to hear from you again. I have been very anxious about my colleagues in Germany. . . . I suppose sterilization has been discontinued in Germany?" Popenoe offered tidbits about various American eugenics luminaries and then sent various eugenics publications. In a separate package, Popenoe sent some cocoa, coffee, and other goodies.

Verschuer wrote back, "Your very friendly letter of 7/25 gave me a great deal of pleasure and you have my heartfelt thanks for it. The letter builds another bridge between your and my scientific work; I hope that this bridge will never again collapse but rather make possible valuable mutual enrichment and stimulation."

Soon, Verschuer again became a respected scientist in Germany and around the world. In 1949, he became a corresponding member of the newly formed American Society of Human Genetics, organized by American eugenicists and geneticists.

In the fall of 1950, the University of Münster offered Verschuer a position at its new Institute of Human Genetics, where he later became a dean. In the early and mid-1950s, Verschuer became an honorary member of numerous prestigious societies, including the Italian Society of Genetics, the Anthropological Society of Vienna, and the Japanese Society for Human Genetics.

Human genetics' genocidal roots in eugenics were ignored by a victorious generation that refused to link itself to the crimes of Nazism and by succeeding generations that never knew the truth of the years leading up to war. Now governors of five states, including California, have issued public apologies to their citizens, past and present, for sterilization and other abuses spawned by the eugenics movement.

Human genetics became an enlightened endeavor in the late twentieth century. Hardworking, devoted scientists finally cracked the human code through the Human Genome Project. Now, every individual can be biologically identified and classified by trait and ancestry. Yet even now, some leading voices in the genetic world are calling for a cleansing of the unwanted among us, and even a master human species.

There is understandable wariness about more ordinary forms of abuse, for example, in denying insurance or employment based on genetic tests. On Oct. 14 [2008], the United States' first genetic antidiscrimination legislation passed the Senate by unanimous vote [and was later signed into law by President George W. Bush]. Yet because genetics research is global, no single nation's law can stop the threats.

4. Why the Nazis Studied American Race Laws for Inspiration

James Q. Whitman

On June 1934, about a year and a half after Adolf Hitler became chancellor of the Reich, the leading lawyers of Nazi Germany gathered at a meeting to plan what would become the Nuremberg Laws, the centerpiece anti-Jewish legislation of the Nazi race regime. The meeting was an important one, and a stenographer was present to take down a verbatim transcript, to be preserved by the ever-diligent Nazi bureaucracy as a record of a crucial moment in the creation of the new race regime.

That transcript reveals a startling fact: the meeting involved lengthy discussions of the law of the United States of America. At its very opening, the minister of justice presented a memorandum on US race law and, as the meeting progressed, the participants turned to the US example repeatedly. They debated whether they should bring Jim Crow segregation to the Third Reich. They engaged in detailed discussion of the statutes from the thirty US states that criminalized racially mixed marriages. They reviewed how the various US states determined who counted as a "Negro" or a "Mongol," and weighed whether they should adopt US techniques in their own approach to determining who counted as a Jew. Throughout the meeting the most ardent supporters of the US model were the most radical Nazis in the room.

The record of that meeting is only one piece of evidence in an unexamined history that is sure to make Americans cringe. Throughout the early 1930s, the years of the making of the Nuremberg Laws, Nazi policymakers

This article originally appeared in *Aeon* as "Why the Nazis Studied American Race Laws for Inspiration" by James Q. Whitman, December 13, 2016. https://aeon .co/ideas/why-the-nazis-studied-american-race-laws-for-inspiration. It has been republished under a Creative Commons License.

looked to US law for inspiration. Hitler himself, in *Mein Kampf* (1925), described the United States as "the one state" that had made progress toward the creation of a healthy racist society, and after the Nazis seized power in 1933 they continued to cite and ponder US models regularly. They saw many things to despise in US constitutional values, to be sure. But they also saw many things to admire in US white supremacy, and when the Nuremberg Laws were promulgated in 1935, it is almost certainly the case that they reflected direct US influence.

This story might seem incredible. Why would the Nazis have felt the need to take lessons in racism from anybody? Why, most especially, would they have looked to the United States? Whatever its failings, after all, the United States is the home of a great liberal and democratic tradition. Moreover, the Jews of the United States—however many obstacles they might have confronted in the early twentieth century—never faced state-sponsored persecution. And, in the end, Americans made immense sacrifices in the struggle to defeat Hitler.

But the reality is that, in the early twentieth century, the United States, with its vigorous and creative legal culture, led the world in racist lawmaking. That was not only true of the Jim Crow South. It was true on the national level as well. The United States had race-based immigration law, admired by racists all over the world; and the Nazis, like their right-wing European successors today (and so many US voters) were obsessed with the dangers posed by immigration.

The United States stood alone in the world for the harshness of its anti-miscegenation laws, which not only prohibited racially mixed marriages but also threatened mixed-race couples with severe criminal punishment. Again, this was not law confined to the South. It was found all over the United States: Nazi lawyers carefully studied the statutes, not only of states such as Virginia, but also of states such as Montana. It is true that the United States did not persecute the Jews—or at least, as one Nazi lawyer remarked in 1936, it had not persecuted the Jews "so far"—but it had created a host of forms of second-class citizenship for other minority groups, including Chinese, Japanese, Filipinos, Puerto Ricans and Native Americans, scattered all over the Union and its colonies. American forms of second-class citizenship were of great interest to Nazi policy makers as they set out to craft their own forms of second-class citizenship for the German Jewry.

Not least, the United States was the greatest economic and cultural power in the world after 1918—dynamic, modern, wealthy. Hitler and other Nazis envied the United States, and wanted to learn how the

Americans did it; it's no great surprise that they believed that what had made America great was American racism.

Of course, however ugly American race law might have been, there was no American model for Nazi extermination camps, even if the Nazis often expressed their admiration for the American conquest of the West, when, as Hitler declared, the settlers had "shot down the millions of Redskins to a few hundred thousand." In any case extermination camps were not the issue during the early 1930s, when the Nuremberg Laws were framed. The Nazis were not yet contemplating mass murder. Their aim at the time was to compel the Jews by whatever means possible to flee Germany, in order to preserve the Third Reich as a pure "Aryan" country.

And here they were indeed convinced that they could identify American models—and some strange American heroes. For a young Nazi lawyer named Heinrich Krieger, for example, who had studied at the University of Arkansas as an exchange student, and whose diligent research on US race law formed the basis for the work of the Nazi Ministry of Justice, the great American heroes were Thomas Jefferson and Abraham Lincoln. Did not Jefferson say, in 1821, that it is certain "that the two races, equally free, cannot live in the same government"? Did not Lincoln often declare, before 1864, that the only real hope of America lay in the resettlement of the black population somewhere else? For a Nazi who believed that Germany's only hope lay in the forced emigration of the Jews, these could seem like shining examples.

None of this is entirely easy to talk about. It is hard to overcome our sense that if we influenced Nazism we have polluted ourselves in ways that can never be cleansed. Nevertheless the evidence is there, and we cannot read it out of either German or American history.

5. Constructing Normalcy

The Bell Curve, the Novel, and the Invention
of the Disabled Body in the Nineteenth Century

Lennard J. Davis

We live in a world of norms. Each of us endeavors to be normal or else deliberately tries to avoid that state. We consider what the average person does, thinks, earns, or consumes. We rank our intelligence, our cholesterol level, our weight, height, sex drive, bodily dimensions along some conceptual line from subnormal to above average. There is probably no area of contemporary life in which some idea of a norm, mean, or average has not been calculated.

To understand the disabled body, one must return to the concept of the norm, the normal body. So much of writing about disability has focused on the disabled person as the object of study, just as the study of race has focused on the person of color. But as with recent scholarship on race, which has turned its attention to whiteness, I would like to focus not so much on the construction of disability as on the construction of normalcy. I do this because the "problem" is not the person with disabilities; the problem is the way that normalcy is constructed to create the "problem" of the disabled person.

A common assumption would be that some concept of the norm must have always existed. After all, people seem to have an inherent desire to compare themselves to others. But the idea of a norm is less a condition of human nature than it is a feature of a certain kind of society. Recent work on the ancient Greeks, on preindustrial Europe, and on tribal peoples, for example, shows that disability was once regarded very differently from the way it is now. As we will see, the social process of disabling arrived with industrialization and with the set of practices and discourses that are linked to late eighteenth- and nineteenth-century notions of nationality, race, gender, criminality, sexual orientation, and so on.

I begin with the rather remarkable fact that the constellation of words describing this concept—"normal," "normalcy," "normality," "norm," "average," "abnormal"—all entered the European languages rather late in human history. The word "normal" as "constituting, conforming to, not deviating or different from, the common type or standard, regular, usual" only enters the English language around 1840. (Previously, the word had meant "perpendicular"; the carpenter's square, called a "norm," provided the root meaning.) Likewise, the word "norm," in the modern sense, has only been in use since around 1855, and "normality" and "normalcy" appeared in 1849 and 1857, respectively. If the lexicographical information is relevant, it is possible to date the coming into consciousness in English of an idea of "the norm" over the period 1840–1860.

If we rethink our assumptions about the universality of the concept of the norm, what we might arrive at is the concept that preceded it: that of the "ideal," a word we find dating from the seventeenth century. Without making too simplistic a division in the historical chronotope, one can nevertheless try to imagine a world in which the hegemony of normalcy does not exist. Rather, what we have is the ideal body, as exemplified in the tradition of nude Venuses, for example. This idea presents a mythopoetic body that is linked to that of the gods (in traditions in which the god's body is visualized). This divine body, then, this ideal body, is not attainable by a human. The notion of an ideal implies that, in this case, the human body as visualized in art or imagination must be composed from the ideal parts of living models. These models individually can never embody the ideal since an ideal, by definition, can never be found in this world. When ideal human bodies occur, they do so in mythology. So Venus or Helen of Troy, for example, would be the embodiment of female physical beauty. . . .

By contrast, the grotesque as a visual form was inversely related to the concept of the ideal and its corollary that all bodies are in some sense disabled. In that mode, the grotesque is a signifier of the people, of common life. As Bakhtin, Stallybrass and White, and others have shown, the use of the grotesque had a life-affirming transgressive quality in its inversion of the political hierarchy. However, the grotesque was not equivalent to the disabled, since, for example, it is impossible to think of people with disabilities now being used as architectural decorations as the grotesque were on the façades of cathedrals throughout Europe. The grotesque permeated culture and signified common humanity, whereas the disabled body, a later concept, was formulated as by definition excluded from culture, society, the norm.

If the concept of the norm or average enters European culture, or at least the European languages, only in the nineteenth century, one has to ask what is the cause of this conceptualization? One of the logical places to turn in trying to understand concepts like "norm" and "average" is that branch of knowledge known as statistics. Statistics begins in the early modern period as "political arithmetic"—a use of data for "promotion of sound, well-informed state policy" (Porter 1986, 18). The word *statistik* was first used in 1749 by Gottfried Achenwall, in the context of compiling information about the state. The concept migrated somewhat from the state to the body when Bisset Hawkins defined medical statistics in 1829 as "the application of numbers to illustrate the natural history of health and disease" (cited in ibid., 24). In France, statistics were mainly used in the area of public health in the early nineteenth century. The connection between the body and industry is tellingly revealed in the fact that the leading members of the first British statistical societies formed in the 1830s and 1840s were industrialists or had close ties to industry (ibid., 32).

It was the French statistician Adolphe Quetelet (1796–1847) who contributed the most to a generalized notion of the normal as an imperative. He noticed that the "law of error," used by astronomers to locate a star by plotting all the sightings and then averaging the errors, could be equally applied to the distribution of human features such as height and weight. He then took a further step of formulating the concept of *"l'homme moyen"* or the average man. Quetelet maintained that this abstract human was the average of all human attributes in a given country. For the average man, Quetelet wrote in 1835, "all things will occur in conformity with the mean results obtained for a society. If one seeks to establish, in some way, the basis of a social physics, it is he whom one should consider" (cited in ibid, 53). Quetelet's average man was a combination of *l'homme moyen physique* and *l'homme moyen morale*, both a physically average and a morally average construct.

The social implications of this idea are central. In formulating the idea of *l'homme moyen*, Quetelet is also providing a justification for *les classes moyens*. With bourgeois hegemony comes scientific justification for moderation and middle-class ideology. The average man, the body of the man in the middle, becomes the exemplar of the middle way of life. Quetelet was apparently influenced by the philosopher Victor Cousin in developing an analogy between the notion of an average man and the *juste milieu*. This term was associated with Louis Philippe's July monarchy—a concept that melded bourgeois hegemony with the constitutional monarchy and celebrated moderation and middleness (ibid., 101). In England too, the middle class as the middle way or mean had been searching for a scientific justification. . . .

Statements of ideology of this kind saw the bourgeoisie as rationally placed in the mean position in the great order of things, developing the kind of science that would then justify the notion of a norm. With such thinking, the average then becomes paradoxically a kind of ideal, a position devoutly to be wished. . . .

The concept of a norm, unlike that of an ideal, implies that the majority of the population must or should somehow be part of the norm. The norm pins down that majority of the population that falls under the arch of the standard bell-shaped curve. This curve, the graph of an exponential function, that was known variously as the astronomer's "error law," the "normal distribution," the "Gaussian density function," or simply "the bell curve," became in its own way a symbol of the tyranny of the norm. Any bell curve will always have at its extremities those characteristics that deviate from the norm. So, with the concept of the norm comes the concept of deviations or extremes. When we think of bodies, in a society where the concept of the norm is operative, then people with disabilities will be thought of as deviants. This, as we have seen, is in contrast to societies with the concept of an ideal, in which all people have a nonideal status.

In England, there was an official and unofficial burst of interest in statistics during the 1830s. . . . The use of statistics began an important movement. . . . The rather amazing fact is that almost all the early statisticians had one thing in common: they were eugenicists. The same is true of key figures in the movement: Sir Francis Galton, Karl Pearson, and R. A. Fisher. While this coincidence seems almost too striking to be true, we must remember that there is a real connection between figuring the statistical measure of humans and then hoping to improve humans so that deviations from the norm diminish—as someone like Quetelet had suggested. Statistics is bound up with eugenics because the central insight of statistics is the idea that a population can be normed. An important consequence of the idea of the norm is that it divides the total population into standard and nonstandard subpopulations. The next step in conceiving of the population as norm and non-norm is for the state to attempt to norm the nonstandard—the aim of eugenics. . . .

MacKenzie asserts that it is not so much that Galton's statistics made possible eugenics but rather that "the needs of eugenics in large part determined the content of Galton's statistical theory" (1981, 52). In any case, a symbiotic relationship exists between statistical science and eugenic concerns. Both bring into society the concept of a norm, particularly a normal body, and thus in effect create the concept of the disabled body.

It is also worth noting the interesting triangulation of eugenicist interests. On the one hand Sir Francis Galton was cousin to Charles Darwin, whose notion of the evolutionary advantage of the fittest lays the foundation for eugenics and also for the idea of a perfectible body undergoing progressive improvement. As one scholar has put it, "Eugenics was in reality applied biology based on the central biological theory of the day, namely the Darwinian theory of evolution" (Farrall 1985, 55). Darwin's ideas serve to place disabled people along the wayside as evolutionary defectives to be surpassed by natural selection. So, eugenics became obsessed with the elimination of "defectives," a category which included the "feebleminded," the deaf, the blind, the physically defective, and so on. . . .

Galton made significant changes in statistical theory that created the concept of the norm. He took what had been called "error theory," a technique by which astronomers attempted to show that one could locate a star by taking into account the variety of sightings. The sightings, all of which could not be correct, if plotted would fall into a bell curve, with most sightings falling into the center, that is to say, the correct location of the star. The errors would fall to the sides of the bell curve. Galton's contribution to statistics was to change the name of the curve from "the law of frequency of error" or "error curve," the term used by Quetelet, to the "normal distribution" curve. (See figure 5.1.)

The significance of these changes relates directly to Galton's eugenicist interests. In an "error curve" the extremes of the curve are the most mistaken in accuracy. But if one is looking at human traits, then the extremes, particularly what Galton saw as positive extremes—tallness, high intelligence, ambitiousness, strength, fertility—would have to be seen as errors. Rather than "errors," Galton wanted to think of the extremes as distributions of a trait. As MacKenzie notes:

> Thus there was a gradual transition from use of the term "probable error" to the term "standard deviation" (which is free of the implication that a deviation is in any sense an error), and from the term "law of error" to the term "normal distribution" (1981, 59).

But even without the idea of error, Galton still faced the problem that in a normal distribution curve that graphed height, for example, both tallness and shortness would be seen as extremes in a continuum where average

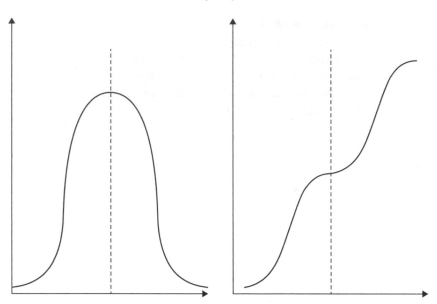

FIGURES 5.1 AND 5.2. *Left,* Adolphe Quetlet's "error curve," which Francis Galton renamed the "normal distribution" curve. *Right,* Francis Galton's "ogive" curve, which used quartiles to rank "desired" traits like tallness and high intelligence above "undesired" deviations. (Figures courtesy of Lennard J. Davis.)

stature would be the norm. The problem for Galton was that, given his desire to perfect the human race, or at least its British segment, tallness was preferable to shortness. How could both extremes be considered equally deviant from the norm? So Galton substituted the idea of ranking for the concept of averaging. That is, he changed the way one might look at the curve from one that used the mean to one that used the median—a significant change in thinking eugenically.

If a trait, say intelligence, is considered by its average, then the majority of people would determine what intelligence should be—and intelligence would be defined by the mediocre middle. Galton, wanting to avoid the middling of desired traits, would prefer to think of intelligence in ranked order. Although high intelligence in a normal distribution would simply be an extreme, under a ranked system it would become the highest ranked trait. Galton divided his curve into quartiles, so that he was able to emphasize ranked orders of intelligence, as we would say that someone was in the first quartile in intelligence (low intelligence) or the fourth quartile (high intelligence). Galton's work led directly to current "intelligence quotient" (IQ) and scholastic achievement tests. In fact, Galton revised Gauss's bell curve to show the superiority of the desired trait (for example, high intelligence). He created what he called an "ogive," which is arranged in quartiles with an ascending curve that features the desired trait as "higher" than the undesirable deviation. . . . (See figure 5.2.)

What these revisions by Galton signify is an attempt to redefine the concept of the "ideal" in relation to the general population. First, the application of the idea of a norm to the human body creates the idea of deviance or a "deviant" body. Second, the idea of a norm pushes the normal variation of the body through a stricter template guiding the way the body "should" be. Third, the revision of the "normal curve of distribution" into quartiles, ranked in order, and so on, creates a new kind of "ideal." This statistical ideal is unlike the classical ideal which contains no imperative to be the ideal. The new ideal of ranked order is powered by the imperative of the norm, and then is supplemented by the notion of progress, human perfectibility, and the elimination of deviance, to create a dominating, hegemonic vision of what the human body should be.

While we tend to associate eugenics with a Nazi-like racial supremacy, it is important to realize that eugenics was not the trade of a fringe group of right-wing, fascist maniacs. Rather, it became the common practice of many, if not most, European and American citizens. When Marx used Quetelet's idea of the average in his formulation of average wage and abstract labor, socialists as well as others embraced eugenic claims, seeing in the perfectibility of the human body a Utopian hope for social improvement. Once people allowed that there were norms and ranks in human physiology, then the idea that we might want to, for example, increase the intelligence of humans, or decrease birth defects, did not seem so farfetched. These ideas were widely influential. . . . The influence of eugenicist ideas persisted well into the twentieth century, so that someone like Emma Goldman could write that unless birth control was encouraged, the state would "legally encourage the increase of paupers, syphilitics, epileptics, dipsomaniacs, cripples, criminals, and degenerates" (Kevles 1985, 90).

The problem for people with disabilities was that eugenicists tended to group together all allegedly "undesirable" traits. So, for example, criminals, the poor, and people with disabilities might be mentioned in the same breath. Take Karl Pearson, a leading figure in the eugenics movement, who defined the "unfit" as follows: "the habitual criminal, the professional tramp, the tuberculous, the insane, the mentally defective, the alcoholic, the diseased from birth or from excess" (cited in ibid., 33). In 1911, Pearson headed the Department of Applied Statistics, which included the Galton and Biometric Laboratories at University College in London. This department gathered eugenic information on the inheritance of physical and mental traits including "scientific, commercial, and legal ability, but also hermaphroditism, hemophilia, cleft palate, harelip, tuberculosis, diabetes, deafmutism, polydactyly (more than five fingers) or brachydactyly (stub fingers), insanity, and mental deficiency" (ibid., 38–39). Here again one sees a strange selection of disabilities merged with other types of human variations. All of these deviations from the norm were regarded in the long run as contributing to the disease of the nation. As one official in the Eugenics Record Office asserted:

> the calculus of correlations is the sole rational and effective method for attacking . . . what makes for, and what mars national fitness. . . . The only way to keep a nation strong mentally and physically is to see that each new generation is derived chiefly from the fitter members of the generation before (ibid., 39–40).

The emphasis on nation and national fitness obviously plays into the metaphor of the body. If individual citizens are not fit, if they do not fit into the nation, then the national body will not be fit. Of course, such arguments are based on a false notion of the body politic—as if a hunchbacked citizenry would make a hunchbacked nation. Nevertheless, the eugenic notion that individual variations would accumulate into a composite national identity was a powerful one. This belief combined with an industrial mentality that saw workers as interchangeable and therefore sought to create a universal worker whose physical characteristics would be uniform, as would the result of their labors—a uniform product.

One of the central foci of eugenics was what was broadly called "feeblemindedness." This term included low intelligence, mental illness, and even "pauperism"; since low income was equated with "relative inefficiency" (ibid., 46). Likewise, certain ethnic groups were associated with feeblemindedness and pauperism. Charles Davenport, an American eugenicist, thought that the influx of European immigrants would make the American population "darker in pigmentation, smaller in stature . . . more given to crimes of larceny, assault, murder, rape, and sex-immorality." In his research, Davenport scrutinized the records of "prisons, hospitals, almshouses, and institutions for the mentally deficient, the deaf, the blind, and the insane" (ibid., 55).

The loose association between what we would now call disability and criminal activity, mental incompetence, sexual license, and so on established a legacy that people with disabilities are still having trouble living down. This equation was so strong that an American journalist writing in the early twentieth century could celebrate "the inspiring, the wonderful, message of the new heredity" as opposed to the sorrow of bearing children who were "diseased or crippled or depraved" (ibid., 67). The conflation of disability with depravity expressed itself in the formulation "defective class." As the president of the University of Wisconsin declared after World War One, "we know enough about eugenics so that if the knowledge were applied, the defective classes would disappear within a generation" (ibid., 68). And it must be reiterated that the eugenics movement was not stocked with eccentrics. Davenport was funded by Averell Harriman's sister Mary Harriman, as well as John D. Rockefeller, prime ministers A.J. Balfour, Neville Chamberlain, and Winston Churchill. President Theodore Roosevelt, H.G. Wells, John Maynard Keynes, and H.J. Laski, among many others, were members of eugenicist organizations. Francis Galton was knighted in 1909 for his work, and in 1910 he received the Copley Medal, the Royal Society's highest honor. A Galton Society met regularly in the American Museum of Natural History in New York City. In 1911 the Oxford University Union moved approval of the main principles behind eugenics by a vote of almost two to one. In Kansas, the 1920 state fair held a contest for "fitter families" based on their eugenic family histories, administered intelligence tests, medical examinations, and venereal disease tests. A brochure for the contest noted about the awards, "this trophy and medal are worth more than livestock sweepstakes.... For health is wealth and a sound mind in a sound body is the most priceless of human possessions" (ibid., 62).

In England, bills were introduced in Parliament to control mentally disabled people, and in 1933 the prestigious scientific magazine *Nature* approved the Nazis' proposal of a bill for "the avoidance of inherited diseases in posterity" by sterilizing the disabled. The magazine editorial said "the Bill, as it reads, will command the appreciative attention of all who are interested in the controlled and deliberate improvement of human stock." The list of disabilities for which sterilization would be appropriate were "congenital feeblemindedness, manic depressive insanity, schizophrenia, hereditary epilepsy, hereditary St. Vitus's dance, hereditary blindness and deafness, hereditary bodily malformation and habitual alcoholism" (cited in MacKenzie 1981, 44).

We have largely forgotten that what Hitler did in developing a hideous policy of eugenics was just to implement the theories of the British and American eugenicists. Hitler's statement in *Mein Kampf* that "the struggle for the daily livelihood [between species] leaves behind, in the ruck, everything that is weak or diseased or wavering" is not qualitatively different from any of the many similar statements we have seen before (cited in Blacker 1952, 143). And even the conclusions Hitler draws are not very different from those of the likes of Galton, Bell, and others....

I have tried to show here that the very term that permeates our contemporary life—the Normal—is a configuration that arises in a particular historical moment. It is part of a notion of progress, of industrialization, and of ideological consolidation of the power of the bourgeoisie. The implications of the hegemony of normalcy are profound and extend into the very heart of cultural production. . . . One of the tasks for a developing consciousness of disability issues is the attempt, then, to reverse the hegemony of the normal and to institute alternative ways of thinking about the abnormal.

REFERENCES

Bell, Alexander Graham. 1969. *Memoir upon the Formation of a Deaf Variety of the Human Race.* Washington, DC: Alexander Graham Bell Association for the Deaf.

Blacker, C. P. 1952. *Eugenics: Galton and After.* Cambridge, MA: Harvard University Press.

Conrad, Joseph. 1924. "An Outpost of Progress." In *Tales of Unrest.* Garden City, NJ: Doubleday, Page.

———. 1924. *Youth.* Garden City, NJ: Doubleday, Page.

———. [1957] 1989. *Under Western Eyes.* London: Penguin.

———. [1968] 1990. *The Secret Agent.* London: Penguin.

———. 1986. *Lord Jim.* London: Penguin.

Defoe, Daniel. 1975. *Robinson Crusoe.* New York: Norton.

Farrall, Lyndsay Andrew. 1985. *The Origin and Growth of the English Eugenics Movement 1865–1925.* New York: Garland.

Flaubert, Gustave. 1965. *Madam Bovary.* Trans. Paul de Man. New York: Norton.

Freud, Sigmund. 1977. *Introductory Lectures on Psychoanalysis.* Trans. James Strachey. New York: Norton.

Kevles, Daniel J. 1985. *In the Name of Eugenics: Genetics and the Uses of Human Heredity.* New York: Alfred A. Knopf.

MacKenzie, Donald. A. 1981. *Statistics in Britain, 1865–1930.* Edinburgh: Edinburgh University Press.

Marx, Karl. 1970. *Capital.* Vol. 1. Trans. Samuel Moore and Edward Aveling. New York: International.

Porter, Theodore M. 1986. *The Rise of Statistical Thinking 1820–1900.* Princeton, NJ: Princeton University Press.

Stallybrass, Peter, and Allon White. 1987. *The Politics of Transgression.* Ithaca, NY: Cornell University Press.

Stigler, Stephen M. 1986. *The History of Statistics: The Measurement of Uncertainty before 1900.* Cambridge, MA: Harvard University Press.

Zola, Emile. 1964. *The Experimental Novel and Other Essays.* Trans. Belle M. Sherman. New York: Haskel House.

———. 1993. *The Masterpiece.* Trans. Thomas Walton. London: Oxford University Press.

6. The Eugenics Legacy of the Nobelist Who Fathered IVF

Osagie K. Obasogie

Robert G. Edwards might not be a household name, but the innovation he pioneered along with Patrick Steptoe certainly is. In vitro fertilization (IVF), the process whereby human eggs are fertilized outside of the body and the resulting embryos implanted in a woman's womb, led to the 1978 birth of Louise Brown—the world's first "test tube baby." To date, an estimated five million children worldwide have been born using this innovation. Edwards received the 2010 Nobel Prize in Physiology or Medicine for this remarkable achievement.

Edwards's passing (in early 2013) prompted an outpouring of praise. He [was] widely described as a maverick researcher disinterested in personal recognition, who simply wanted to give babies to those who couldn't make them on their own. The *New York Times* quoted Edwards's former collaborator, Barry Bavister, as saying "Dr. Edwards's motivation—his passion, in fact—was not fame or fortune but rather helping infertile women." Bavister continued, "He believed with all his heart that it was the right thing to do."

But Edwards's views on the technology he created and the uses to which it should be put may be more complicated than this portrayal. One detail omitted from the obituaries published around the world was that Edwards was a member in good standing of the Eugenics Society in Britain for much of his career. Recently uncovered documents show that Edwards served on the organization's council—its leadership body—as a trustee on three

Adapted with permission. Copyright © 2013. Scientific American, a division of Nature America, Inc. All rights reserved. Originally published by Osagie K. Obasogie, "The Eugenics Legacy of the Nobelist Who Fathered IVF. " *Scientific American.* October 4, 2013. https://www.scientificamerican.com/article/eugenic-legacy-nobel-ivf/.

73

separate occasions: from 1968 to 1970, from 1971 to 1973, and once again from 1995 to 1997 after the group euphemistically renamed itself "The Galton Institute" for the founder of the eugenics movement, Francis Galton. As we consider Edwards's legacy in light of his passing, it is important to think critically about the relationship between Edwards's development of IVF and his participation in an organization that was dedicated to promoting one of the most dangerous ideas in human history: that science should be used to control human reproduction in order to breed preferred types of people.

Coined by Galton in the late 1800s to mean "wellborn," eugenics became a dominant aspect of Western intellectual life and social policy during the first half of the twentieth century. It started with the seemingly simple proposition that one's social position is rooted in heritable qualities of character and intellect.

Eugenicists of that era also believed that people with what they considered the least desirable traits tend to have the most children, precipitating what they saw as an inevitable decline in a society's intellectual and physical vigor. Taking their cue from livestock breeders, eugenicists argued that socially disadvantageous characteristics could be bred out of human populations through policies that limited the reproduction of "the unfit"—the "feebleminded," the poor, and the weak. Many eugenicists considered these qualities to be more prevalent among racial and ethnic minorities.

Eugenics has had disturbing implications since its inception. Galton characterized it in 1883 as the "science of improving stock . . . to give to the more suitable races or strains of blood a better chance of prevailing speedily over the less suitable than they otherwise would have had." Framed in this manner, eugenics has held an appeal to dictatorial and authoritarian regimes seeking to eradicate or discourage the growth of disfavored groups. The Nazis invoked eugenics to justify the extermination of people with disabilities, Jews, and other marginalized populations. And as a social philosophy eugenics also received approbation and financial support from the wealthy and other elites—particularly in the United States. It was followed by the likes of philanthropists John D. Rockefeller and Nobel Prize–winning scientists such as William Shockley and Alexis Carrel. It even gained the support of women's health advocates such as Margaret Sanger.

Eugenics was duly stigmatized after World War II, when the world saw the horrors of its most infamous implementation, the Holocaust. But many scientists continued to believe in its core tenet: that social problems have fundamental biological underpinnings that can be eliminated through scientific control of human reproduction. Whereas prewar eugenics focused on "negative" applications—weeding out undesirables through practices such as forced sterilization and genocide—eugenicists in the postwar years emphasized so-called positive eugenics, or strategically breeding in favorable traits. As early as the 1960s, and continuing right up to the present, some scientists and others—including a number of well-respected scholars at prestigious institutions—have tried to resurrect eugenic projects in one guise or another. These enthusiasts promote the troubling idea that eugenic principles can be separated from the inequality, intolerance, and oppression that gave rise to the movement in the late nineteenth century.

No one questions Edwards's clinical concern and goodwill in treating his patients, as documented by obituary writers. Yet he also seemed comfortable with placing IVF and other reproductive and genetic technologies in an intellectual framework that would have been eminently recognizable to Galton and his early disciples. Edwards believed that increased control over human reproduction could not only treat the infertile but also allow for socially favored characteristics to be selected and bred into the population. Edwards himself hinted at the link between IVF and eugenics when reflecting on the twenty-fifth anniversary of Louise Brown's birth in 1993, saying that developing IVF "was about more than infertility. . . . I wanted to find out exactly who was in charge, whether it was god himself or whether it was scientists in the laboratory." Edwards's conclusion?—"It was us."

Although there is nothing inherently eugenic about IVF, being able to manipulate human conception outside of the womb is an essential platform technology for any modern eugenic goal. Edwards implicitly acknowledged this link in 1999 when he said, "Soon it will be a sin of parents to have a child that carries the heavy burden of genetic disease. We are entering a world where we have to consider the quality of our children." The implications of this statement suggest not simply using new technologies to treat disease, but also to prevent the births or reconfigure the traits of individuals thought to be of low value.

But Edwards didn't stop there. He also supported the use of modern selection technologies for predetermining nonmedical traits, such as having a boy instead of a girl, that are viewed as more desirable in some societies. Edwards's technology, IVF, combined with preimplantation genetic diagnosis—the ability to screen embryos for a particular trait before implanting them—are vitally important steps toward being able to select the features of future generations much like we currently configure the details for a new car. Edwards fully supported using sex selection technologies for social and not just medical reasons, saying, "Go ahead and use it. Those parents have to raise those children."

It is not simply that Edwards believed in the permissive implementation of reproductive and genetic technologies in a manner that happened to coincide with eugenic aspirations. Rather, he embraced eugenics as morally justifiable—save for the brief period when a presumably pure science was politicized by fanatics. Edwards said as much in 2004 during testimony before the Parliamentary Committee on Science and Technology in the British House of Commons. When asked whether a line could be drawn between new reproductive and genetic technologies and eugenics, Edwards—in a style reminiscent of former Pres. Bill Clinton—said, "It depends what we mean by eugenics." He continued, "Eugenics was started in the 1870s by an English geneticist who had the welfare of mankind in his mind. The work became degraded after 1930 caused by the Nazis ... [and] various other things where people were found not to be behaving themselves correctly. So, the word became degraded. . . . "

Whereas Edwards noted that these technologies can go forward without using the specific term "eugenics," his statement is quite revealing if only for its stunning inaccuracy. No respectable historian of this period would characterize Galton's ambitions so charitably nor would any suggest that the Nazis were the first or only group to brutally implement eugenics as law and public policy. (The world's first eugenics law—providing for forced sterilizations—was passed in Indiana in 1907. Many of these practices continued in other states until the 1970s.) Edwards's thinly veiled attempt at rehabilitating eugenics as not only a defensible concept but somehow as a social movement with a misunderstood history is deeply problematic.

Some try to distinguish (and therefore obscure and excuse) postwar eugenic advocates like Edwards from their prewar predecessors by focusing on the modern emphasis on individuals. By no means did Edwards or his contemporaries advocate the state-enforced eugenic policies of the 1920s and 1930s. They continue to believe in markets and choice; individuals should have access to technologies that allow them to control and enhance human traits. The new eugenicists hope that the market dynamics surrounding the availability of selection and enhancement technologies will mirror other consumer pressures—akin to upgrading to the newest iPhone—to elicit broad-based societal demand that facilitates the goal of population-wide improvements in the "quality" of the human genetic stock.

But this emphasis on individual choice as a distinguishing factor is illusory at best. Past state-enforced eugenics and the prospects for what some call a new free-market eugenics are connected by an incessant desire to link social problems to human biology rather than to political and economic conditions that underlie apparent disparities in abilities among different social groups. A troubling yet consistent theme for eugenics is that only certain types of people—the smart, the beautiful and those without disability—deserve to exist, as those who are less than ideal are simply too burdensome.

It is this notion that science can perfect the human condition and technology can solve social problems that leads otherwise intelligent and well-meaning people to adhere to eugenics' disastrous social policies and scientific projects. Despite attempts to rehabilitate the term, what connects old-school eugenics with the postwar and twenty-first-century versions is the hubris embedded in an underlying attitude that scientists and other elites should take control of evolution to predetermine—or redesign—humanity's biological landscape. Edwards articulated this perspective in 1974 when speculating on the possibility of cloning humans—a quintessentially eugenic pursuit that envisions replicating individuals with genetic endowments thought to be ideal. Although acknowledging possible limitations, he broadly stated that "any [scientific] method of potential value in raising human standards should be considered, and [human reproductive] cloning might contribute towards this end by providing pools of talent."

We can all celebrate the happiness that IVF has brought to many. And, with sufficient oversight of an increasingly sprawling and unregulated fertility industry, IVF need not lead to a parade of horribles. But as we consider the legacy of Robert G. Edwards we must put into perspective the views he held that bear clear links to ignoble attitudes and social policies: the breeding of humans to fit certain predefined specifications that conform to social norms that are all too easily biased. In that sense Edwards's eugenic aspirations should be part of his legacy. The arrogance and cruelty embedded in these beliefs should serve as a clear warning against revisiting a path that science and society should never again travel.

Bioethics and Its Discontents

． ． ． ． ．

As the introduction to this volume notes, abuses and scandals during the twentieth century revealed that doctors and biomedical researchers were capable of inflicting great harm and that oversight was urgently required. Bioethics emerged from this recognition during the postwar period and began to establish itself as a professional and academic field by the late 1960s and early 1970s. As advanced medical technologies developed, bioethics' mission expanded, and it came to address both the practical and philosophical implications of end-of-life decision making, organ donation, human experimentation, and related biomedical issues.

Bioethics' strong suits include the light it shines on the importance of informed consent and the protections it offers to people in the vulnerable roles of patient and research subject. These are no small matters, especially in an era of increasingly high-tech biomedical and pharmaceutical interventions that are developed, marketed, and used in lucrative commercial contexts.

But from its beginnings, and increasingly in recent years, bioethics has come to emphasize an individualistic approach. As a field, it often fails to seriously engage the social and political meanings of reproductive, genetic, and biomedical technologies, and it lacks a social justice and public interest understanding of their development. And bioethics often underestimates the influence of dominant sociocultural values regarding gender, race, class, and ability that continue to shape and skew the work of doctors and researchers—as well as bioethicists themselves. As in other fields, the need for continual critical reflection is evident.

In recent years, some bioethicists' financial and professional ties to the biotechnology and pharmaceutical industries have further compromised their roles. The contributions in this section call attention to these and other limitations of mainstream bioethics, including a critical analysis of the ways in which this standard approach obscures the interests and experiences of women, people with disabilities, and racial minorities.

The first article, by sociologist John Evans, examines how and why a small set of principles, which others have criticized for narrowing the focus of bioethics, has become so widely accepted in the professions. Using the sociology of knowledge to analyze bioethical "principlism," Evans examines the social forces and consequences that have led to the dominance of this approach.

Margaret Olivia Little's article, "Why a Feminist Approach to Bioethics?" is a call for "'feminist-inspired' moral theory." Little argues that "feminist reflection" is important to bioethics because it can reveal gendered reasoning. This can bias which policies are adopted and what counts in the first place as an important question to examine, as well as basic moral concepts.

Gregor Wolbring's "Disability Rights Approach toward Bioethics?" points out that bioethics often uses the term *disabilities* as a synonym for defects, diseases, and subnormal abilities while framing these differences as "medical problems" in need of medical solutions. This view within mainstream bioethics diverges fundamentally from the disability rights perspective, which situates these issues within a social justice framework rather than a medical one.

Catherine Myser argues that there has been inadequate attention to and questioning of "the dominance and normativity of whiteness in the cultural construction of bioethics in the United States." The result, she says, is that bioethics risks reproducing white privilege and white supremacy in its theory, method, and practices.

In "Bioethical Silence and Black Lives," Derek Ayeh contrasts the bioethics community's rapt consideration of high-tech developments like cloning and gene editing with its silence about police killings in communities of color. He observes, "Bioethics often overlooks poverty, unemployment, and gun violence, leaving them for other healthcare professionals." But Ayeh holds out hope that bioethics can become "more than it is today," that it has the potential to probe "the foundations of why people believe what they believe," and thereby "find resolutions through reasonable discourse."

"The Ethicists," a chapter from Carl Elliott's 2010 book *White Coat, Black Hat,* is a hard-hitting examination of the limits and distortions of bioethics as it is now practiced. Elliott begins by pointing out that mainstream bioethicists often put "informed consent" at the center of their analyses without considering the conditions under which people agree to participate in a drug study or medical experiment. If justice and exploitation are taken into account, he argues, "it makes a crucial difference if a subject is sleeping on a hot-air grate and foraging for meals from a dumpster." Elliott then traces the emergence of industry-funded bioethics and the resulting conflicts of interest. The biggest problem, he concludes, is that bioethicists now "seek to become trusted advisers, rather than gadflies or watchdogs."

7. A Sociological Account of the Growth of Principlism

John H. Evans

The three principles of respect for persons, beneficence, and justice first articulated in the Belmont Report and the four similar principles published in Beauchamp and Childress's *Principles of Biomedical Ethics* have structured decision making in bioethics.[1] In this paper I do not address whether these principles are correct, or apply them to problems. Nor do I explain why certain principles have become institutionalized rather than others. Rather, I engage in an investigation of principlism from the perspective of the sociology of knowledge: Why has the use of a small set of principles—whichever they may be—become common in bioethics?

The sociology of knowledge assumes that decision-making systems, such as principlism, do not become influential because they are "the best" or "correct," but rather because the social conditions are right for those promoting the system to defeat the champions of competing ideas. That an idea is the "most coherent," for example, is important in this competition only if the people in authority to judge the legitimacy of the ideas agree that "coherence" is important.

Everyone recognizes that to understand the dominance of principlism we must go back in history, perhaps to the Nuremberg trials. I propose that to understand the social determinants of principlism we must go back much farther, to 1494, when the first textbook for double-entry bookkeeping was written. This form of bookkeeping is the tabulation that any department chair is familiar with: What are our costs for the next year? What is our

Original publication by John Evans. "A Sociological Account of the Growth of Principilism." *Hastings Center Report* 30, no. 5 (2000): 31–38. Reprinted by permission of John Wiley and Sons. Due to space limitations, portions of this chapter have been deleted or edited from the original.

income? And more specifically, are the costs associated with this component of the business generating returns that justify the costs? This process is taken so much for granted it is hard to imagine an alternative. Before the invention of this system, however, accounts from businesses were basically a "rambling story with numbers" that served to "assist the memory of the businessman," but not help with evaluation of the businessman's actions.[2] . . .

Double-entry bookkeeping was a major innovation in economic history. Two changes in the accounts system also transformed it into a procedure that allowed for calculability, efficiency, and predictability in human action, paving the way for modern capitalism. The first change was that the new system was a means of discarding information deemed to be extraneous to decision making. . . . The new accounting just had numbers, with all other information removed.

The second change was that these numbers took on a new degree of calculability. Instead of putting proceeds on one list (the account above) and costs on another, these two were translated into a common metric called "profit," making an evaluation of each action much more precise. With the previous, "rambling narrative" style of bookkeeping, owners could not readily determine whether an action (such as delivering to the abbot) was efficient at maximizing their end (profit). With accounting that combined information about the costs and proceeds for dealings with the abbot, the efficacy of selling to him could be calculated.

A similar evolution, I suggest, has happened in bioethics, and revealing the similarity lets us tap into a long history of scholarship on these types of transformations. The Babel of information formerly thought to be relevant to an ethical decision has been whittled down to a much more manageable level through the use of principles, and the principles give us a commensurable unit—akin to "profit" in bookkeeping—that also allows for much simpler decisions. Weber called the calculable logic of the new accounting system "formally rational," and he believed that it would spread to other institutions, making a formally rational social life "no mere possibility, but the inescapable fate of the modern world."[3] From this stance of resignation in the face of large social changes, the ethics of science and medicine was simply one of the last holdouts against the forces of rationalization.

PRINCIPLISM AS A FORMALLY RATIONAL INSTITUTION

. . . K. Danner Clouser described medical ethics in the 1960s as "a mixture of religion, whimsy, exhortation, legal precedents, various traditions, philosophies of life, miscellaneous moral rules, and epithets."[4] With a system such as this, how an ethical decision would be made was not calculable and predictable to anyone but the person making the decision, a point to which I will return below.

Principlism does in fact offer the lure of calculability and predictability not offered by the jumble described by Clouser. This calculability and predictability—this formal rationality—can be summarized by the notion of commensuration. Commensuration is a method of "measuring different properties normally represented by different units with a single, common standard or unit."[5] Philosophers will immediately recognize utility as one such commensurable metric. Scientists will recognize risk-benefit analysis, which translates all of the information of a situation into a universal commensurable metric of pain and pleasure. And all of us of course recognize money, the most common commensurable metric of all. Commensuration is essentially a method for discarding information in order to make decision making easier by ignoring aspects of the problem that cannot be translated to the common metric.[6]

Principlism is a form of commensuration, although not as pure a form as money or utility, and not as commensurable as some critics would like. If money or utility represents one commensurable scale, the principles as articulated by Beauchamp and Childress represent four metrics, with no agreed on method for deciding what to do when the metrics on the four scales point in different directions.

Yet the principles are a system of commensuration nonetheless. They are a method that takes the complexity of actually lived moral life and translates this information into four scales by discarding information that resists translation. To see the principles as commensuration, we can ask, Why are there four principles in Beauchamp and Childress and not ten or perhaps twenty? Why did a member of the National Commission complain that there were "too many principles" in an early draft of the Belmont Report (there were seven), and that the list was not "crisp enough?"[7] The answer is that the principles were created to enhance calculability or, in more common language, to simplify bioethical decision making. For example, Beauchamp says that the principles "provide frameworks of general guidelines that condensed morality to its central elements and gave people from diverse fields an easily grasped set of moral standards."[8]

This calculability or simplicity is largely gained by discarding information about deeper epistemological or theoretical commitments. Although Beauchamp is a professed "rule utilitarian" and Childress a professed "rule deontologist," this common metric of principles allows for ethical decision making they can both agree to, despite the massive amount of information about their ethical inclinations represented by the phrases "rule utilitarian" and "rule deontologist." They find that "many forms of rule utilitarianism and rule deontology lead to identical rules and actions,"[9] that is, that there is a commensurable metric between the two. Yet "rule utilitarianism" must provide some information about how to make a decision or the authors would no longer need such labels. Thus despite longstanding differences between utilitarians and deontologists, decision making can be more efficiently calculated with the principles, aptly described by Albert Jonsen as "the common coin of moral discourse." . . .

WHY HAVE THE PRINCIPLES PROSPERED?

Principlism was not first used in debating the issue of human genetic engineering, but rather in the context of experimentation with human subjects. It is important to go back to this issue to observe the early growth of principlism. Later, I will discuss why principlism, in the words of one bioethicist, "grew from the principles underlying the conduct of research into the basic principles of bioethics."[10]

The "rambling narrative" era of bioethics, before the commensurable system of principles was created, was an era when the first ethicists were attempting to influence individuals directly. For example, Henry Beecher's 1966 whistle-blowing article, "Ethics and Clinical Research," appealed to the medical profession to change its ways and suggested a collective mechanism whereby only articles that discussed how the authors obtained the consent of the research subjects would be published.[11] Beecher made no mention of government action or the involvement of citizens in this reform.

When the individual appeal to scientists and physicians fell on deaf ears, and the public simultaneously became more concerned about the ethics of science and medicine, citizens called on the state to protect them. David Rothman reports that it was scandals in human experimentation, such as the Tuskegee experiment, that provided the final impetus for long-term government intervention in the ethics of researchers.[12] The state in turn created commissions and promulgated ethical regulations with which all researchers who used federal money had to comply. It is here that the relevance of fifteenth-century bookkeeping to bioethics is found. The literature on the rise of formal rationality points to the state as one of the foremost proponents of formal rationality. The critical point is that it was people making decisions on behalf of the state and other bureaucratic institutions who found the commensurable set of ethical decision-making principles to be useful, and it is this usefulness that explains the rise and success of principlism.

The birth of principlism itself is intertwined with the advent of state intervention in ethics. Warren Reich, in discussing the early days of the Kennedy Institute (which was later the birthplace of principlism), says that in the early 1970s "there was a political urgency to many of the biomedical issues," and that "the media craved the biomedical controversies and federal and state policymakers wanted answers."[13] Principlism was "evidently found more congenial to the educational and policymaking purposes being pursued" than what he saw as its competitor. It was "open to a broad range of concepts and arguments, in particular those that favored clarification in the public forum of policymaking." Reich goes on to describe how the founder of the Kennedy Institute "marshaled hitherto untapped federal and private funding" and "fostered the need for medical bioethics in the government agencies."

It is also well known that the Belmont principles were created at the urging of the state, and enacted as regulations, with the help of Kennedy Institute members who were simultaneously writing *Principles of Biomedical Ethics*. The original congressional mandate to the National Commission included the command to "identify the ethical principles which should underlie the conduct of biomedical and behavioral research with human subjects and develop guidelines that should be followed in such research."[14] Jonsen later concluded that the principles, which had become part of public law, had "met the need of public-policy makers for a clear and simple statement of the ethical basis for regulation of research."[15]

It is critical for the purposes of this essay to ask why policy makers needed a "clear and simple statement of the ethical basis of research." Put in the terms used in this essay, why did policy makers need a commensurable ethical scale that discarded information? There are two primary reasons. First, government authority in Western liberal democracies, such as the United States, needs to be transparent to citizens. Second, this government is bureaucratic in form.

In liberal democracies, decision making must be transparent. Unlike decrees coming from the subjective perspective of the European sovereign, the US political system was partly founded "on the idea that politics is transparent, that political agents, political actions, and political power can be viewed."[16] This would not be a problem in the city-state described by Plato, where all decision makers knew each other, but in representative democracies this becomes a critical issue. The surface manifestation of this impulse is that government decision-making proceedings, like bioethics commissions, have open meetings, allowing the public to view the decision-making process undertaken on its behalf by its elected officials (and appointed officials, like commissioners). The subtler manifestation of the need for transparency is that government decision making tends toward methods that purport to be objectively transparent.

Historian Theodore Porter explains the popularity of commensurable scales in government decision making, such as cost-benefit analysis, as the result of government officials needing to appear not really to be exercising judgment, but rather to be following both transparent and objective laws or rules. As he puts it, in other countries government officials are "trusted to exercise judgment wisely and fairly. In the United States, they are expected to follow rules."[17] This is because, put simply, it is part of US political culture not to trust authority, especially government authority, and the authority of bureaucrats in particular. "In a country where mistrust of government is rife, the temptation to substitute supposedly impersonal calculation for personal, responsible decisions . . . cannot but be exceedingly strong."[18]

If Americans do not trust government officials to exercise their discretion over where to build a dam (one of Porter's examples), they certainly do not trust the government to make what are construed as moral decisions. Thus an institutional review board (IRB), for example, which is indirectly representing the government, cannot approve of research simply because IRB members think that it is ethical; rather they must "show" or, better yet, "prove" with objective, transparent methods, that the research is ethical.[19] They must show their reasoning in a manner that the public can judge. "Weighing" principles such as beneficence and nonmaleficence (risks and benefits)—notice the simple decision-making rule—purports to be transparent. Like the allure of cost-benefit analysis, the public can feel that it understands the decision being made on its behalf, giving the decision legitimacy.

The commensurable principles have allure for a further reason: the parts of the government that are supposed to protect us ethically are bureaucracies. Bureaucracies evolved to control larger and larger institutions, and part of this ability to control far-flung activities rests on creating standardized rules.[20] Standardized rules, even in ethics, are desirable from a bureaucratic perspective.

Consider the IRB system, which in classic bureaucratic form has "guidelines" that are to be used by recipients of federally funded research involving human subjects. Each institution must set up a committee to make ethical decisions for the government, and the government mandates that certain ethical principles be used and certain records be kept. Imagine the bureaucratic chaos if each IRB were to determine its own principles. The key problem would be that the NIH would not be able to demonstrate to its overseers that proper ethics were actually in use.

The bureaucracy has become even more important since the 1960s. Not only does the Office for Human Research Protection (formerly the Office for Protection from Research Risks) supervise the IRBs, but ethics committees spread throughout the federal bureaucracy, such as the Recombinant DNA Advisory Committee, also oversee human experimentation. Each of these must demonstrate that it is making transparent decisions, and the principles are useful for this purpose. . . .

The underlying link between the two reasons for the rise of principlism is the need to legitimate decisions to people whom decision makers do not know. The greater the need to legitimate decisions, whether because decision makers represent citizens or because they function in a bureaucracy, the more likely it is that principlism will be used. . . .

THE SPREAD OF PRINCIPLISM

Part of the explanation for the dominance of principlism is that it spread from human experimentation to other debates in science and medicine. One reason for this expansion is the increasing number of issues with which the state or bureaucracies are involved. In Callahan's terms, an increasing number of issues call for "clinical," "regulatory," or "health policy" bioethical work. At first our elected officials intervened in the ethics of human experimentation. Having succeeded in this task, more and more issues that somehow seemed related also seemed to cry out for state intervention. A blunt indicator of this spread in state interest are the titles of the first and second government bioethics commissions. The "National Commission" was "for the Protection of Human Subjects of Biomedical and Behavioral Research." The "President's Commission" was "for the Study of Ethical Problems in Medicine and Biomedical and Behavioral Research." "Ethical problems" suggests a much more expansive task than research on human subjects, and the reports of the latter commission reflected this expansiveness.

As I have repeatedly stated, states and bureaucracies prefer formally rational decision-making systems. However, principlism has spread beyond issues with which the state or bureaucracy is involved into what Callahan calls "cultural" bioethics. For example, while the President's Commission's deliberations over human genetic engineering sharply exemplify principlism, many people have used the principles to write about human genetic engineering outside of state contexts. The reason for this further spread of principlism is that principlism itself has become an institution. . . .

Principlism has similarly taken on a life of its own, independent of the conditions that encouraged its early growth, although the continued appetite of the state and bureaucracies for bioethical decisions continues to encourage principlism directly. This independent life of principlism has been encouraged by the rise of bioethics as a profession.

The dominant view in the study of professions is that they are not defined by the existence of an association, nor by having a specialized degree. Rather, they are defined by having a distinct system of knowledge that they use to solve the problems in their jurisdiction.[21] This system is taught by the elite to the average members of the profession—and the dominant system in bioethics is principlism. Of course, all professions have an elite that debates the details of the system of knowledge—academic theologians maintain the system for teaching parish clergy, and psychology professors maintain the system for practicing psychologists. Elite bioethicists likewise perform a maintenance role (which Callahan calls "foundational" bioethics), and are probably more questioning of the inherited system than the average person making bioethical decisions. However, the average member of an ethics committee is unlikely to question the system—who has the time?—and is going to open his or her copy of Beauchamp and Childress because it sets out a legitimate method for making decisions.

. . . The principles have been replicated so many times—printed in so many textbooks, spoken in so many meetings—that it becomes hard to imagine any other way of making decisions. The fact that principlism is also the legally required decision-making system for recipients of federal research funds also encourages this process. . . .

Principlism is rigidly applied not because newcomers to bioethics are lazy, but rather because they are smart. They know what a legitimate decision is, they know what the standards for publishing are, and they know that to continue as a bioethicist they cannot invent their own ethical system (unless they are lucky enough to be elites working on "foundational" bioethics). As they are asked to write on or decide issues far removed from the first bioethical issue, human experimentation, the tendency to use principlism will be quite understandable.

Weber was ambivalent about formal rationality, which he felt was making the world a cold, alienating place, spreading the sense of disenchantment. Many scholars feel similarly about the spread of formally rational commensurable metrics in bioethics. . . . It is hard to imagine going back to the "pastiche" of methods that reportedly comprised decision making in medical ethics before the principles. Yet many feel that the use of principles does not capture the moral life quite properly.[22] By obtaining a better understanding of the social forces that lead to the principles having such a hold on us we can, I hope, discern better ways in which we as a society can make our ethical decisions in medicine and science.

NOTES

1. National Commission for the Protection of Human Subjects of Biomedical and Behavioral Research, *The Belmont Report: Ethical Principles and Guidelines for the Protection of Human Subjects of Research* (Washington, DC: GPO, 1978); T.L. Beauchamp and J.F Childress, *Principles of Biomedical Ethics,* 3rd ed. (New York: Oxford University Press, 1989).

2. B.G. Carruthers and W. Nelson Espeland, "Accounting for Rationality: Double-Entry Bookkeeping and the Rhetoric of Economic Rationality," *American Journal of Sociology* 97, no. 1 (1991): 40.

3. R. Brubaker, *The Limits of Rationality: An Essay on the Social and Moral Thought of Max Weber* (Boston: George Allen and Unwin, 1984), 44.

4. K.D. Clouser, "Bioethics and Philosophy," Hastings Center Report 23, no. 6 (1993): S10.

5. W. Nelson Espeland, *The Struggle for Water: Politics, Rationality and Identity in the American Southwest* (Chicago: University of Chicago Press, 1998), 24. . . .

6. Ibid., 25.

7. A.R. Jonsen, *The Birth of Bioethics* (New York: Oxford University Press, 1998), 103.

8. T.L. Beauchamp, "Principlism and Its Alleged Competitors," *Kennedy Institute of Ethics Journal* 5, no. 3 (1995): 181.

9. Jonsen, *Birth of Bioethics,* 332.

10. Ibid., 104.

11. H.K. Beecher, "Ethics and Clinical Research," *New England Journal of Medicine* 274 (1966): 1354–60.

12. D.J. Rothman, *Strangers at the Bed-side: A History of How Law and Bioethics Transformed Medical Decision Making* (New York: Basic Books, 1991), 182–89.

13. W.T. Reich, "The Word 'Bioethics': The Struggle over Its Earliest Meanings," *Kennedy Institute of Ethics Journal* 5, no. 1 (1995): 19–33.

14. Jonsen, *Birth of Bioethics,* 102.

15. A.R. Jonsen, foreword to *A Matter of Principles? Ferment in U.S. Bioethics,* ed. E.R. Du Bose, R.P. Hamel, and L.J. O'Connell (Valley Forge, PA: Trinity Press International, 1994), xvi.

16. Y. Ezrahi, *The Descent of Icarus: Science and the Transformation of Contemporary Democracy* (Cambridge, MA: Harvard University Press, 1990), 69.

17. T.M. Porter, *Trust in Numbers: The Pursuit of Objectivity in Science and Public Life* (Princeton, NJ.: Princeton University Press, 1995), 195.

18. R. Hammond, "Convention and Limitation in Benefit-Cost Analysis," *Natural Resources Journal* 6 (1966): 222.

19. Ezrahi, *Descent of Icarus.*

20. C. Perrow, *Complex Organizations: A Critical Essay,* 3rd ed. (New York: Random House, 1986); M. Weber, *Economy and Society* (Berkeley: University of California Press, 1968), 220–23.

21. A. Abbott, *The System of Professions: An Essay on the Division of Expert Labor* (Chicago: University of Chicago Press, 1988).

22. E.R Du Bose, R.P. Hamel, and L.J. O'Connell, *A Matter of Principles? Ferment in U.S. Bioethics* (Valley Forge, PA: Trinity Press International, 1994).

8. Why a Feminist Approach to Bioethics?

Margaret Olivia Little

Those who work in feminist bioethics are all too familiar with the question, "Why think that feminism offers a distinctive contribution to bioethics?" When asked respectfully, I take it to be a fair question. After all, even if we were to stipulate that the tenets of feminism are profound and wise, it would not guarantee that they offer substantial illumination in every subject matter. However, while it is a good question to ask, it also has a good answer. In this essay, I outline why it is, and how it is, that feminist insights provide such a valuable theoretical aid to the study of bioethics. . . .

At its most general, feminist theory can be thought of as an attempt to uncover the ways in which conceptions of gender distort people's view of the world and to articulate the ways in which these distortions, which are hurtful to all, are particularly constraining to women. These efforts involve theory—and not merely benign protestations of women's value or equality—because the assumptions at issue are often so subtle or so familiar as to be invisible, and, crucially, because the assumptions about gender have shaped not only the ways in which we think about men and women, but also the contours of certain fundamental concepts—from "motherhood" to "rationality"—that constitute the working tools of theoretical analyses. According to feminist theory, that is, distorted and harmful conceptions of gender have come to affect the very ways in which we frame our vision of the world, affecting what we notice, what we value, and how we conceptualize what does come to attention.

Little, Margaret Olivia. "Why a Feminist Approach to Bioethics?" *Kennedy Institute of Ethics Journal* 6:1 (1996), 1–18. © 1996 Johns Hopkins University Press. Reprinted with permission of Johns Hopkins University Press and authorized adaptations from author. Due to space limitations, portions of this chapter have been deleted or edited from the original.

If these claims are correct, then feminist theory will be useful to disciplines whose subject matter or methods are appreciably affected by such distortions—and it will be useful in ways that far outstrip the particular policy recommendations that feminists might give to some standard checklist of topics. For one thing, feminist reflection may change the checklist—altering what questions people think to ask, what topics they regard as important, what strikes them as a puzzle in need of resolution. Or, again, such reflection may change the analyses underlying policy recommendations—altering which assumptions are given uncontested status, which moves feel persuasive, what elements stand in need of explanation, and how substantive concepts are understood and deployed. If such reflections sometimes yield policies similar to those offered by nonfeminists, the differences in approach can still matter, and matter greatly, by influencing what precedent one takes oneself to have set, what dangers one is alerted to watch for, what would later count as reason to abandon or rethink the policy. And if such reflections are sometimes followed by diverse policy recommendations, we should not be surprised, much less frustrated; for the diagnostic work that forms the core enterprise of feminist theory leads to policy recommendations only in combination with commitments on a variety of other fronts, from economic theory to the empirical facts of the case, about which feminists will understandably disagree.

This, however, is so far rather abstract. To give a more concrete sense of how feminist theory might contribute to bioethics, we need to dip into the theory itself. . . .

ANDROCENTRISM

One of the central themes of feminist theory is that human society, to put it broadly, tends to be androcentric, or male-centered. Under androcentrism, man is treated as the tacit standard for human: he is the measuring stick, the unstated point of reference, for what is paradigmatic of or normal for humans. To start with an obvious example, man is used as the supposedly generic representative of humanity. That is, when we want to refer to humans independently of gender, it is man that is cast for the job: in language ("Man does not live by bread alone"), in examples (such as the classic illustration of syllogistic reasoning, "All men are mortal, Socrates is a man, therefore Socrates is mortal"); in pictorial representations (according to the familiar depiction of evolution—still used in current biology texts—the indeterminate primate, gradually rising to bipedalism, is inevitably revealed in the last frame to be a man).

This depiction of "human" arguably places man in an unfairly privileged position, since he is not only a constituent, but the representative, of all humanity. But much deeper problems than this are at issue, for these supposedly neutral uses of man are not actually neutral. They are false generics, as revealed in our tendency to drop the so-called gender-neutral "he" in favor of "she" when speaking of professions (such as nanny) that are held mostly by women, or again by our difficulty in imagining the logic professor saying, "All men are mortal, Sally is a man (woman?), therefore Sally is mortal."

The first problem resulting from this hidden bias is that androcentrism has a disturbing cumulative effect on our understanding of "human." . . . Certain features of men—their experiences, their bodies, their values— have subconsciously come to be regarded as constituting the human norm. . . .

Second, treating man as the human norm affects, in subtle but deep ways, our concept of "woman." Males and females obviously differ from one another in various ways. "Different from" is a relation, of course, and a symmetrical one at that: if x is different from y, it is just as true that y is different from x. Under androcentrism, however, we tend to anchor man as the reference point and view woman's nature as a departure from his. A subtle but powerful message is communicated when we always anchor one side of what is logically a symmetrical relation as the fixed point of reference: the anchored point gains the status of the center; the other receives the status of the margin. Because man has been fixed as the reference point for so long, part of our very conception of woman has become the conception of "other"—she is, as Simone de Beauvoir put it [in 1952], the second sex. Instead of thinking that men differ from women who differ from men, a subtle conceptual shift occurs, and we begin to think of women as simply [intrinsically] "different." . . . In the end, it is a short step to regarding aspects of woman's distinct nature as vaguely deviant.

Further, woman becomes closely defined by the content of her departure from man. The fundamental ways in which women and men differ are, of course, in certain biological features. But when man's body is regarded as the neutral "human" body, woman's biological sex becomes highlighted in such a way that, in the end, awareness of woman very often is awareness of her sex. The phenomenon is akin to one that occurs with race. In white-dominated societies, being white gets anchored as the tacit reference point; over time, the fact that whites have a race tends to fade from consciousness, while people of color are seen as somehow more intrinsically raced. . . . In a similar way, woman's sex comes to be seen as more essential to her nature than man's sex is to his. We are more likely to see woman as ruled by the whims of her reproductive system than man is by his; more subtly, if no less dangerously, we are simply more likely to think of and be concerned with reproductive issues when thinking of women than of men.

Finally, under androcentrism, woman is more easily viewed in instrumental terms—in terms, that is, of her relation to others and the functions she can serve them. . . . Awareness of these general androcentric themes will give new food for thought on any number of topics in bioethics. The medicalization of childbirth, for instance—too often packaged as a tiresome debate between those generically loyal to and those generically suspicious of technology—takes on more suggestive tones when we consider it in light of the historical tendency to regard women as "other" or deviant and hence in need of control. Certain patterns of research on women and AIDS emerge with greater clarity when viewed against our proclivity to view women instrumentally: until very recently such research focused almost entirely on women as transmitters of the disease to their fetuses, rather than on how the disease manifests itself, and might be treated, in the women themselves. Let me develop in slightly more detail, though, an example that brings to bear the full range of androcentric themes outlined above.

Many people were taken by surprise when a 1990 U.S. Government Accounting Office report indicated that women seemed to be underrepresented in clinical trials. To give a few now-famous examples, the Physicians Health Study, which concluded in 1988 that an aspirin a day may help decrease the risk of heart disease, studied twenty-two thousand men and no women; the Baltimore Longitudinal Study, one of the largest projects ever to study the natural processes of aging, included no women at its inception in 1958 and still had no data on women by 1984, although women constitute 60 percent of the population over age sixty-five in the United States. It is difficult to be precise about women's overall representation in medical research because information on participants' sex often is not gathered; but there does seem to be legitimate cause for concern. . . .

The possibility of significant underrepresentation has raised concerns that women are being denied equal opportunity to participate in something they may regard as valuable and that women may face compromised safety or efficacy in the drugs and procedures they receive (for instance, the difference in the average weights of women and men raises questions about the effects on women of drugs that are highly dosage-sensitive). Now, determining what policy we should advocate with respect to women's inclusion in medical research is a complicated matter—if only because adding sex as a variable in research protocols can significantly increase the cost of research. What is clear, though, is that awareness of various androcentric motifs can highlight important issues that might otherwise remain hidden or camouflaged. Without the perspective of feminist theory, that is, certain concerns are likely not even to make it to the table to be factored in when policy questions arise. . . .

One argument against the inclusion of women commonly offered by those running clinical trials is that women's hormones represent a "complication": the cyclicity of women's hormonal patterns introduces a variable that can make it harder to discern the effects of the drug or procedure being studied. Now this is an interesting argument, for acknowledging the causal power of women's hormonal cyclicity might also suggest the very reason that it might be important to include women in studies, namely, the possibility that the cyclicity affects the underlying action of the drug or procedure. Medicine has only begun to consider and study this possibility in earnest. Early results include preliminary evidence that surgical treatment for breast cancer is more effective if done in the second, rather than the first, half of a woman's menstrual cycle, and that the effectiveness of antidepressants varies across a woman's menstrual cycle, suggesting that women currently receive too much for one half of the month and too little for the other. Trust in all-male studies seems to reflect a broad confidence in the neutrality of treating the male body as the human norm and a familiar tendency to regard that which is distinct to woman as a distortion. . . .

Another reason often given for the underrepresentation of women by those running clinical trials is that women are harder to find and to keep in studies. There is an important element of truth here: questionnaires reveal that women report greater problems navigating the logistics of participating in drug trials—they find it more difficult, for instance, to arrange for transportation and child care (Cotton 1993; Laurence and Weinhouse 1994, 70–71). But if it is currently harder for women to participate than for men, it is not because of some natural or neutral ordering of things; it is in large part because drug trials are currently organized to accommodate the logistical structure and hassles of men's lives. . . .

A different concern lay behind the now-defunct FDA guidelines barring women of childbearing potential from early clinical trials. Here the explicit rationale was fetal protection: the drugs women would be exposed to might harm fetuses they knowingly or unknowingly carried. A closer look, however, once again reveals the subtle presence of androcentrism: granting society's interest in fetal health, protective measures are applied quite differently to men and women. The guidelines in essence barred all fertile women from early trials—including single women not planning to have intercourse, women using reliable birth control, and women whose partners had had vasectomies (Merton 1994). In contrast, when trials were conducted on drugs suspected of increasing birth defects by affecting men's sperm (a possibility often forgotten), fertile men were simply required to sign a form promising to wear condoms during the trial (Laurence and Weinhouse 1994, 72–73). The regulation was able to think of men under guises separate from their reproductive capacities ... one might argue that respect for parental autonomy argues in favor of allowing the individual to decide whether participation is worth the risk. But when respect for parental autonomy conflicts with protection for fetuses or children, society is much more willing to intrude on the autonomy if it belongs to a woman than to a man. ...

GENDERED CONCEPTS

A second core theme of feminist theory maintains that assumptions about gender have, in subtle but important ways, distorted some of the broad conceptual tools that philosophers use. Certain key philosophical concepts, such as reason and emotion or mind and body, seem in part to be gendered concepts—that is, concepts whose interpretations have been substantively shaped by their rich historical associations with certain narrow conceptions of male and female.

One such distortion stems from the fact that, historically, that which is tightly and consistently associated with woman tends to become devalued. Throughout history, woman has been regarded as a deficient human: as a group, at least, she does not measure up to the standard set by man. ... Aristotle defined woman as "a mutilated male," placing her just above slaves in the natural hierarchy. In post-Darwinian Victorian society, when a theory emerged according to which "lower forms" of human remained closer to embryonic type, a flurry of studies claimed to demonstrate the childlike aspects of woman's anatomy. ... Against this background, those things associated with woman can gradually inherit a depreciated status. "Womanly" attributes, or aspects of the world regarded as somehow "feminine," become devalued. ... To give just one illustration, think of the associations we carry about voice types and authority. A resonant baritone carries a psychological authority missing in a high squeaky voice. This is often cited as a reason women have trouble being viewed as authority figures; but it is also worth asking why authority came to be associated with a baritone rather than a soprano in the first place. Clearly, the association both reflects a prior conception of man as naturally more authoritative and reinforces that commitment, as women's voices then stand in the way of their meeting the "neutral" standard of authority.

Another common distortion stems from the fact that pairs of concepts whose members are associated with man and woman, respectively, tend to become interpreted in particularly dualistic ways. For much of Western history, but especially since the Scientific Revolution, men and women have been understood as having different appropriate spheres of function. Man's central role was in the public sphere—economics, politics, religion, culture; woman's central role was in the private sphere—the domestic realm of caretaking for the most natural, embodied, and personal aspects of humans. This separation of spheres was understood to constitute a complementary system in which each contributed something of value that, when combined, made an ideal whole—the marriage unit. Of course, given the devaluation of that which is associated with woman, it is not surprising that woman's sphere was regarded as less intrinsically valuable: it is man, and what is accomplished in the public sphere, that represents the human ideal (a view reflected in history books, which are histories of wars and political upheavals, not of hearth and home). In any event, because the division was understood as grounded in the natures of man and woman, the separation was a rigid one; the idea that either side of the division could offer something useful to the other's realm would simply not emerge as a possibility. This dualistic picture of the nature and function of women and men, with its subtle devaluing of women, can bleed over to concepts that have been tightly associated with the sexes. When abstract concepts, such as, say, mind and body, come to be paired with the concepts of male and female consistently enough, their substantive interpretations often become tainted with the dualism that characterizes the understanding of those latter concepts. The nature of each comes to be understood largely in opposition to the other, and, while the pair is understood as forming a complementary whole, the functions of the components are regarded as rigidly separated, and the one that is regarded as "male"—here, mind—is held in higher philosophical esteem.

These themes are mirrored in the interpretation of certain central philosophical concepts. An important instance is the traditional conception of reason and emotion, which plays a large role in moral philosophy. For all the hotly disputed debates in the history of ideas, one theme that emerges with remarkable consistency is an association of women with emotion and men with reason. According to Aristotle, women have rationality "but without authority"; Rousseau (1979, 386) gives Sophie a different education from Emile because "the search for abstract and speculative truths, principles and axioms in the sciences, for everything that tends to general ideas, is not within the competence of women"; and according to Kant

(1960, 79), "women's philosophy is not to reason but to sense." Science has contributed its support—for example, tracing woman's supposedly greater proclivity toward volatile emotions to disorders of the womb (hence "uterus" as the root of "hysteria") and her restricted intellect to the "hormonal hurricanes" of her menstrual cycle. As James Allan wrote, "In intellectual labor, man has surpassed, does now, and always will surpass woman for the obvious reason that nature does not periodically interrupt his thought in application" (Russett 1989, 30). . . .

The conception of reason and emotion found in much of traditional ethical theory bears the mark of these entrenched associations. There is a tendency to regard reason and emotion as having completely separate functions and to regard emotion, at best, as irrelevant to the moral enterprise and, at worst, as something that infects, renders impure, and constantly threatens to disrupt moral efforts. Emotion is conceptualized as something more to do with the body we have as animals than the mind we have as humans; it is viewed as a faculty of blind urges, akin to pains and tickles, rather than as responses that reflect evaluations of the world, and that hence can be "tutored" or developed into mature stances.

Thus, most traditional moral epistemology stresses that the stance appropriate to moral wisdom is a dispassionate one. To make considered, sound moral judgments, we are told to abstract from our emotions, feelings, and sentiments. Emotions are not part of the equipment needed to discern moral answers; indeed, only trouble can come of their intrusion into deliberations about what to do, for they "cloud" our judgment and "bias" our reasoning. To be objective is to be detached; to be clear-sighted is to achieve distance; to be careful in deliberation is to be cool and calm. Further, the tradition tends to discount the idea that experiencing appropriate emotion is an integral part of being moral. . . .

Feminists argue that these presuppositions may not survive their gendered origins. . . . While our passions and inclinations can mislead us and distort our perceptions, they can also guide them. . . . Distance, that is, does not always clarify. Sometimes truth is better revealed, the landscape most clearly seen, from a position that has been called "loving perception" or "sympathetic thinking" (Lugones 1987, Jaggar 1989, Walker 1992b). And again, emotion arguably forms an integral part of being moral. Simply to perform a required action—while certainly better than nothing—is often not enough. Being moral frequently involves feeling appropriate emotions, including anger, indignation, and especially caring. The friend who only ever helps one out of a sense of duty rather than a feeling of generous reciprocity is not in the end a good friend; the citizen who gives money to the poor, but is devoid of any empathy, is not as moral as the one whose help flows from felt concern. . . .

In another important instance, that which is associated with the private or domestic sphere is given short shrift in moral theory. Relations in the private sphere, such as parent-child relations, are marked by intimacies and dependencies, appropriate kinds of partiality, and positive but unchosen obligations that cannot be modeled as "contracts between equals." Furthermore, few would imagine that deliberations about how to handle such relations could be settled by some list of codified rules—wisdom here requires skills of discernment and judgment, not the internalization of set principles. But traditional moral theory tends to concentrate on moral questions that adjudicate relations between equal and self-sufficient strangers, to stress impartiality, to acknowledge obligations beyond duties of noninterference only when they are incurred by voluntary contract, and to emphasize a search for algorithmic moral principles or "policies" that one could apply to any situation to derive right action (Walker 1992a, Baier 1987).

This tendency to subsume all moral questions under a public "juridical" model tends, for one thing, to restrict the issues that will be acknowledged as important to those cast in terms of rights. "The" moral question about abortion, for instance, is often automatically cast as a battle between maternal and fetal rights, to the exclusion of, say, difficult and nuanced questions about whether and what distinctly maternal responsibilities might accompany pregnancy. And it often does violence to our considered sensibilities about the morality of relations involving dependencies and involuntary positive obligations. For instance, in considering what it is to respect patient autonomy, many seem to feel forced into a narrow consumer-provider model of the issue, in which the alternative to simply informing and then carrying out the patient's wishes must be regarded as paternalism. While such a model may be appropriate to, say, business relations between self-sufficient equals, it seems highly impoverished as a model for relations marked by the unequal vulnerabilities inherent in physician-patient relations. . . .

Finally, when ethical approaches more characteristic of the private sphere do make it onto the radar screen, there is still a tendency to segregate these approaches from those we take to the public sphere. That is, in stark contrast to the tendency to subsume the morality of intimates into the morality of strangers, rarely do we ask how the moral lessons garnered from reflecting on private relations might shed light on moral issues that arise outside of the purely domestic context. To give just one example, patients often feel a deep sense of abandonment when their surgeons do not personally display a caring attitude toward them: the caring they may receive from other health care professionals, welcome as it may be, seems unable to compensate for this loss. This phenomenon will seem less puzzling if, borrowing a concept from the private realm, we realize that surgery involves a special kind of intimacy, as the surgeon dips into the patient's body. Seen under this guise, the patient's need becomes more understandable—and the surgeon's nontransferable duty to care clearer—for reflection on more familiar, domestic intimacies, such as those involved in sexual interactions, reminds us that intimacy followed by a vacuum of care can constitute a kind of abandonment.

In summary, then, reflection in feminist theory is important to bioethics in at least two distinct ways. First, it can reveal androcentric reasoning present in analyses of substantive bioethical issues—reasoning that can bias not only which policies are adopted, but what gets counted as an important question or persuasive argument. Second, it can help bioethicists to rethink the very conceptual tools used in bioethics—specifically, helping to identify where assumptions about gender have distorted the concepts commonly invoked in moral theory and, in doing so, clearing the way for the development of what might best be called "feminist-inspired" moral theory.

REFERENCES

Baier, Annette. 1987. "The Need for More Than Justice." In *Science, Morality and Feminist Theory*, edited by Marsha Hanen and Kai Nielsen, 41–56. Calgary: University of Calgary Press.

Beauvoir, Simone de. 1952. *The Second Sex*. New York: Alfred A. Knopf.

Bem, Sandra L. 1993. *The Lenses of Gender*. New Haven, CT: Yale University Press.

Bird, Chloe E. 1994. "Women's Representation as Subjects in Clinical Studies: A Pilot Study of Research Published in JAMA in 1990 and 1992." In *Women and Health Research: Ethical and Legal Issues of Including Women in Clinical Studies*. Vol. 2, Institute of Medicine, 151–73. Washington, DC: National Academy Press.

Blackstone, William. [1765–1769] 1979. *Commentaries on the Laws of England*. Chicago: University of Chicago Press.

Bordo, Susan. 1986. "The Cartesian Masculinization of Thought." *Signs* 11 (3): 439–56.

Broverman, Inge K., Donald M. Broverman, and Frank E. Clarkson, et al. 1970. "Sex-Role Stereotypes and Clinical Judgments of Mental Health." *Journal of Consulting and Clinical Psychology* 34 (1): 1–7.

Carse, Alisa L., and Hilde Lindemann Nelson. 1996. "Rehabilitating Care." *Kennedy Institute of Ethics Journal* 6:19–35.

Cotton, Paul. 1990. "Examples Abound of Gaps in Medical Knowledge because of Groups Excluded from Scientific Study." *Journal of the American Medical Association* 263:1051, 1055.

——. 1993. "FDA Lifts Ban on Women in Early Drug Tests." *Journal of the American Medical Association* 269:2067.

Daniels, Cynthia. 1993. *At Women's Expense: State Power and the Politics of Fetal Rights*. Cambridge, MA: Harvard University Press.

DeBruin, Debra A. 1994. "Justice and the Inclusion of Women in Clinical Studies: An Argument for Further Reform." *Kennedy Institute of Ethics Journal* 4:117–46.

Faden, Ruth, N. Kass, and D. McGraw. Forthcoming. "Women as Vessels and Vectors: Lessons from the HIV Epidemic." In *Feminism and Bioethics: Beyond Reproduction*, edited by Susan Wolf. New York: Oxford University Press.

Fausto-Sterling, Anne. 1992. *Myths of Gender: Biological Theories about Women and Men*. New York: Harper Collins.

Gatens, Moira. 1991. *Feminism and Philosophy*. Bloomington: Indiana University Press.

Hamilton, Jean, and Barbara Parry. 1983. "Sex-Related Differences in Clinical Drug Response: Implications for Women's Health." *JAMWA* 38 (5): 126–32.

Institute of Medicine: Committee on the Ethical and Legal Issues Relating to the Inclusion of Women in Clinical Trials. 1994. *Women and Health Research*. Vols. 1 and 2. Washington, DC: National Academy Press.

Jaggar, Alison. 1989. "Love and Knowledge: Emotion in Feminist Epistemology." In *Women, Knowledge, and Reality*, edited by Ann Garry and Marilyn Pearsall, 129–55. Boston: Unwin Hyman.

Kant, Immanuel. 1960. "Observations on the Feeling of the Beautiful and Sublime." Section Three. In *Of the Distinction of the Beautiful and Sublime in the Interrelations of the Two Sexes*. Berkeley: University of California Press.

Laurence, Leslie, and Beth Weinhouse. 1994. *Outrageous Practices: The Alarming Truth about How Medicine Mistreats Women*. New York: Fawcett Columbine.

Little, Margaret Olivia. 1995. "Seeing and Caring: The Role of Affect in Feminist Moral Epistemology." *Hypatia* 10 (3): 117–37.

Lloyd, Genevieve. 1983. "Reason, Gender, and Morality in the History of Philosophy." *Social Research* 50:490–513.

——. 1984. *The Man of Reason: Male and Female in Western Philosophy*. London: Methuen.

Lugones, Maria. 1987. "Playfulness, World-Traveling, and Loving Perception." *Hypatia* 2 (2): 3–19.

MacKinnon, Catharine A. 1987. *Feminism Unmodified: Discourses on Life and Law*. Cambridge, MA: Harvard University Press.

Merton, Vanessa. 1994. "Impact of Current Federal Regulations on the Inclusion of Female Subjects in Clinical Studies." In *Women and Health Research: Ethical and Legal Issues of Including Women in Clinical Studies.* Vol. 2. Institute of Medicine, 65–83. Washington, DC: National Academy Press.

Minow, Martha. 1990. *Making All the Difference: Inclusion, Exclusion, and American Law.* Ithaca, NY: Cornell University Press.

Okin, Susan Miller. 1979. *Women in Western Political Thought.* Princeton, NJ: Princeton University Press.

Pateman, Carole. 1989. *The Disorder of Women.* Stanford, CA: Stanford University Press.

Rosenberg, Charles. 1976. *No Other Gods: On Science and American Social Thought.* Baltimore: Johns Hopkins University Press.

Rothman, Barbara Katz. 1982. *In Labor: Women and Power in the Birthplace.* New York: Norton.

Rousseau, Emile. 1979. *Emile, or On Education.* New York: Basic Books.

Russett, Cynthia E. 1989. *Sexual Science: The Victorian Construction of Womanhood.* Cambridge, MA: Harvard University Press.

Smith-Rosenberg, Carroll. 1972. "The Hysterical Woman: Sex Roles in 19th Century America." *Social Research* 39:652–78.

Tuana, Nancy. 1992. *Woman and the History of Philosophy.* New York: Paragon House.

US General Accounting Office. 1990. *National Institutes of Health: Problems in Instituting Policy on Women in Study Populations.* Washington, DC: GAO.

Walker, Margaret Urban. 1992a. "Feminism, Ethics, and the Question of Theory." *Hypatia* 7 (3): 23–38.

———. 1992b. "Moral Understandings: Alternative Epistemology for a Feminist Ethics." In *Explorations in Feminist Ethics,* edited by Eve Browning Cole and Susan Coultrap-McQuinn, 165–75. Bloomington: Indiana University Press.

9. Disability Rights Approach toward Bioethics?

Gregor Wolbring

IS THERE AN ABSOLUTE OBJECTIVE ETHICAL APPROACH TO GOVERN SCIENCE AND TECHNOLOGY?

Throughout history, science and technology (S&T) have had, and in the future will have, positive and negative consequences for humankind. It has been said that S&T is value-neutral and that inanimate technological inventions cannot harbor values (The European Commission 2003; Sundloff 2000). These claims are false, perhaps simplistic or perhaps simply beside the point. As the result of human activity, S&T is imbued with intention and purpose. The goals for which S&T are advanced are value-laden, reflecting the cultural, economical, ethical, spiritual, and moral frameworks of society. Technology follows social norms. S&T embodies the perspectives, purposes, prejudices, and objectives of society and of powerful social groups within society. It is believed that many negative consequences of S&T for humankind could be avoided by using ethical principles to govern them. How do we come up with these ethical principles? There are religious-based ethics and secular-based ethics. For the sake of this article, I will focus on secular-based ethics and the academic field of bioethics. . . .

Within the academic debate over bioethical issues, certain ethical principles are put forward time after time, namely, the *principles of autonomy, beneficence, nonmaleficence, and justice.* However, different philosophies and approaches to bioethics interpret the concept and boundaries of autonomy, beneficence, nonmaleficence, and justice in different ways and come up with

Gregor Wolbring. "Disability Rights Approach Toward Bioethics?" *Journal of Disability Policy Studies* 14, no. 3, pp. 174–180, copyright © 2003 by SAGE. Reprinted by Permission of SAGE Publications, Inc. Due to space limitations, portions of this chapter have been deleted or edited from the original.

additional principles to define ethical behavior. These varieties in philosophies and principles give rise to different possibilities for governing S&T. How do we decide which philosophy, which ethics, to use? It has been assumed that the academic development of ethical principles is free from political interventions, but such is not the case. It also is not free from prejudices or judgments. . . . In the same way S&T is shaped by societal perspectives, so is ethics. Ethics also embodies the perspectives, purposes, prejudices, and objectives of society and of powerful social groups within society. Because the academic discourse is not free of politics and prejudice, by itself this discourse cannot lead us to an ethics with which we could govern S&T for the good of everyone in society.

BIOETHICS ISSUES AND DISABILITIES: INFLUENCING THE DEBATE

Various parties use the characteristic *disability* to justify and promote their views of particular bioethics issues. Biogenetic technology, for example, is promoted as a tool for fixing disabilities, whereby *disability* is often a synonym for impairments, diseases, defects, and "subnormal" abilities. Biogenetic technology is seen as a tool for diminishing suffering and as having the potential to free us from the "confinement of our genes" (Wolbring 2003). Nanotechnology is promoted as offering the potential to free us from the "confinement of our biological bodies" (Wolbring 2003) through the use of nonbiological "assistive" bionic technologies such as prosthetic limbs that adjust to the changes in the body, more biocompatible implants, artificial retinas or ears, neural prostheses, and the "spinal patch," to name just a few. Euthanasia is sold with the slogan of increasing the autonomy of the terminally ill and of individuals with irreversible disabilities, allowing them to decide their time of death and a dignified death. Mercy killing/infanticide is justified with the argument that it prevents the suffering of the so-called disabled. As *disability* is one battleground for so many bioethics issues, one might expect that people with disabilities would shape the debates or at least be dominant players in them. Sadly, such is not the case. The bioethics issues debate happens on many levels—academia, policy making, governmental, and within civil society—and a disability rights angle is often excluded from and rejected by all of these. . . .

ONE EXAMPLE: THE SEX-SELECTION DEBATE

I will now discuss one example of how the debate surrounding a bioethics issue plays out for individuals with disabilities. The practice of sex selection is as old as humankind. For the greater part of our history, this selection of one gender over another (mostly male over female) took place after birth and led to infanticide or neglect of the newborn of the "wrong" sex (mostly female). Infanticide and neglect of the newborn increasingly is being replaced by new technologies such as (a) prenatal sex diagnosis and termination of pregnancies or (b) preimplantation sex diagnosis and selection of the embryo of the "right" sex for implantation into the womb of the mother or by the use of sperm sorting. The demand for these new technologies has been generated not only in countries such as China and India, where the preference for boys is still strong, but also in countries such as the United States, Australia, and Germany, where the prejudice against women has somewhat diminished. The reasons might be different between the first two cases (prejudice against girls) and last three cases (a sex-balanced family), but the acceptance of and demand for using prenatal diagnostic procedures to have a child of a desired sex seems to be equally high (Wertz and Fletcher 1998).

Taking into account the high demand for sex selection worldwide, how does the legal system deal with the demand for sex-selection procedures? A variety of laws and legal proposals that prohibit sex selection for "nonmedical reasons" exist. Furthermore, at least three international documents demand the prohibition of sex selection for nonmedical reasons, as follows:

1. Article 14 (Non-Selection of Sex) of the *European Convention for the Protection of Human Rights and Dignity of the Human Being with Regard to the Application of Biology and Medicine: Convention on Human Rights and Biomedicine* states, "The use of techniques of medically assisted procreation shall not be allowed for the purpose of choosing a future child's sex, except where serious hereditary sex-related disease is to be avoided" (Council of Europe 1997).

2. Paragraph 21 of the World Health Organization (WHO) *Draft Guidelines on Bioethics* states that "sex is not a disease" (1999).

3. Paragraph 71 of the *Draft Report on Pre-implantation Genetic Diagnosis and Germ-Line Intervention* of the Working Group of the International Bioethics Committee of the United Nations Educational, Scientific, and Cultural Organization (UNESCO) states, "Destruction of embryos for non-medical reasons or termination of pregnancies because of a specific gender are not 'counterbalanced' by avoiding later suffering by a severe disease. Sex selection by PGD or PD is therefore considered as unethical" (2002).

The debate over the prohibition of sex selection leads to a few questions for people with disabilities:

1. Are the arguments used to demand the prohibition of sex selection also valid for demanding the prohibition of "ability selection" and disability deselection?

2. Does the demand for prohibiting sex selection also lead to a demand for prohibiting ability selection/disability selection?

3. Does this line of argument lead to another discriminatory approach toward the characteristic of disability?

Are the arguments used to demand the prohibition of sex selection also valid for demanding the prohibition of ability selection and disability deselection?

One line of argument says that sex selection poses significant threats to the well-being of children, the children's sense of self-worth, and the attitude of unconditional acceptance of a new child by parents, which is so psychologically crucial to parenting. However, is this a specific argument for the prohibition of sex selection? Could the argument not read as follows: "One line of argument says that ability selection/disability deselection poses significant threats to the well-being of children, the children's sense of self-worth, and the attitude of unconditional acceptance of a new child by parents, so psychologically crucial to parenting." Would not this argument also justify the prohibition of ability selection/disability deselection? Is not unconditional acceptance also endangered if parents choose their child based on wanted abilities or unwanted disabilities? People can debate whether the above argument is valid, but if it is a valid one to use in demanding the prohibition of sex selection, then it is also valid to use in demanding the prohibition of ability selection/disability deselection and, in essence, the prohibition of selection for any reason. This thus is not a sex-selection–specific argument.

A second line of arguments justifies sex selection prohibition by pointing out the negative consequences for the unwanted sex. . . . Still other researchers see sex selection leading to the enhancement of sex stereotypes, which means that people will have certain expectations of individuals of one sex versus the other. Would not the following version of the above also hold true? People explain that ability selection/disability deselection is leading to the oppression of the persons with unwanted disabilities, which in turn is leading to social injustice. Other persons see ability selection/disability deselection as a form of disability discrimination. Still other individuals see ability selection/disability deselection leading to the enhancement of ability/disability stereotypes, which means that people will have certain expectations of people with one ability/disability or another.

Indeed, these arguments were, and still are, used by people with disabilities and others when they look at the consequences of gene technology for individuals with disabilities. Some persons connect the use of prenatal testing for Down syndrome with the diminishing rights of people with that syndrome. Andrew Brown of Amnesty International stated, "If society regards the presence of such disease as an acceptable reason for aborting a fetus, this makes it harder to preserve equality of respect for those already born. One might argue that their human worth, if not their human rights, have been diminished" (1998, 19).

A third line of argument sees sex selection leading to "designer babies," trivializing the selection procedure and leading to selection based on "cosmetic reasons" (Human Fertilization and Embryology Authority 1993). However, one could ask, What is *cosmetic*? Cosmetic is something based on established norms. Is it cosmetic to have no legs, to be shorter, to be obese, to have black hair, to be intelligent, or to have blue eyes? Is the term *cosmetic* another synonym for characteristics not affecting abilities? This leads to the following questions: Which abilities are needed that still fit within acceptable variations from the norm, and which don't? Who decides what are cosmetics? The same questions have to be raised for the usage of the term *designer baby*.

A fourth line of argument says that sex selection is wrong because it is not a disease. This argument is not an ethical argument but rather a "hierarchy" argument. Someone decides, based on his or her prejudices and power, that testing for one characteristic is all right but testing for another characteristic is not. Furthermore the term *disease* is rather undefined and by itself is a social construct (Wolbring 2000).

What are the reactions if people with disabilities employ the arguments used to demand the prohibition of sex selection to demand a prohibition of disability deselection?

The reaction is one of nonacceptance, and some of the arguments used to justify this nonacceptance are as follows:

1. The deselection of characteristics labeled a disability is already happening with other methods without the consequence of neglecting people with disabilities (Knoppers 1993).

2. We prevent the birth of nondisabled people without diminishing the worth of the ones who are living (Nelson 1999).

3. People with disabilities are not necessarily stigmatized because there are people with disabilities who have high self-esteem but who do not want their child to have the same disability they have (Tannsjorn 1998).

4. People with disabilities cannot be seen simply as a human variation but only as an aberration because their numbers are not great enough. A larger number of people with disabilities within society would lead to less support for them (Wertz 2000).

5. Sex is not a disease (WHO 1999).

The question here is whether these arguments are valid and, if they are, whether they are also valid for denouncing the arguments used to demand the prohibition of sex selection.

The first argument is obviously not true, as people with disabilities feel neglected within society. However, it is difficult to prove that the situation of a particular group worsened after tests were created to eliminate the characteristic that defines the group. For example, did the situation of people with polio worsen after the introduction of the polio vaccine? Some people actually would say "yes," but no statistics exist to prove or disprove this statement. Furthermore, independent of the truth of the above statement, it is obvious that the argument can be used just as well to negate the prohibition of sex selection because no studies exist to prove that the introduction of sex-selection diagnostics led to a decrease in quality of life for women. The link seems to be logical to us because if we feel so bad about a characteristic that we do not think anyone in our family should have it, this cannot improve the perception of the people with the characteristic. However, it is hard to draw a direct link between availability of tests and a decrease in quality of life.

The second [and the third arguments] would also support opposition to a sex-selection prohibition. [The former] because, obviously, abortion of unwanted children takes place in countries where sex selection occurs. [The latter] . . . because there are strong women who still support sex selection. . . . People with disabilities could use this counterargument also because most countries adhere to a culture of "ableism" and oppress individuals with disabilities. The fourth argument begs the question of whether the statement would be accepted if it would read as follows: "A greater number of women within society leads to less support for them." . . .

Again this is not a moral argument but rather a power argument. If this is the level of debate, then we have a new type of bioethics philosophy—an "animal farm" bioethics (some are more equal than others).

The fifth argument also is not an ethical argument but an argument based on preference. Why should none of the ethical arguments used to demand prohibition of sex selection be applicable to the prohibition of disease deselection, which in essence often also means defect deselection and disability deselection, because the arguments are being applied to characteristics labeled as diseases, defects, and disabilities? This does sound like an animal farm philosophy. In addition, this argument raises a few more questions. Obviously, a huge variety of characteristics are labeled as diseases, defects, and disabilities. Does the above argument include every disease, disability, and defect or just the "severe" ones, as is often claimed? In the end, there is no way to distinguish among different disabilities, diseases, and defects (Disabled People's International Europe 2000). . . .

If we can't make distinctions among diseases because every single disease is perceived differently based on societal settings, would the same not be true for any characteristic? Which characteristics parents can cope with in general depends on their cultural, societal, economical, and other settings. Indeed, some bioethicists have used this argument to demand that parents have the right to select for whatever characteristic they wish. Furthermore, would not that statement also extend to the fact that people perceive different characteristics as diseases? Diseases are a societal construct. In different cultural or societal settings, different characteristics will be seen as diseases. One prime example is that homosexuality is seen in some settings as a disease and in others as a trait or lifestyle.

THE CONSEQUENCE

It seems that every argument used to justify a sex-selection prohibition could also be used to demand disability deselection prohibition. Furthermore, any arguments used to denounce the demand for the prohibition of disability deselection can be used just as well to denounce the demand for the prohibition of sex selection.

The only possible way to justify disability deselection but not gender selection is by arbitrarily defining disabilities, defects, and diseases as medical problems in need of medical solutions and seeing them in a different moral light based on that medical label. This approach in turn would lead to the development of a double system of morality/ethics—an animal farm philosophy. One would merely have to define something as a medical problem and the acceptable actions would broaden. The argument that "It is a medical problem" trumps all other arguments.

The animal farm philosophy is evident in the debates over many bioethics issues, such as end of life, antigenetic discrimination laws, genetic enhancement, nongenetic cures and enhancements, wrongful life suits, wrongful birth suits, bionics, transhumanism, personhood, mercy killing, infanticide, and treatment of neonatals. In each of these issues, the animal farm philosophy depends on disabilities, diseases, and defects being seen as medical problems.

This philosophy is strengthened by two facts: (a) a disability rights approach is mostly excluded from the discourse—academic or otherwise—of bioethics issues and the development of bioethics theories, and (b) a disability rights approach toward bioethics issues and the development of bioethics theories can be ignored without much fear of repercussion because of the marginalization of people with disabilities.

The animal farm philosophy leads to chasms among different groups as defined by their characteristics because it is based on a hierarchy of applied ethics, that is, the one group that is the most powerful will draw a line between itself and all other groups in regards to the ethical usage of genetic technology. Because this philosophy makes a distinction between medical reasons and social reasons, people with disabilities will find it impossible to establish ableism within the same human rights framework as racism, sexism, ageism, homophobia, and other "isms." . . .

Furthermore, antidisability discrimination laws such as the Americans with Disabilities Act, might only cover people with a particular disability until a medical or technological cure for that disability is found. They might not be seen as having a disability if medical or technological "fixes" are available that might rob them of any legal protection and also possibly force them into using the "cure," even if they don't want to do so. These individuals might be forced into the medical understanding of their characteristics. This scenario is on the horizon after the United States Supreme Court ruled, in *Sutton v. United States* (1998), that "the Americans With Disabilities Act does not cover people whose disabilities can be sufficiently corrected with medicine, eyeglasses or other measures" (see also *Murphy v. United Parcel Service*, 1998). In short, the whole atmosphere of the debate makes it very difficult to gain acceptance for a social justice view of disability, something with which the disability movement identifies.

REFERENCES

Alderson, P. 2001. "Prenatal Screening: Past, Present and Future." In *Before Birth*, edited by E. E. Hore. London: Ashgate.

Ärzte Zeitung. 2001. *Absage an Gentests, PID und Pränataldiagnostik,* March 3. Accessed January 26, 2003. www.aerztezeitung.de/docs/2001/03/22 /054a0602.asp.

Asch, A., and E. Parens. 1999. "The Disability Rights Critique of Prenatal Genetic Testing." *Hastings Center Report* 29 (suppl.): S1–S25.

Beauchamp, T. L., and J. F. Childress 1978. *Principles of Biomedical Ethics.* New York: Oxford University Press.

Birnbacher, D. (1999). *Kongreß für Philosophie, Konstanz, 4.-8.10.99, Referat am 07.10.99 Dieter Birnbacher, Selektion am Lebensbeginn—ethische Aspekte.* Accessed January 26, 2003. www.netlink.de/gen/Zeitung/1999 /991007e.htm.

Brown, A. 1998. "Amnesty's Latest Fear: How Our Genes May Determine Our Fate, *Independent,* February 18, 19.

Calaça, C., and A. Akin. 1995. "The Issue of Sex Selection in Turkey." *Human Reproduction* 10:1631–32.

Council of Europe. 1997. *European Convention for the Protection of Human Rights and Dignity of the Human Being with Regard to the Application of Biology and Medicine: Convention on Human Rights and Biomedicine* (Article 14: non-selection of sex). http://conventions.coe.int/treaty/en /Treaties/Html/164.htm.

Diederich, N., and D. Maroger 2001. *Les personnes handicapées face au diagnostic prénatal: Eliminer avant la naissance ou accompagner?* Ramonville, Ste Ann, France: Editions ERES.

Disabled Peoples' International Europe. 2000. *The Right to Live and Be Different* (Solihull Declaration). Accessed January 26, 2003. www.johnnypops.demon .co.uk/bioethicsdeclaration/index.htm.

DOK Zürich Schweiz. 1998. Diskriminierung behinderter Menschen in der Schweiz. Benachteiligungen und Maßnahmen zu deren Behebung. Hg.: *Dachorganisationenkonferenz der privaten Behindertenhilfe.* Accessed January 26, 2003. http://216.239.37.100/search?q = cache:vgGZEQAGfx8C: www.disabilityresearch.ch/d/shared/dokumente/Diskb98d.pdf+Diskriminie rung+behinderter+Menschen+in+der+Schweiz&hl = en&ie = UTF-8.

Down Syndrome Network of Germany. 1994. *Eltern fordern Lebensrecht und Unterstützung ohne Einschränkung.* Accessed January 26, 2003. www .down-syndrom.org/ak204b.htm.

Federal Organization for Physically and Multiply Disabled People. 2003. *Netzwerk gegen Selektion durch Pränataldiagnostik.* Accessed January 26, 2003. www.bvkm.de/netzwerk/.

German Ministry for Health and Social Affairs. 1990. *Gesetz zum Schutz von Embryonen* (Embryonenschutzgesetz—ESchG) Vom 13. December 1990 BGBl. I 1990 S. 2746–2748 (BGBl III 453–19) '3 Verbotene Geschlechtswahl. Accessed January 26, 2003. www.bmgesundheit.de/rechts/genfpm/embryo /embryo.htm.

Harris, J. 2000. "Is There a Coherent Social Conception of Disability?" *Journal of Medical Ethics* 26:95–100. Accessed January 26, 2003. www.ncbi.nlm.nih

.gov:80/entrez/query.fcgi?cmd = Retrieve&db = PubMed&list_uids = 10786318&dopt = Abstract.

Hennen, L., T. Petermann, and A. Sauter. 2000. *Stand und Perspektiven der genetischen Diagnostik* [Genetic diagnostics: Status and prospects] (Report No. 66). Germany: Institute for Technology Assessment and Systems Analysis at the Karlsruhe Research Center, Office of Technology and Assessment at the German Parliament. Accessed January 26, 2003. www.tab.fzk.de/de/projekt/zusammenfassung/Textab66.htm.

House of Commons of Canada. 2002. *An Act Respecting Assisted Human Reproduction.* Bill C-13. First reading, October 9, 2002. Accessed January 26, 2003. www.parl.gc.ca/37/2/parlbus/chambus/house/bills/government/C-13/C-13_1/C-13_cover-E.html.

Hubbard, R. 1990. *The Politics of Women's Biology.* New Brunswick, NJ: Rutgers University Press.

Human Fertilization and Embryology Authority. 1993. *Sex Selection* (Public Consultation Document).

India Ministry of Human Resource Development, Department of Women & Child Development. 1994. *Convention on Elimination of All Forms of Discrimination against Women (CEDAW): India's First Report.* Accessed January 26, 2003. wcd.nic.in/CEDAW4.htm.

International Centre for Bioethics, Culture, and Disability. 2003a. *Rationale for the Establishment of an International Center for Bioethics, Culture and Disability.* Accessed January 26, 2003. www.bioethicsanddisability.org/Centerrational.html.

International Centre for Bioethics Culture and Disability. 2003b. *Sex Selection.* Accessed January 26, 2003. www.bioethicsanddisability.org/sexselection.html.

Kaplan, D. 1993. "Prenatal Screening and Its Impact on Persons with Disabilities." *Clinical Obstetrics and Gynecology* 36:605–12. Accessed January 26, 2003. www.ncbi.nlm.nih.gov/entrez/query.fcgi?cmd = Retrieve&db = PubMed&list_uids = 8403607&dopt = Abstract.

Kirschner, K.L., K.E. Ormond, and C.J. Gill. 2000. "The Impact of Genetic Technologies on Perception of Disability." *Quality Management in Health Care* 8 (3): 19–26. Accessed January 26, 2003. fwww.ncbi.nlm.nih.gov/entrez/query.fcgi?cmd = Retrieve&db = PubMed&list_uids = 10947381&dopt = Abstract.

Knoppers, B.M. 1993. "Human Genetics: Parental, Professional and Political Responsibility (1992 Picard Lecture in Health Law)." *Health Law Journal* 1:14–23.

Kolata, G. 2001. "Fertility Ethics Authority Approves Sex Selection." *New York Times,* September 28, A16. Accessed January 26, 2003. www.genetics-and-society.org/resources/items/20010928_nytimes_kolata.html.

Law on Maternal and Infant Health Care. 1994. *Renmin Ribao* (Peoples Daily, China), October 28.

Lippman, A. 1991. "Prenatal Genetic Testing and Screening: Constructing Needs and Reinforcing Inequities." *American Journal of Law and Medicine* 17 (1/2):

15–50. Accessed January 26, 2003. www.ncbi.nlm.nih.gov:80/entrez/query
.fcgi?cmd = Retrieve&db = PubMed&list_uids = 1877608&dopt = Abstract.

Little People of America. 1997. *What Is LPA's Position on the Implications of
These Discoveries in Genetics?* www.lpaonline.org/ resources_faq.html.

Mallik, R. 2002. *A Less Valued Life: Population Policy and Sex Selection in
India.* Takoma Park, MD: Center for Health and Gender Equity. Accessed
January 26, 2003. www.genderhealth.org/pubs/MallikSexSelectionIndia
Oct2002.pdf.

Medical Consultation and Judgment. 1989. 18 PA. CONS. STAT. § 3204(c), as
amended November 17, 1989.

Middleton, A., J. Hewison, and R. Mueller. 2001. "Prenatal Diagnosis for
Inherited Deafness—What Is the Potential Demand?" *Journal of Genetic
Counseling* 10 (2): 121–31. Accessed January 26, 2003. www.ncbi.nlm.nih
.gov/entrez/query.fcgi?cmd = Retrieve&db = PubMed&list_uids =
11767801&dopt = Abstract.

Murphy v. United Parcel Service. 1998. Accessed January 28, 2003. www
.washingtonpost.com/wp-srv/national/longterm/supcourt/1998–99
/murphy.htm.

Nelson, J. 1999. "Meaning of the Act." In "The Disability Rights Critique of
Prenatal Genetic Testing." *Hastings Center Report* (suppl.), September
/October, S3.

Pembrey, M. 1998. "In the Light of Preimplantation Genetic Diagnosis: Some
Ethical Issues in Medical Genetics Revisited." *European Journal of Human
Genetics* 6:4–11.

Retsinas, J. 1991. "Impact of Prenatal Technology on Attitudes toward Disabled
Infants." In *Research in the Sociology of Healthcare,* edited by D. Wertz,
89–90. Westport, CT: JAI Press.

Rivera y Carlo, R. 2002. "Targeting the Disabled." *Boundless Magazine.*
Accessed January 26, 2003. www.boundless.org/2002_2003/features
/a0000685.html.

Savulescu, J. 1999. "Sex Selection: The Case For." *Medical Journal of Australia*
171:373–75. Accessed January 26, 2003. www.mja.com.au/public
/issues/171_7_041099/savulescu/savulescu.html.

Sherwin, S. 1992. *No Longer Patient: Feminist Ethics and Health Care.*
Philadelphia: Temple University Press.

Singer, P. 2001. "Response to Mark Kuczewski." *American Journal of Bioethics*
1 (3): 55–57.

Skene, L. 1993. "Why Prenatal Screening Is Not Eugenics. *Age,* A12.

Sutton v. United States. 1998. Accessed January 28, 2003. www.washingtonpost
.com/wp-srv/national/longterm/supcourt/1998–99/sutton.htm.

Tannsjorn, T. (1998). "Compulsory Sterilisation in Sweden." *Bioethics* 12 (3):
236–49.

Toynbee, P. 2001. "Rights Are for the Living." *Guardian* (London), August 24.
Accessed January 26, 2003. www.guardian.co.uk/comment/story/0,3604,
541665,00.html.

United Nations Educational, Scientific, and Cultural Organization. Working Group of the International Bioethics Committee. 2002. *Draft Report on Pre-implantation Genetic Diagnosis and Germ-Line Intervention.* Accessed January 26, 2003. www.unesco.org/ibc/en/actes/s9/ibc9draftreportPGD.pdf.

US President's Commission for the Study of Ethical Problems in Medicine and Biomedical and Behavioral Research. 1983. *Screening and Counseling for Genetic Conditions.* Washington, DC: US Government Printing Office.

Wendling, M. 2001. "UK Authorities Look to Tighten Sex Selection Laws." *CNSNews,* November 5. Accessed January 26, 2003. www.cnsnews.com /ViewForeignBureaus.asp?Page=/ForeignBureaus/archive/200111 /FOR20011105h.html.

Wertz, D. C. 2000. "Drawing Lines for Policymakers." In *Prenatal Testing and Disability Rights,* edited by E. Parens and A. Asch. Washington, DC: Georgetown University Press.

Wertz, D. C., and J. C. Fletcher. 1989. *Ethics and Human Genetics: A Crosscultural Perspective.* New York: Springer Verlag.

———. 1993. *A Critique of Some Feminist Challenges to Prenatal Diagnosis.* Accessed January 26, 2003. www.shriver.org/Research/SocialScience/Staff /Wertz/critique.htm.

———. 1998. "Ethical and Social Issues in Prenatal Sex Selection: A Survey of Geneticists in 37 Nations." *Social Science and Medicine* 46:255–73. Accessed January 26, 2003. www.shriver.org/Research/SocialScience/Staff/Wertz /sexselect.htm.

Wolbring, G. (2000). *Science and the Disadvantaged* (Occasional Paper of the Edmonds Institute). Accessed January 26, 2003. www.edmonds-institute .org/wolbring.html.

———. 2001. *Folgen der Anwendung genetischer Diagnostik fuer behinderte Menschen* [Consequences of the application of genetic diagnostics for disa-bled people] (Expert opinion for the Study Commission on the Law and Ethics of Modern Medicine of the German Bundestag). Accessed January 26, 2003. http://www.bundestag.de/gremien/medi/medi_gut_wol.pdf.

———. 2003. "Confined to Your Legs." In *Living with the Genie: Essays on Technology and the Quest for Human Mastery,* edited by A. Lightman, D. Sarewitz, and C. Dresser. Washington, DC: Island Press.

World Health Organization. 1999. *Draft Guidelines on Bioethics 1999.* http:// helix.nature.com/wcs/b23a.html.

10. Differences from Somewhere

The Normativity of Whiteness in Bioethics in the United States

Catherine Myser

THE DUTIES OF SELF-REFLECTION AND SELF-REMEDY IN BIOETHICS

When examining the cultural practice of academics . . . we cannot separate questions about method, theory, and knowledge construction from questions about origins and standpoints based on, for example, ethnicity, class, gender, sexual orientation, physical abilities or disabilities, religion, nation, region, language, and/or academic discipline. In bioethics, however . . . we have not paid as much attention as we should have to the origins and standpoints of dominant theories and methods in the field. I say should here because I do think that such self-reflection and self-remedy are an obligation, even an ethical imperative, especially for those working in the field of ethics. This is important for individual academics . . . in relation to their own work, but it is particularly important for the majority of "the field" that constructs the dominant, mainstream theories and methods. Such self-reflection and self-remedy are particularly salient as we consider at this point in the evolution of the field of bioethics questions of origin and knowledge such as Where does medical ethics come from? Which medicine? Whose ethics? Is there an implicit or tacit ethic to medicine? Who knows medical ethics? Is biomedical ethics a matter of common sense and common knowledge?

This publication is derived, in part, from an article by Catherine Myser, "Differences from Somewhere: The Normativity of Whiteness in Bioethics in the United States." *The American Journal of Bioethics* 3.2 (Spring 2003): 1–11. doi: 10.1162/152651603766436072. Reproduced by permission of Taylor and Francis Group, LLC, a division of Informa plc. (visit http://www.informaworld.com). Due to space limitations, portions of this chapter have been deleted or edited from the original.

More specifically then . . . we have inadequately noticed and questioned the dominance and normativity of whiteness in the cultural construction of bioethics in the United States. By thus allowing the whiteness of bioethics to go unmarked, we risk repeatedly reinscribing white privilege—white supremacy even—into the very theoretical structures and methods we create as tools to identify and manage ethical issues in biomedicine. In other words, by not seeing or locating the whiteness in bioethics, by theorizing from this unself-reflective white standpoint and by extending its cultural capital into bioethics policies and practices, we risk functioning as cultural colonizers who do violence to social-justice concerns related to race and class.

What I am writing about here may seem controversial, but it will seem even more so if it is misunderstood. To head off misunderstandings, I would like to specify what I am not doing and what I am doing in this article. Although the 2001–2002 American Society of Bioethics and Humanities member survey identified that responding members working in the field of bioethics are overwhelmingly white/Caucasian, and although I regard this as a serious concern for the field, I am not talking about skin color, or white people, or "white Anglo-Saxon Protestant" people per se. That would be a very different endeavor. Rather, by talking about whiteness I am talking about a marker of location or position within a social, and here racial, hierarchy—to which privilege and power attach and from which they are wielded—and how this is complicated by a forgetting of the history of whiteness in the United States and by its current invisibility. Furthermore, by focusing on the dominance of whiteness in the cultural construction of bioethics in the United States, I do not wish to imply that whiteness does not operate in bioethics in other countries. Clearly the issue of whiteness is not simply a problem in the United States. As such, the question of whiteness concerns not only those who are shaping bioethics theory and policy but also those who will be affected by this theory and policy in the United States and elsewhere.

In making this claim, I would like to acknowledge that feminist bioethics has made significant efforts to address social-justice concerns—primarily related to gender, but increasingly to race and class—and accordingly to revise mainstream bioethics. However, given legitimate ongoing critiques by women of color regarding the dominance and unconscious positioning of the views of conservative, white, North American women in feminism in general, whiteness is a subject that feminists also need to interrogate in greater depth.

My own goal is to begin marking the unmarked marker status of whiteness in the history and practice of bioethics in the United States and thus to begin to color the seeming invisibility of white epistemologies and performance in its academic corpus. First, I will define what I mean by whiteness in the specific context of the relevant legal and social history of the United States as a nation. In this context—in which whiteness and American-ness are inextricably linked—I will recast Fox's already revealing work on the sociology of bioethics in the United States, in which she tags its American-ness as an important initial marking of its whiteness. Next, I will briefly consider the attempts of social scientists and others to highlight sociocultural "diversity" as a corrective in bioethics. To explore and analyze such diversity work at greater length would be a useful future endeavor, but it is beyond the scope of this article.

Thus, I will merely point out a key gap to date in the work of these diversity researchers: because these scholars fail to problematize white dominance and normativity and white-other dualism when they study and describe the "unique" and "varying" standpoints of African American, Asian American, and Native American others, their work merely inoculates difference in bioethics.

DEFINING WHITENESS: THE US CONTEXT

When I lament the fact that the vast majority of those working in bioethics in the United States are white/Caucasian, I do not lament the color of that bioethics so much as the apparent invisibility—and thus unreflective acceptance—of its dominance and normativity to most in and outside of the field. To understand what I am problematizing by marking the whiteness of bioethics in the United States, it is instructive to recall that moment in its history when "the first Congress . . . under [its] Constitution voted in 1790 to require that a person be 'white' in order to become a naturalized citizen." In other words, "the very claiming of [American-ness] involved . . . a claiming of whiteness" (Roediger 1994), which referred not to color but to a relational and hierarchical position or location conferring social and legal status, rights, privileges, and power.

The social standpoint in question was that of the dominant and normative white Anglo-Saxon Protestant (WASP), who was at that moment in US history engaged in declaring the British American white as white—as the real American self. Against this center of whiteness and American-ness, cultured or ethnic "others"—from Blacks to Irish to Jews—were differentiated and thus denied such status and privileges, although the latter two groups eventually won the status and rights of "new white" ethnics. WASP dominance and normativity was accorded actual legal status, "[converting] whiteness from privileged identity to a vested [property] interest" and relationship. That is, "[the] law's construction of whiteness defined and affirmed critical aspects of identity (who is white); of privilege (what benefits accrue to that status); and of property (what legal entitlements arise from that status)" (Harris 1998).

Accordingly, DuBois could validly assert three decades later [in 1920]: "The discovery of personal whiteness among the world's peoples is a very modern thing—a nineteenth and twentieth century matter . . . whiteness is the ownership of the earth forever and ever." Much more recently Baldwin reminded us that "No one was white before he/she came to America. It took generations, and a vast amount of coercion. . . . White men—from Norway, for example, where they were Norwegians—became white . . . and we—who were not Black before we got here either, were defined as Black by the slave trade" (Roediger 1994). Thus WASP whiteness and American-ness operated interchangeably in the history of this nation as locations of cultural—and specifically racial—dominance and normativity, and to become American, one had to assimilate WASP whiteness, culture, values, and practices.

MARKING WHITENESS IN THE CULTURAL CONSTRUCTION OF BIOETHICS IN THE UNITED STATES

If white academics in particular forget this all too pertinent and ever-unfolding US history and operate in a race- and power-evasive manner in our construction of bioethical theories, knowledge, methods, and policies, we risk reproducing white privilege and supremacy in our own cultural practice. That is why I am arguing that we need to be more vigilant and self-reflective regarding our production of knowledge and our cultural practice within the largely racially homogenous (i.e., Caucasian) zone of bioethics in the United States. Such vigilance will enable us to enter into a more complex, dynamic, and historically situated analysis of our own positioning—particularly as white academics in bioethics—in a racially hierarchical society. . . .

SOCIOCULTURAL DIVERSITY AND DIFFERENCE IN UNITED STATES BIOETHICS: GENUINE CORRECTIVE OR MERE INOCULATION?

. . . A growing number of social scientists and other professionals working in US bioethics have sought to introduce sociocultural diversity and difference as further correctives in the field, thus going beyond [mere] . . . increased disciplinary diversity. The hope is that "ethicists [will be] made to rethink their agendas" (DeVries 1990) and adjust their theorizing and practice by understanding and incorporating the "unique" and "varying" sociocultural contexts and standpoints, values, beliefs, and practices of, for example, African American, Asian American, Hispanic American, and Native American "others." I support the goal of increased sociocultural diversity in bioethics. However, I argue that unless researchers of diversity and difference additionally problematize white dominance and normativity and the white-other dualism when they study and describe the beliefs and practices of other ethnic groups, their work merely legitimates and maintains "minoritized spaces" in bioethics, and these minority spaces will remain marginal in relation to the unmarked, dominant "majority space" of white theory and practice in US bioethics. . . .

In other words, introducing sociocultural diversity and difference without recognizing, highlighting, and problematizing the dominance and normativity of whiteness in US bioethics merely inoculates difference in bioethics, ignoring the question of against whose invisible and seemingly neutral norms such "difference" is defined; ignoring the need to subject whiteness to the same kind of surveillance and exploration; and thus not only failing to decenter whiteness, but effectively maintaining its dominance. By invoking Barthes's (1972) concept of "inoculation," I mean to argue that these forms of apparently critical or oppositional thinking are ineffective in the end, maintaining and enabling rather than displacing white normativity and its dominant, dualistic stance in relation to the other. Accordingly, rather than challenging mainstream bioethics theory and practice as intended, it merely stimulates a kind of immune response, leaving the main body of traditional bioethics intact. This is true because researchers of diversity and difference fix our gaze on the other, allowing whiteness to stand as the unmarked or neutral category (unrecognized absence) in relation to which the "difference" of "others" (recognized presence) emerges in high relief. Thus, the dominant gaze is never focused "behind the mirror," and white theorizing and practice in bioethics remain wholly unaffected and unchanged by all this apparent diversity and difference, thereby remaining deceptively neutral and invisibly normative. This dominant white center must be problematized, displaced, and relocated for diversity work to make a difference in determining what counts as an ethical issue and to adjust or revise dominant bioethics values (e.g., hyperindividualism and truth-telling) and concepts (e.g., autonomy). . . .

I want to be clear that by making these critiques of existing sociocultural diversity research in US bioethics, I am not arguing that such research is unwelcome and could never be useful. Rather, I am arguing that such diversity research cannot be truly effective until it is accompanied by equal efforts to recognize and decenter unself-reflective white theorizing and practice in the field, against whose norms such sociocultural "difference" is currently defined and constructed. Indeed, it is through this very discursive process that whiteness actually maintains and reinscribes itself in bioethics.

POSTCOLONIAL BIOETHICS DISCOURSE, THEORY, AND PRACTICE: A CHALLENGE FOR A MORE SELF-REFLECTIVE AND INCLUSIVE BODY BIOETHIC

White academics in particular must find effective ways to explore their ethnicity and class (e.g., as academic elites), as well as other social positionings and norms operating in the cultural construction of bioethics in the United States and elsewhere. All who work in the field of bioethics need to recognize and name whiteness and carefully consider its implications for the theories and practices of the field—to decenter "whiteness as ownership of the world forever and ever" (DuBois 1920) and to enable bioethics better to serve a richly pluralistic society in the United States and elsewhere.

For example, those of us who participate in bioethics research, publishing, and presenting at conferences need to notice not only who is in the field—seeking and creating ever greater sociocultural diversity among such professionals—but also, as [bell] hooks (1997) urges a different audience, "when the usual arrangements of white supremacist hierarchy [are] mirrored in terms of who [is] speaking, who [is] in the audience, what voices [are] deemed worthy to speak and be heard, [and] how bodies [are] arranged on the stage" and seek to remedy this. Those among us who are medical educators need to see more clearly as well as study and challenge "the varied practices that constitute and reproduce medical professions in the United States, [the] set of institutional routines and 'white cultural practices' . . . evident in establishing and maintaining privileges generally associated with being white" (Hartigan 1997).

We must also find or create more democratic methods by which to engage equal collaborators from the broad range of voices and visions currently suppressed, silenced, or excluded in United States bioethics. "Other" or "different" voices should not be sought merely to be maintained in . . . minoritized spaces, to be interpreted or translated in and through the majority space. Rather, such voices and visions should be sought fully to participate in creating and revising "mainstream" bioethics issues, concepts, theories, paradigms, policies, and practices. For example, a method that engages bioethics researchers and participants (from marginalized communities in particular) as equal collaborators—such as community-based participatory research (CBPR)—will be invaluable. The CBPR method can open a space of contact between minoritized spaces and majority space, consciously undermining and erasing such hierarchical positionings. Through CBPR, greater democracy in bioethics theorizing and policy making can be achieved because "researchers" and "participants" are interactively and interdependently linked, without a hierarchical relationship serving as a means of cultural control. Accordingly, social meanings and values, as well as bioethics issues, standards, and policies, are relationally negotiated, and bioethics knowledge is socially created and constructed.

I do not assume that the above endeavors will be easy or that they will lead to quick fixes of bioethics theory and practice. To the contrary, such endeavors pose extremely complex challenges. McIntyre (1997), for example, describes and warns of "white talk" as a means of "actively subvert(ing) the language white people need to decenter whiteness as a dominant ideology." White talk can insulate white people from examining their own individual and collective roles in racism and from related self examination and critique, so the "dilemmatic nature of white talk" is such that "white people [need to be engaged] in conversations about whiteness while simultaneously being cognizant of the strategies we use to derail those discussions." Such challenges, and the ongoing self-reflective vigilance they require of us, make it all the more important that we engage in them routinely. This is not work that ends but rather human history that goes on forever as categories such as whiteness and cultural practices such as the production of bioethics knowledge and standards evolve, are performed and interrogated, and continue to evolve. Keeping in mind that "theory is always written from some 'where'" (Clifford 1989), we need to decolonize our minds, imaginations, theories, concepts, and practices. This article is only a beginning.

REFERENCES

Baldwin, J. 1984. "On Being White . . . and Other Lies." In *Black on White: Black Writers on What It Means to Be White*, edited by D. R. Roediger. New York: Schocken Books.

Barthes, R. 1972. *Mythologies*. Translated by A. Lavers. New York: Hill and Wang.

Braun, K. L., and R. Nichols. 1997. "Death and Dying in Four Asian American Cultures: A Descriptive Study." *Death Studies* 21 (4): 327–59.

Brodkin, K. 1998. *How Jews Became White folks and What That Says about Race in America*. New Brunswick, NJ: Rutgers University Press.

Brookhiser, R. 1997. "Others and the WASP World They Aspired To." In *Critical White Studies: Looking behind the Mirror,* edited by R. Delgado and J. Stefancic, 362–67. Philadelphia: Temple University Press.

Browner, C.H., et al. 1999. "Ethnicity, Bioethics, and Prenatal Diagnosis: The Amniocentesis Decisions of Mexican-Origin Women and Their Partners." *American Journal of Public Health* 89 (11): 1658–66.

Carrese, J.A., and L.A. Rhodes. 1995. "Western Bioethics on the Navajo Reservation." *JAMA* 274 (10): 826–45.

Clifford, J. 1989. "Notes on Travel and Theory." In *Inscriptions 5: Traveling Theories, Traveling Theorists,* edited by J. Clifford and V. Dhareshwar. Santa Cruz: UCSC Center for Cultural Studies.

DeVries, R. 1990. Review of *Social Science Perspectives on Medical Ethics,* edited by G. Weisz. *Social Science and Medicine* 33.

DuBois, W.E.B. 1920. "The Souls of White Folk (from *Darkwater*)." In *W.E.B. DuBois: Writings,* edited by N. Huggins. New York: Library of America, 1986.

Dula, A., and S. Goering, eds. 1994. *It Just Ain't Fair: The Ethics of Health Care for African Americans.* Westport, CT: Praeger.

Eagleton, T., et al. 1990. *Nationalism, Colonialism and Literature.* Minneapolis: University of Minnesota Press.

Flack, H.E., and E.D. Pellegrino, eds. 1992. *African-American Perspectives on Biomedical Ethics.* Washington: Georgetown University Press.

Fox, R. 1990. "The Evolution of American Bioethics." In *Social Science Perspectives on Medical Ethics,* edited by G. Weisz. Boston: Kluwer Academic.

Frankenberg, R. 1993. *The Social Construction of "Whiteness": White Women, Race Matters.* Minneapolis: University of Minnesota Press.

———. 1994. "Whiteness and Americanness: Examining Constructions of Race, Culture, and Nation in White Women's Life Narratives." In *Race,* edited by S. Gregory and R. Sanjek. New Brunswick, NJ: Rutgers University Press.

———. 1997. "White Women, Race Matters: The Social Construction of Whiteness." In *Critical White Studies: Looking behind the Mirror,* edited by R. Delgado and J. Stefancic. Philadelphia: Temple University Press.

Fussell, P. 1983. *Class: A Guide through the American Status System.* New York: Touchstone.

Giordano, J., and M. McGoldrick. 1996. "European Families: An Overview." In *Ethnicity and Family Therapy,* edited by M. McGoldrick, et al. 2nd ed. New York: Guilford Press.

Harris, C. 1998. "Whiteness as Property." In *Black on White: Black Writers on What It Means to Be White,* edited by D.R. Roediger. New York: Schocken Books.

Hartigan, J. 1997. "Establishing the Fact of Whiteness." *American Anthropologist* 99 (3): 495–506.

Hern, H.E. Jr., et al. 1998. "The Difference That Culture Can Make in End-of-Life Decisionmaking." *Cambridge Quarterly of Healthcare Ethics* 7 (1): 27–40.

Hoffmaster, B. 1992. "Can Ethnography Save the Life of Bioethics?" *Social Science and Medicine* 35 (12): 1421–31.

hooks, b. 1990. *Yearnings: Race, Gender, and Cultural Politics.* Boston: South End Press.

———. 1997. "Representing Whiteness in the Black Imagination." In *Displacing Whiteness: Essays in Social and Cultural Criticism,* edited by R. Frankenberg. Durham, NC: Duke University Press.

———. 2000. *Where We Stand: Class Matters.* New York: Routledge.

Ignatiev, N. 1995. *How the Irish Became White.* New York: Routledge.

Israel, B.A., et al. 1998. Review of *Community-Based Research: Assessing Partnership Approaches to Improve Public Health. Annual Review of Public Health* 19:173–202.

Jecker, N., et al. 1995. "Caring for Patients in Cross-Cultural Settings." *Hastings Center Report* 25 (1): 6–14.

Kaufert, J., et al. 1991. "The Cultural and Political Context of Informed Consent for Native Canadians." *Arctic Medical Research* (suppl.): 81–184.

King, P.A., and L.E. Wolf. 1998. "Lessons for Physician Assisted Suicide from the African-American Experience." In *Physician Assisted Suicide: Expanding the Debate,* edited by P. Battin, et al. New York: Routledge.

Laguerre, M. 1999. *Minoritized Space: An Inquiry into the Spatial Order of Things.* Berkeley: University of California, Institute of Governmental Studies Press.

Lopez, Ian F. 1996. *White by Law: The Legal Construction of Race.* New York: New York University Press.

Marshall, P.A. 1992. "Anthropology and Bioethics." *Medical Anthropology Quarterly* 6 (1): 49–73.

———. 2001. "The Relevance of Culture for Informed Consent in U.S.-Funded International Health Research." Commissioned Paper. In *Ethical and Policy Issues in International Research: Clinical Trials in Developing Countries.* Vol. II. Bethesda, MD: National Bioethics Advisory Commission.

Marshall, P.A., et al. 1998. "Multiculturalism, Bioethics, and End-of-Life Care: Case Narratives of Latino Cancer Patients." In *Health Care Ethics: Critical Issues for the 21st Century,* edited by J. Monagle and D. Thomasma, 421–31. Gaithersburg, MD: Aspen.

McGill, D., and J.K. Pearce. 1982. "British Families." In *Ethnicity and Family Therapy,* edited by M. McGoldrick, et al. New York: Guilford Press.

McIntyre, A. 1997. *Making Meaning of Whiteness: Exploring Racial Identity with White Teachers.* Albany: State University of New York Press.

Muller, J.H. 1994. "Anthropology, Bioethics, and Medicine: A Provocative Trilogy." *Medical Anthropology Quarterly* 8 (4): 448–67.

Murray, R.F. 1992. "Minority Perspectives on Biomedical Ethics." In *Transcultural Dimensions in Medical Ethics,* edited by E. Pellegrino, et al., 35–42. Frederick, MD: University Publishing Group.

Orona, C., et al. 1994. "Cultural Aspects of Nondisclosure." *Cambridge Quarterly of Healthcare Ethics* 3 (3): 338–46.

Perkins, H.S. et al. 1993. "Autopsy Decisions: The Possibility of Conflicting Cultural Attitudes." *Journal of Clinical Ethics* 4 (2): 145–54.

Roediger, D. 1994. *Towards the Abolition of Whiteness.* New York: Verso.

Sacks, K. 1998. "How Jews Became White." In *Race, Class, and Gender in the United States: An Integrated Study,* edited by P.S. Rothenberg, 100–114. New York: St. Martin's Press.

Said, E. 1978. *Orientalism.* New York: Random House Vintage Books.

Sanders, C. 1992. "Surrogate Motherhood and Reproductive Technologies: An African American Perspective." *Creighton Law Review* 25 (5): 1707–18.

———. 1994. "European-American Ethos and Principlism: An African-American Challenge." In *A Matter of Principles? Ferment in U.S. Bioethics,* edited by Edwin R. DuBose, et al., 148–63. Valley Forge, PA: Trinity Press International.

Tatum, B.D. 1997. *Why Are All the Black Kids Sitting Together in the Cafeteria?* New York: Basic Books.

11. Bioethical Silence and Black Lives

Derek Ayeh

When confirmation was released that researchers from China had genetically modified human embryos for the first time ever [in spring 2015], there was a sudden explosion of activity on the Web from the bioethics community. Physicians, academics, and anyone else who could claim some affiliation to the field wrote articles for magazines discussing the ethical dimensions of the issue. After all, human enhancement and genetic modification are staples of bioethical discourse. Who wouldn't want to add their two cents and take part in such an important discussion?

Conversely, when the news of Freddie Gray's death became public [a young Black man who died at the hands of Baltimore police officers in April 2015], I was greeted by a surprising but familiar bioethical silence. Surprising because I thought that the relationship between Freddie Gray's death and bioethics was rather obvious: here was a man who requested healthcare numerous times but was refused it—the justification being that he was a criminal and either faking his pain or self-inflicting it. While there are likely numerous reasons why Freddie Gray died, do bystanders have moral responsibilities when they witness an injured person? There's often debate about whether a bystander has a moral responsibility to intervene. However, as public servants, police officers surely have some ethical responsibility to ensure that even criminals receive medical treatment when badly injured.

It is ethically troubling that individuals charged with protecting the public ignored a man who was begging for and needed immediate medical

Originally appeared as "Bioethical Silence and Black Lives." *Voices in Bioethics: An Online Journal.* August 3, 2015. www.voicesinbioethics.net/opeds/2015/08/03/bioethical-silence-black-lives#_edn1.

treatment.[1] The dimensions of the situation also intrigued me: even incarcerated individuals are entitled to receive "adequate" healthcare, so on what moral grounds does a police officer stand when he/she ignores the cries for treatment of someone who has been seriously injured? While other aspects of the case concern me, these were questions I was able to ask purely as a student of bioethics—questions I thought bioethicists should have opinions about and be interested in discussing. Yet, while bioethicists have had no issue condemning genetic experimentation, they seem to sew their mouths shut on the matter of black lives.

The lack of any analysis or statement from bioethicists on Freddie Gray's case is familiar. Just last year, I was greeted by a similar silence from my field in response to Eric Garner [a black man who similarly died at the hands of New York City police officers in July 2014]. It is even easier to claim his case is within the realms of bioethical inquiry—those squeamish about discussing race could ask: Why was Garner refused CPR, the standard of care? While some claimed that Garner was still breathing (despite his now famous last words, "I can't breathe"), the union president for EMTs and paramedics, Israel Miranda, was quoted widely as saying that the emergency medical team that arrived at the scene ignored the state protocol of supplying oxygen to an individual having difficulty breathing.[2] The ethics of the case are fairly easy—the standard of care exists for a reason, and the refusal to uphold it should have been condemned.

Even the . . . death of Sandra Bland overlaps with bioethical inquiry. Though there is currently dispute over whether her death was a suicide, let's assume that everything that Waller County Jail has told the public is true. Her intake forms indicate that she attempted suicide in the previous year.[3] There is a standard of medical care for inmates who may be actively suicidal or have exhibited past suicidal behavior. Waller County Jail failed to remove the plastic bag from her cell as a potential tool of self-harm and did not keep her under close surveillance.[4] If it was truly a suicide, Sandra Bland's death was the result of the jail withholding the standard of care. As bioethicists we know that disregarding the principle of justice is akin to asking for a health care scandal. We learn about the Tuskegee syphilis study so we can recognize how racism, inequality, and poverty can affect what type of health care an individual receives. Denouncing medical injustice doesn't end because the victim isn't a patient and the crime scene isn't a hospital.

Why hasn't bioethics spoken up about the "Black Lives Matter" movement, especially when the health profession at large has contributed widely to the discussion? Medical students have hosted "white coat die-ins" to show that they stand in solidarity with the protesters in Ferguson and

Baltimore. Public health officials like Dr. Mary Bassett, New York City's health commissioner, have taken this opportunity to try and educate the public about the connections between health and racism. The health professions have been entrenched in our nation's conversation over the value of black lives and the problems that persons of color face when they come into contact with the criminal justice system. These professionals didn't wait for an invitation to speak their minds on these issues, as there was never any question that the health professions belong in this conversation. Still, bioethics shies away.

Leigh Turner, associate professor for the Center for Bioethics at the University of Minnesota, has criticized bioethics for exactly this tendency. He believes that the field is obsessed with the "cutting-edge."[5] Anytime a new technology or innovation that impacts human health springs up, so do the bioethicists. We are even drawn to ethical issues that are still decades away, while questions of race or inequality fall just outside of our purview. With just a quick search, hundreds of papers come up covering human enhancement and cloning. However, bioethics often overlooks poverty, unemployment, and gun violence, leaving them for other health care professionals. The academics in our field are fixated on high technology, which, as Turner points out, paints a picture of bioethical inquiry as only being useful for addressing the concerns of the upper and middle classes. He likely sees us as the epitome of elitist scholars commenting on the world from our ivory tower.

However, I see our field's silence as a problem for a very different reason than Turner's. The late Adrienne Asch once wrote, "Bioethics is at its best when people don't merely ask each other what their views are, but really take the time to find out what is behind those views."[6] She believed that bioethical inquiry was a profound tool that could transcend the drudgery of political polarization and get at the foundations of why people believe what they believe. The ideal bioethicist not only articulates his or her own views but also understands what perceptions and life experiences shape those views and can see what values lie behind the views of others. To Asch, bioethicists are capable of thinking far beyond simply identifying with the political "left" or "right," probing further to find out why people think differently in order to stimulate reasonable discourse.

I want to believe that Adrienne Asch is right and that I study bioethics because of its potential to find resolutions through reasonable discourse. Her vision of bioethics is truly striking and certainly not what the field always is, but what it should constantly aspire to be. Most importantly, it is Asch's bioethics that would be truly useful in our country's current debate over racism and black lives. It's hard to even call it a debate—we constantly spend our time talking over one another and raising our voices louder in hopes that someone will hear our views. Our country is completely divided on this issue. Where some see criminals and rioters, others see disenfranchised individuals and suffering communities.

I am not saying that a handful of bioethicists writing about these issues will correct the gap between our perceptions, nor do I think that the field should stop talking about technological advancements. Nevertheless, I do believe that we have a place in this important conversation and hope that those in the field with far more experience than I will seize the opportunity to make bioethics more than it is today.

NOTES

1. Ed Payne, "We Failed to Get Freddie Gray Timely Medical Care after Arrest," CNN, April 24, 2015, www.cnn.com/2015/04/24/us/baltimore-freddie-gray-death/.

2. Benjamin Mueller, "Medical Workers Face Scrutiny after Man's Death in Police Custody," *New York Times*, July 21, 2014, www.nytimes.com/2014/07/22/nyregion/medical-workers-face-scrutiny-after-mans-death-in-police-custody.html?_r=0.

3. Ben Mathis-Lilley, "Sandra Bland Reportedly Told Jail Staff She'd Previously Attempted Suicide," *Slate*, July 22, 2015, www.slate.com/blogs/the_slatest/2015/07/22/sandra_bland_previous_suicide_attempt_jail_intake_form_disclosed_attempt.html.

4. Dana Liebelson, "A Texas Jail Failed Sandra Bland, Even if It's Telling the Truth about Her Death," *Huffington Post*, July 21, 2015, www.huffingtonpost.com/entry/sandra-bland-jail-death_55ae9f12e4b07af29d569875.

5. Leigh Turner, "Bioethics, Social Class, and the Sociological Imagination," *Cambridge Quarterly of Healthcare Ethics* 14, no. 4 (2005): 374–78.

6. A. Asch, "Big Tent Bioethics: Toward an Inclusive and Reasonable Bioethics," *Hastings Center Report* 35, no. 6 (2005): 11–12.

12. The Ethicists

Carl Elliott

. . . Bioethics has always had an ambivalent relationship with power. The field emerged in the 1960s, when protesting against authority was practically a generational requirement for those of a certain age, and many ethicists see it as their role to speak for the powerless—for patients against doctors, for research subjects against sponsors, for the medically underserved against the insurers and the bureaucrats. Like any new field, bioethics has struggled to establish its legitimacy; bioethicists must be taken seriously if their work is going to have any effect. As the field has come of age, however, many bioethicists have grown uneasy with their rising status. It is one thing to be a social critic and another thing entirely to be the voice of moral authority. Some scholars have recoiled, emphatically rejecting the notion that their voices should count more than others' on ethical affairs. A few cringe at the term bioethicist, which carries a slightly self-important air of professionalism and expertise. Yet others display their moral authority like a newly promoted captain showing off his stripes. For them, bioethics is a professional service that deserves social respect, market compensation, and the occasional appearance on a television news program.

A handful of bioethicists have started their own businesses. When Glenn McGee was unexpectedly dismissed as director of the Alden March Bioethics Center at Albany Medical School, he registered a for-profit consulting business called BENE, or the Bioethics Education Network, LLC. He told the *Business Review* that he was "trying to build what will become at least a $500,000 business in Albany."[1] David Perlman used to work as an ethicist for GlaxoSmithKline; today he works at the University of Pennsylvania, where he has founded Eclipse Ethics Education Enterprises, LLC. (He is also identified as "the inventor of the Crucial Choices learning format," which is based on his experience at GlaxoSmithKline developing "ethics cases wrapped around a Jeopardy-type game format for our senior leadership teams."[2]) Bruce Weinstein, who markets himself as the "Ethics Guy," was a faculty member in bioethics at the West Virginia University Health Sciences Center before he started a for-profit service called Ethics at Work. He has published a series of self-help ethics books offering moral advice on everything from dating to personal hygiene. You can measure your "ethics IQ" on his website.[3]

Other scholars have entered the consultation market in a more modest way. In the late 1990s, when issues such as human cloning and embryonic stem cell research first became topics of heated ethical discussion, many biotechnology companies began setting up bioethics advisory boards. These boards meet periodically to consider issues and offer the companies advice. A list of those reported to have served on such advisory boards reads like a who's who of bioethics: Nancy Dubler of Montefiore Medical for DNA Sciences; Ronald Green of Dartmouth for Advanced Cell Technology; Arthur Caplan of the University of Pennsylvania for Celera Genomics and DuPont; Karen Lebacqz of the Pacific School of Religion, and Laurie Zoloth of Northwestern for Geron Corporation.[4] Consultation fees vary considerably. According to the *New York Times*, in 2001 Advanced Cell Technology was paying its advisory board members a mere two hundred dollars per meeting while Geron paid a thousand dollars, Celera compensated Arthur Caplan annually in stock options; he converted them to cash and donated the money to the Center for Bioethics at the University of Pennsylvania, which he directs. One year that stock was worth more than a hundred thousand dollars.

Not surprisingly, these industry payments began to generate ethical questions of their own. Barbara Koenig, then at Stanford University, said that she was paid two thousand dollars per meeting for one corporate ethics board, an amount that troubled her. "I realized that the only reason I was going to this was because I was getting paid," she said. "And that raised a red flag, and I immediately resigned."[5] But many bioethicists insist that they are learning from their industry relationships and shaping company policy for the better. Over the years, the list of bioethics advisers to industry has grown: Jonathan Moreno of the University of Pennsylvania for GlaxoSmithKline, James Childress of Virginia for Johnson and Johnson,

Tom Beauchamp of Georgetown, and Robert Levine of Yale for Eli Lilly.[6] A task force commissioned by the two major American professional bioethics bodies, the American Society for Bioethics and Humanities and the American Society of Law, Medicine and Ethics, concluded that private corporations should be encouraged to seek out paid bioethics consultants, because "bioethics will have an impact on that [corporate] activity only if bioethicists can be part of the dialogue." In fact, the task force did not merely encourage bioethics consultation; it suggested that bioethicists might consider advertising their services. (Interestingly, the task force refused to disclose the names of the companies for which its members had consulted).[7]

Defenders of corporate consultation often bristle at the suggestion that accepting money from industry compromises their impartiality or makes them any less objective as moral critics. "Objectivity is a myth," DeRenzo told me, marshaling arguments from feminist philosophy to bolster her cause. "I don't think there is a person alive who is engaged in an activity who has absolutely no interest in how it will turn out." Thomas Donaldson, director of the ethics program at the Wharton School, has compared ethics consultants to the external accounting firms often employed by corporations to audit their financial records.[8] Like accountants, ethicists may be paid by the very industries they are assessing, but they are kept honest by the need to maintain a reputation for integrity.

But ethical analysis does not really resemble a financial audit. If a company is cooking its books and the accountant closes his eyes to this fact in his audit, the accountant's wrongdoing can be reliably detected and verified by outside monitors. It is not so easy with an ethics consultant. Ethicists have widely divergent views. They come from different religious standpoints, use different theoretical frameworks, and profess different political philosophies. They are also free to change their minds at any point. How do you tell the difference between an ethics consultant who has changed her mind for legitimate reasons and one who has changed her mind for money? What distinguishes the consultant who has been hired for her integrity from the one who has been hired because her moral viewpoint lines up nicely with that of company executives?

In *Merchants of Immortality*, a book that examines the history of the biotechnology industry, Stephen Hall offers little evidence that the teams of bioethicists contracted by two biotechnology companies, Geron and Advanced Cell Technology, played any meaningful role in shaping company policy. In one case, Hall calls the ethical review a "midwife to fund-raising." His assessment is reinforced by Michael West, the biologist-entrepreneur who founded Geron and who later became CEO of Advanced Cell Technology. When questioned about the ethicists he recruited to advise Advanced Cell Technology on their ethically controversial research, West answered, "In the field of ethics, there are no ground rules, so it's just one ethicist's opinion versus another ethicist's opinion. . . . You're not getting whether something is right or wrong, because it all depends on who you pick."[9] Few executives will pick an ethicist who wants to put the brakes on the corporate mission.

West's comment highlights a problem with ethics consultation. Any ethical problem can be approached from many different perspectives, each of which will come with its own subtleties and nuances and compromises. It is entirely possible that puzzled executives may want to hire an ethicist to guide them through some perilous terrain. However, they might also simply want a congenial, like-minded ethicist to provide cover for what they plan to do anyway. And to the ethicist who is hired, this will not feel like a moral compromise. It will feel like working with an ally.

Consider the case of Eli Lilly, which found itself in the media spotlight in 1996 when the *Wall Street Journal* reported that for two decades the company had been paying homeless alcoholics to test drugs at its Phase I clinic in Indianapolis.[10] Company officials insisted that the homeless subjects were driven by altruism. "These individuals want to help society," said Dr. Dwight McKinney, the executive director of clinical pharmacology. The subjects themselves told a different story. "The only reason I came here is to do a study so I can buy me a car and a new pair of shoes," said a twenty-three-year-old former crack addict who had heard about the Lilly clinic on the streets in Nashville. "I'll get a case of Miller and an escort girl and have sex," another subject told the *Journal.* "The girl will cost me $200 an hour."

Unlike many pharmaceutical companies, which contract with university hospitals or contract research organizations to conduct Phase I trials, Lilly had been operating its own testing clinic since 1926. Dr. Leigh Thompson, a former chief scientific officer for Lilly, told the *Journal* that Lilly was already using homeless people as subjects when he arrived at the company in 1982. "We were constantly talking about whether we were exploiting the homeless," he recalled. But he and others felt that the company was offering the subjects a decent bargain. "Providing them with a nice warm bed and good medical care and sending them out drug- and alcohol-free was a positive thing to do." However, the *Journal* noted that Lilly paid subjects the lowest per diem in the business and recruited them from a homeless shelter supported by the Lilly Endowment, a charity funded by Lilly stock. Many of the subjects had alarming stories to tell about their time in the clinic. One said that a Lilly drug had given him a heart problem so bad "they had to put things on my chest to start my heart up again."

Research on human subjects has generated libraries of scholarly commentary. The dominant stream in bioethics sees the crucial issue as informed consent. Can the subject make a free and informed choice about whether to take part in a drug study? Does he know what he is agreeing to do? But a minority opinion sees the issue as one of justice and exploitation. To these scholars, it makes a crucial difference if a subject is sleeping on a hot-air grate and foraging for meals from a dumpster. Poverty, illness, and addiction render subjects vulnerable to exploitation by powerful corporations. And of course, a pharmaceutical company seeking honest moral counsel about testing drugs on homeless alcoholics would get very different answers depending on which ethicists they asked.

After the *Wall Street Journal* story broke, Lilly hired a team of bioethics consultants that included Tom Beauchamp of Georgetown, the co-author of *Principles of Biomedical Ethics*, perhaps the best-known academic textbook on the subject, and Robert Levine of Yale, the editor of *IRB*, a prominent research ethics journal. Both are scholars squarely in the bioethics mainstream. Beauchamp, in fact, is the co-author of a notable book on informed consent. What the bioethicists privately advised Lilly to do is unknown, but given their previous writing and ideologies, their published take on the issue was not surprising. In an article in the *Journal of Medicine and Philosophy*, they and their co-authors concluded, "It is not unethical or exploitative to use homeless people in Phase I studies if the system of subject selection is fair, consents are well informed and bona fide, and the risks are not exceptional for the pharmaceutical industry."[11]

It is hard to know if these ethicists were compromised by their payments from Lilly, since their published answer to the problem is consistent with their past writings. There was no dramatic reversal after the Lilly consultation, no inexplicable turnaround. Equally hard to know is whether Lilly picked these particular ethicists because they seemed likely to endorse what the company wanted to do. Yet if a particular ethicist's answer is thoroughly predictable, what exactly is the purpose of hiring the ethicist to say it? Why not simply read what he or she has written?

For critics of corporate consultation, the main reason companies are setting up advisory boards and hiring consultants is damage control. Ethically controversial actions can be defended by saying "We ran it by our ethicists." When scientists at the Jones Institute of Reproductive Medicine found themselves being scrutinized in 2001 after announcing that they had created embryos for research purposes, they told the media they had consulted with three panels of ethics experts before proceeding.[12] (They would not name the ethicists, however.) The fact that the approval of an ethics board counts as a meaningful justification for controversial actions is testimony to how far the field of bioethics has risen. What is unclear is whether hired ethicists actually have the power to stop unethical actions. Do they modify company policy in a meaningful way, or are they hired to make selling easier?

．　．　．

Historians quibble about when and where the field of bioethics emerged, but many observers agree that bioethics is rooted in scandal. In the United Kingdom the catalyzing event was Maurice Pappworth's 1967 book *Human Guinea Pigs*, which cataloged over two hundred cases of brazen research abuse, such as drilling holes in patients' skulls in order to study the physiology and injecting patients with malaria parasites and cancer cells, often without the subjects' knowledge.[13] In New Zealand, the scandal was the "unfortunate experiment" in the 1960s and 1970s at Auckland Women's Hospital, in which women with precancerous cervical abnormalities were studied but not treated for cancer, despite the availability of treatment.[14] Many of these women subsequently died from cervical cancer. The United States saw an entire catalog of research scandals emerge during this period, the two most notorious of which occurred at the Willowbrook State Hospital in New York, where mentally disabled children were injected with the hepatitis virus, and in Tuskegee, Alabama, where black men with syphilis went untreated for years after treatment was available.[15]

The 1960s and 1970s were also a time of deeply puzzling moral dilemmas, many of them raised by new medical technologies: organ transplantation, ventilators, the artificial heart, genetic engineering, and in-vitro fertilization. One of the earliest controversies, documented in a *Life* cover story by Shana Alexander in 1960, concerned the distribution of dialysis to patients with failing kidneys at a hospital at the University of Washington. The decision about which patients should receive dialysis was made by what came to be known as the "God committee"—a group of community members who decided, as Alexander put it, who shall live and who shall die.[16] The development of dialysis itself was not nearly as ethically controversial as the process by which the decision to allocate it was made, especially the appeal to the "social worth" of candidates.[17] The "God committee" had explicitly recommended that patients deemed more valuable to society should be given preference for dialysis.

In the early days of bioethics, most of the academics who were drawn to these issues came from outside clinical medicine. Many were theologians or scholars in religious studies, such as Paul Ramsey, Joseph Fletcher, and Richard McCormick. Few had any links to organized medicine. The first bioethics think tank, the Hastings Center, was independent of any university or medical school when it was established in 1969 by Dan Callahan, a philosopher and Willard Gaylin, a psychiatrist. The Kennedy Institute of Ethics was founded at Georgetown University soon afterward, but it was not connected to the medical school. Many of the earliest conversations about bioethics included physicians, but often these were physicians deeply worried about the direction that their profession was taking. In fact, many were sharp critics of medicine—concerned about the way technology was shaping society, skeptical of the research imperative, distrustful of medical authority.

Gradually, however, the shape of bioethics began to change. As it gained a measure of legitimacy, bioethics started to become more tightly incorporated within the structures of medicine itself. Partly this was driven by academic necessity: in the early days many philosophers, lawyers, and religious studies scholars found they had little sense of what actually went on in hospitals and research labs. To write about a critical care unit or an operating room without ever having set foot in one seemed a little like an anthropologist writing about the Trobriand Islanders without ever having left campus. But allowing ethicists on the inside made sense for academic health centers as well. Physicians concerned about the path that medicine was taking felt that ethicists could change the profession's course by teaching ethics to medical students and residents and by making ethics part of hospital culture. And for administrators worried about the threat of external regulation and oversight due to ethical scandals, hiring bioethicists seemed like a way to demonstrate their good intentions.

One notable sign of this change was the emergence in the 1980s of "clinical ethics."[18] Clinical ethicists offered a practical, hands-on service intended to improve patient care. The signature issue of clinical ethics was end-of-life care, and its intellectual center was the University of Chicago, which established its Center for Clinical Medical Ethics in 1984. Whereas academic bioethics had been dominated by theologians and philosophers with the occasional lawyer in the mix, clinical ethics was dominated by physicians. From the start, clinical ethics positioned itself within hospitals. No longer would doctors be forced to listen to the carping and criticism of outsiders; they could do bioethics themselves. And they could do it in the way that they practiced medicine: wearing a white coat, carrying a pager, writing consultation notes in patients' charts. Clinical ethicists set up ethics consultation services on hospital wards, offering moral advice the way a consulting neurosurgeon might recommend a lumbar puncture or a dermatologist might suggest a biopsy.

Another important push came from the federal government. The Human Genome Project was a massive $3 billion program founded in 1990 and headquartered in the Department of Energy and the National Institutes of Health. Headed by the Nobel laureate James Watson (of Watson and Crick double-helix fame), the Human Genome Project funded scientific efforts to map the genetic constitution of the human species. Fortunately for bioethicists, the project also set aside a percent of its budget for ethical and legal issues. The ELSI (Ethical, Legal and Social Implications) Research Program proved to be a financial windfall. (Arthur Caplan famously called it the "full employment act for bioethicists.") ELSI brought a much wider range of scholars into bioethics: not just geneticists interested in ethics, but also public-health researchers, epidemiologists, and medical sociologists. Indirectly, it also allowed university administrators to fund bioethics programs in the same way as other medical research centers, requiring faculty members to support their own salaries through external grants. . . .

In the early days, the primary centers for bioethics were state universities and newly established medical schools. By the 1990s, the elite medical centers had started to join in, establishing their own bioethics programs. Bioethics also started migrating into other centers of medical power. The American Medical Association established a bioethics program. So did the Veterans Administration, NASA, and the National Institutes of Health. President Clinton appointed a National Bioethics Advisory Commission, which was followed by the President's Council on Bioethics. Some ethicists began working in the private sector, in businesses such as pharmaceutical and biotechnology companies, and at commercial IRBs. As the status of bioethics began to rise in the outside world, universities began to take notice. University administrators made more high-profile bioethics appointments, sometimes with endowed chairs. Often these appointments were determined not on the basis of the ethicist's scholarship but because he or she had played a prominent role for a professional body or a federal bioethics commission.

By the early 2000s, the figure of the bioethicist had changed. It had become possible for a person to be employed as a bioethicist without ever having worked as a professor, a doctor, a lawyer, a minister, or anything else. The bioethicist was invested with a kind of social authority, partly because of his or her specialized education, but also because of the individual's distinct place in an institution's bureaucracy. The clinical ethicist, for instance, was given authority simply by virtue of the fact that he or she occupied the

institutional position of "clinical ethicist," which came with certain trappings (a white coat, an office, a hospital ID, responsibility for certain committees) and a certain amount of deference within the organization. Today
many physicians and nurses working in hospitals feel they cannot just
ignore the moral advice of the clinical ethicist, even if they believe it is
wrong.

Bioethics scholarship also began to change. No longer was the default
position of bioethicists a suspicion of medical technology. Human cloning,
gene therapy, embryonic stem cell research, cosmetic psychopharmacology:
a new generation of ethicists saw the same kind of utopian promise in biotechnology that others saw in the digital revolution. Many ethicists began
to press to increase funding for medical research. Some offered wildly
enthusiastic predictions about our technology-enhanced future. (In 2003,
bioethicist Glenn McGee told the *Philadelphia Inquirer* that a device the
size of an iPod would soon give us daily downloads about what to eat and
let us have drugs tailor-made for our personal use. "I'm talking five years,"
he said.[19] In his book *Beyond Genetics*, McGee predicted that "No middle-
class suburbanite will be without home DNA analysis in 2010." In 2004,
James Hughes forecast that in two decades "transsexuals will be able to
have new, fully functional genitalia cloned, grown and transplanted." He
also suggested it will be possible to genetically engineer elves, unicorns, and
centaurs, although he did not give a timeline for when the technology
would be developed.[20]) To some outsiders, it appeared that bioethics had
been co-opted by the very institutions it was intended to study. Many
observers agreed with sociologist Jonathan Imber when he called bioethics
"the public relations division of modern medicine."[21]

One striking example of this transformation came after one of the most
notorious research scandals of the 1990s. Jesse Gelsinger was only eighteen
years old when he flew from Tucson to Philadelphia to take part in a gene
therapy study at the University of Pennsylvania. Gelsinger had been born
with ornithine transcarbamylase (OTC) deficiency, a rare metabolic illness
that makes it impossible for the liver to clear ammonia from the body. OTC
deficiency kills most children shortly after they are born. Gelsinger, however, was a relatively healthy young man. He had a mild form of the disorder, which he had been able to control with drugs and diet. The gene therapy study at Pennsylvania used a modified cold virus to carry corrective
genes to the liver, in the hope that they would fix the metabolic deficiency.
When the Penn researchers injected Gelsinger with the virus, however, he
rapidly went into multisystem organ failure. Within only a few days, he
was dead.

Gelsinger's death stunned scientists who had seen gene therapy as an extraordinarily promising area of research. It also shocked many bioethicists. An FDA panel found that Gelsinger should never have been considered for the trial because of the condition of his liver.[22] A lawsuit charged that the Penn researchers had not been completely honest with Gelsinger about the potential risks of the study.[23] The study was also compromised by serious conflicts of interest: the researcher in charge, James Wilson, held a 30 percent controlling interest in Genovo, the biotechnology company that stood to profit from the study. Genovo also provided four million dollars a year to the research program Wilson directed, the Institute for Human Gene Therapy. Still another ethical controversy concerned the design of the trial. Most potentially risky trials of new drugs for serious illnesses (such as cancer) are done on the patients most severely affected by the disease and whose prognosis is very poor without any treatment. But because the severely affected patients in the case of OTC deficiency are newborns, unable to give informed consent, the Penn researchers had decided to test the gene therapy on adult patients with mild forms of the disorder, such as Gelsinger. This meant that healthy patients with long lives ahead of them were exposed to the risk of serious injury and death.

Shortly after Gelsinger died, two bioethicists at the University of Pennsylvania, Arthur Caplan and David Magnus, published a newspaper op-ed about the study. Remarkably, they did not criticize the study in any way. Rather, they lamented the fact that Gelsinger's death might slow the pace of medical research by leading to tighter regulation. "We do a disservice to Jesse Gelsinger and others who have been hurt or killed in medical research by simply adding layers of bureaucracy in the path of clinical research," the bioethicists wrote. "If we are not careful we may wind up allowing our collective grief over the death of a young research subject to justify the imposition of bad public policy governing the future of gene therapy."[24]

The op-ed was not widely noticed, yet it represented a remarkable transformation for bioethics. In the 1970s and 1980s bioethicists were often deeply resented by doctors, who regarded them as naive and arrogant outsiders intruding on the doctors' professional turf. By the mid-1990s, however, bioethicists were demonstrating that they not only did not threaten medicine but also could be loyal members of the medical team. In 1996, Benjamin Freedman, a colleague of mine in the Biomedical Ethics Unit at McGill University, wrote a controversial article called "Where Are the Heroes of Bioethics?"[25] If bioethicists were genuinely speaking their minds about ethical problems in hospitals and research labs, Freedman thought, surely there would be times when they would be punished. Bioethicists critical of unethical practices would lose their jobs, find themselves censured by hospital authorities, and have their tenure applications denied. Yet there seemed to be remarkably few cases, if any, where this had happened. Freedman wondered, what does this say about the direction that bioethics is traveling?

. . .

. . . In March 2009, the US Government Accountability Office (GAO) announced the successful conclusion of a clandestine "sting" operation.[26] The targets of the sting were institutional review boards (IRBs), the committees responsible for overseeing the ethics of medical research in the United States. Posing as a bogus corporation called Device Med-Systems, the GAO asked three commercial IRBs to review a study of a fictitious substance called Adhesiabloc gel. The supposed purpose of the study was to find out whether Adhesiabloc would prevent surgical adhesions, a side effect of surgery that occurs when scar tissue causes internal organs to bind together. The GAO designed the Adhesiabloc protocol to appear so risky that any reasonable IRB would turn it down. It called for a liter of Adhesiabloc gel to be poured into the abdominal cavity of a patient after surgery, where it would remain for up to five months.

The GAO intentionally kept the three IRBs in the dark about the protocol. It did not give them the results of any animal studies. It did not tell them where the experimental substance was manufactured. It did not reveal where the proposed surgery would take place, or who would perform it. Although the GAO noted that the experimental substance was 2.5 percent propylene glycol, it did not provide any information about the substance's other 97.5 percent. The principal investigator for the study was given a forged medical license from Virginia dated 1990, which meant that even if the license had been legitimate, it would have already expired. The only contact information the GAO included was a P.O. box and a cell phone number. "It's the worst thing I've ever seen," said a reviewer for one of the IRBs. "Doing a major surgery on a patient, and then a mystery guy comes in and dumps a solution in the body? Where's the safety for the patient?" Yet one commercial IRB approved the protocol within a week.[27]

That was Coast IRB, based in Colorado Springs, Colorado. The GAO had good reason to suspect that Coast IRB was not the most rigorous board in the business. Coast had offered pharmaceutical companies coupons, inviting them to "take us for a free test drive" and to "coast through your next study." Of the 356 protocols that Coast reviewed over a five-year period, it rejected only one. It had also posted an advertisement on YouTube featuring a cartoon schooner. People speaking in phony English accents apparently meant to convey sailors' voices narrated the ad: "Cap'n, we've got to find our way out of this fog!" says one. "Ah, thar she blows! The Coast Independent Review Board! With speed, quality, and service, guided by the light of ethics."[28] Despite these tactics (or perhaps because of them), annual revenue for Coast IRB doubled in only four years, reaching $9.3 million in 2008.[29]

On March 26, 2009, the Energy and Commerce Committee of the US House of Representatives conducted a hearing on the GAO sting operation. One after another, angry members of Congress waved liter bottles of the fake Adhesiabloc, professing outrage that such a bizarre study had been given a stamp of ethical approval. But Dan Dueber, the CEO of Coast IRB, remained defiant. Refusing to admit any flaws whatsoever in the fake Adhesiabloc study, Dueber maintained that the protocol had been thoroughly vetted by his expert reviewers. In fact, it was Coast IRB that had been victimized, Dueber claimed. "Innocent citizens of this country cannot be lawfully defrauded by the government," he told the congressional committee, noting that he had reported the GAO fraud to the appropriate law enforcement officials.[30]

Representatives from the FDA and the Office for Human Research Protections also attended the congressional hearing. To the surprise of everyone there, however, they refused to criticize Coast IRB. Their refusal to pass judgment appeared to make members of Congress even angrier. "I'm so mad at the company I can hardly be civil," said Rep. Joe Barton, a Republican from Texas, "but I'm almost as upset with our government folks who are supposed to oversee these IRBs. This company has gotten four or five notice letters in the last two to three years and yet they're still in business. And they have the gall to come here and threaten to sue the government!"[31]

IRBs have a unique place in the bioethics universe. They are the closest things to a regulatory arm for which bioethicists can claim any credit. Established in the 1970s, IRBs are supposed to provide independent ethical review before a medical research study proceeds. IRBs were set up as an institutional response to the research scandals of the 1960s and 1970s, such as the ones at Tuskegee and Willowbrook. Today it is a rare bioethics course that does not include the ethics of human research, and many bioethicists serve on IRBs at their own institutions. Over the years, research ethics has evolved into the most bureaucratic, legalistic subfield of bioethics, spawning its own unique subculture with members that include not just academics but also industry executives, IRB members, and government functionaries, all of whom gather each year at a conference called Public Responsibility in Medicine and Research (PRIM&R).

In the early days, IRBs were mainly volunteer committees made up of clinicians and researchers working in the hospitals and medical schools where studies were being conducted. During the 1990s however, as drug studies migrated out of academic health centers and into the private sector, a new type of IRB emerged: independent, for-profit boards that reviewed studies for pharmaceutical companies and other research sponsors in exchange for a fee. This arrangement would prove profitable for both parties. By 2002, more than forty commercial IRBs were operating in the United States with over sixty million dollars in annual revenue.[32] In 2004, Chesapeake IRB, one of the largest commercial IRBs, was named by Deloitte as one of the fastest-growing tech companies in America.[33]

Commercial IRBs have proven controversial, for the obvious reasons. Since they are paid by the companies whose protocols they review, commercial IRBs have a financial interest in keeping their clients happy. They market themselves by promising industry-friendly service and a quick turnaround for reviews. No research sponsor is obligated to stick with a single IRB; if a study is rejected by one commercial IRB as too risky or deceptive, the sponsor can simply send it to other IRBs until it is approved. Proponents of commercial IRBs argue that it is in the interest of research sponsors to obtain a strict review, if only to head off potential litigation later. (In fact, lawsuits about clinical trials are still relatively rare.) However, the costs of delaying a study because of a slow, rigorous ethical review can be considerable. In its "lost schooner" advertisement, Coast IRE told potential customers that a delay of a single day for a fifty-center, Phase III trial could cost a sponsor six million dollars.

Surprisingly, bioethicists have given commercial IRBs a fairly easy ride. Some bioethicists even work for commercial IRBs as reviewers. Goodwyn IRB in Ohio, for example, lists a number of distinguished university-based bioethicists on its review board, even as it markets itself as a "good friend to pharmaceutical manufacturers."[34] Dr. Ezekiel Emanuel, formerly the chief of bioethics at the National Institutes of Health, has been a vocal supporter of commercial IRBs, especially Western IRB in Olympia, Washington. "I think there are a lot of reasons that make Western and a few of the others very good," Emanuel told the President's Council on Bioethics a few years ago. "One is certainly leadership and dedication to doing the right thing, and believing that by doing the right thing, you'll be successful, and in their case profitable."[35]

IRBs themselves are not carefully overseen. There is an accreditation body for IRBs, but accreditation is voluntary; the federal Office for Human Research Protections maintains an IRB registry, but registration for an IRB does not exactly guarantee quality. In fact, the GAO sting operation that sank Coast IRB also targeted the OHRP registration process. GAO staffers were able to register a fake IRB located in Chetesville, Arizona, chaired by a three-legged German shepherd called Truper Dawg, with members named April Phuls, Alan Ruse, and Timothy Wittless.[36] . . .

Defenders of commercial IRBs point out that whatever their financial conflicts of interest, their members are not mixed up in the kind of personal and administrative entanglements that make university IRBs so fraught. Commercial IRBs may be paid by research sponsors, but they are often located thousands of miles away from research sites, and their members are usually total strangers to any investigator whose research they review. A review from a commercial IRB can be as faceless and anonymous as a passport application. Anonymity is no guarantee of quality, of course, as the GAO sting of Coast IRB showed. But some commercial IRBs have developed a reputation for serious review. . . .

. . .

On the website of the Washington Speakers Bureau is a video clip of Arthur Caplan, the director of the Center for Bioethics at the University of Pennsylvania.[37] Caplan has developed a unique niche in bioethics; he is a critic of pharmaceutical industry gifts and payments to doctors but works as a paid consultant to industry himself. "Even small gifts influence behavior," Caplan told the *Minneapolis Star Tribune* in 2008.[38] Industry payments to doctors are "too damn lucrative to believe anyone can resist," Caplan said to the *New York Times,* noting that payments will distort people's judgment.[39] Yet Caplan has consulted for companies such as Pfizer, DuPont, and Celera, and the University of Pennsylvania Center for Bioethics has disclosed funding from dozens of pharmaceutical and healthcare corporations.

In the video Caplan describes his work as a consultant for Pfizer in the 1990s, when the company was deciding how to market Viagra, its drug for erectile dysfunction. "Pfizer was terrified," says Caplan. "They thought, 'If we come out with this pill, people are going to attack us for promoting sex.'" Caplan points out that Pfizer had not been in the business of reproductive medicine or sexual health. "They were terrified that this drug could get them in trouble on ethics grounds—that people would call them from churches and other organizations and say, 'Why are you promoting this kind of thing?'" They needed to know "How do you advertise this thing so that you can say to people, 'Look this treats a serious problem. It's not just a joke, or a lark.'"

Caplan goes on to explain how he helped Pfizer as a bioethicist. "At meeting after meeting we were trying to figure out who might step forward and say: they had a problem in the sexual function area; they were happy to have the pill help them; it was a disease; erectile dysfunction would be seen as a disease; and say that this is something they used within their marriage, and it was a good thing," says Caplan. "I suggested that what they needed was a spokesperson who would stand for integrity, and who would never be confused for someone who was just there as a sex object. And so I invented Bob Dole."

Bob Dole, of course, is the former Kansas senator and Republican presidential candidate who endorsed Viagra in television advertisements during the late 1990s. Dole's image as an elderly, serious-minded public figure—the polar opposite of the sexually adventurous demographic that came to embrace Viagra a few years later—helped brand the drug as a legitimate treatment for a medical disorder. Caplan praises Pfizer for blending good bioethics with good marketing. "So that's how they handled the ethical challenge," he says. "They thought about it ahead of time, it was all a part of their marketing campaign; they never made a move without being sure they were talking about the disease and what was going on."

Caplan's work for Pfizer is emblematic of a certain vision of bioethics: the bioethicist's desire to be the power behind the throne, the one whispering in the king's ear. First you must establish a presence in the corridors of power, then you work carefully and judiciously to shape the opinions of the people in charge. It is not an unreasonable aspiration, of course. Yet sometimes the king turns out to be venal, greedy, or corrupt, and when that happens, the line between advice and complicity is blurred. When Caplan advises Pfizer to hire a figure like Bob Dole, for example, it sounds less like sorting out an ethical dilemma and more like an effort to devise a better marketing campaign. What does it say about bioethics when the ethicist is indistinguishable from the public relations counsel?

Critics of industry-funded bioethics usually describe the issue as a conflict of interest, pointing out that bioethicists have other roles in which they are presumed to have a measure of distance from the pharmaceutical industry. Bioethicists teach ethics to college students, for example, who generally do not suspect that the professor may be getting a paycheck from the very corporations whose actions and policies they are discussing in class. But a larger question concerns the direction of bioethics as a field. As bioethics has matured, its practitioners have aspired to a kind of professional expertise. Bioethicists have gained recognition largely by carving out roles as trusted advisers. But embracing the role of trusted adviser means foregoing other potential roles, such as that of the critic. It means giving up on pressuring institutions from the outside, in the manner of investigative reporters. As bioethicists seek to become trusted advisers, rather than gadflies or watchdogs, it will not be surprising if they slowly come to resemble the people they are trusted to advise. And when that happens, moral compromise will be unnecessary, because there will be little left to compromise.

Ken De Ville, an attorney and historian of medicine at East Carolina University, wonders whether bioethics will continue to be an enterprise worth pursuing once it is thoroughly infused with corporate money. "If ethicists are transformed into a bunch of corporate shills who exist only to serve the machine," he asks, "where is the honor in taking part?" Of course, De Ville's comment presumes that there is a distinction between honor and serving the machine. Once the very discipline of bioethics is itself a part of the machine, service is an honor. Laurie Zoloth, the former president of the American Society for Bioethics and Humanities, has written that the real temptations of industry associations are not financial but in the honor and status of corporate consultancies. If she is right and advising a corporation is an honor, then bioethicists have already made the shift from outsider to insider, from critic of the machine to loyal servant.

In recent years, the reputation of the pharmaceutical industry has taken a beating. Profits have fallen off; litigation has risen; and Senator Grassley's investigations have ensured a constant stream of poor publicity. From 1997 to 2005, according to Harris Polls, the public approval rating of the pharmaceutical industry dropped from 80 percent to 9 percent, putting its reputation below every other major business except oil and tobacco companies.[40] Prominent physicians have relentlessly attacked industry practices; several former thought leaders have publicly recanted their work. Yet it is still unclear whether bioethicists are willing to cut their industry ties. If anything, many top bioethicists appear to be ramping up their consulting work.

A few years ago I was invited to give a talk about the pharmaceutical industry at a prominent academic health center. At that point I had written several disapproving articles about pharma-funded bioethics, and my presentation included some pointed criticism. The colleague who invited me appeared to share my views. As I was standing next to the lectern, waiting to be introduced, he leaned over to me and whispered, "Do you mind if I thank Janssen Pharmaceuticals for sponsoring your presentation?"

NOTES

1. Adam Sichko, "In Transition, McGee Turns to Growing a Bioethics Business for Profit," *Business Review* (Albany), August 29, 2008, http://albany.bizjournals .com/albany/stories/2008/09/01/story6.html?b = 1220241600^1692509.

2. See www.e-four.org/Home/about-crucial-choices.

3. See Weinstein's Web page at www.theethicsguy.com.

4. Carl Elliott, "Pharma Buys a Conscience," *American Prospect* 12 (2001): 16–20; Sheryl Stolberg, "Bioethicists Fall under Familiar Scrutiny," *New York Times,* August 2, 2001.

5. Stolberg, "Bioethicists Fall."

6. For Moreno, see his Web page at the University of Pennsylvania: http://hss .sas.upenn.edu/mtstatic/faculty/department_faculty/jonathan_moreno_phd_ professor.php. For Beauchamp and Levine, see the Eli Lilly Corporate Citizenship Report 2005–06, www.socialfunds.com/csr/reports/Lilly_20052006_Corporate_ Citizenship_Report.pdf. For Childress, see the *University of Virginia Center for Bioethics Ethics Annual Report,* 2002–03.

7. Baruch Brody and Nancy Dubler, et al., "Bioethics Consultation in the Private Sector," *Hastings Center Report* 32 (2002): 14–20, http://repository.upenn.edu/cgi /viewcontent.cgi?article1024&context = bioethics_papers.

8. Thomas Donaldson, "The Business Ethics of Bioethics Consulting," *Hastings Center Report* 31 (2001): 12–14.

9. Stephen Hall, *Merchants of Immortality: Chasing the Dream of Human Life Extension* (Boston: Houghton Mifflin, 2003), 321, 323.

10. Laurie P. Cohen, "Stuck for Money," *Wall Street Journal,* November 14, 1996.

11. Tom Beauchamp, Bruce Jennings, Eleanor Kinney, and Robert Levine, "Pharmaceutical Research Involving the Homeless," *Journal of Medicine and Philosophy* 27 (2002): 547–64.

12. Stolberg, "Bioethicists Fall."

13. Maurice Pappworth, *Human Guinea Pigs* (Boston: Beacon Press, 1967).

14. Sandra Coney and Phillida Bunkle, "An 'Unfortunate Experiment' at National Women's," *Metro* (1987): 47–65; Alastair V. Campbell, "A Report from New Zealand: An 'Unfortunate Experiment,'" *Bioethics* 3 (1989): 59–66.

15. David Rothman, "Were Tuskegee and Willowbrook 'Studies in Nature'?" *Hastings Center Report* (1982): 5–7; James Jones, *Bad Blood: The Tuskegee Syphilis Experiment* (New York: Free Press, 1981).

16. Shana Alexander, "They Decide Who Lives, Who Dies," *Life* 53 (1962): 102–25.

17. Albert Jonsen, *The Birth of Bioethics* (New York: Oxford University Press, 1998), 212.

18. Mark Siegler, "Ethics Committees: Decisions by Bureaucracy," *Hastings Center Report* 16 (1986): 22–24; Mark Siegler and Peter Singer, "Clinical Ethics Consultation: Godsend or 'God Squad'?" *American Journal of Medicine* 85 (1988): 759–60.

19. Eils Lotozo, "Bioethicist Foresees a Wild Frontier in Genetics Field," *Philadelphia Inquirer,* October 28, 2005.

20. Glenn McGee, *Beyond Genetics* (New York: HarperCollins, 2005), 6.

21. James Hughes, *Citizen Cyborg* (New York: Westview Press, 2004), 21, 92.

22. Jonathan Imber, "Medical Publicity before Bioethics: Nineteenth-Century Illustrations of Twentieth-Century Dilemmas," in *Bioethics and Society: Constructing the Ethical Enterprise,* ed. Raymond DeVries and Janardan Subedi (New York: Prentice Hall, 1998), 30.

23. Sheryl Gay Stolberg, "FDA Officials Fault Penn Team in Gene Therapy Death," *New York Times,* December 9, 1999, http://partners.nytimes.com/library /national/science/health/120999hth-gene-therapy.html,

24. Joanne Silberner, "A Gene Therapy Death," *Hastings Center Report* 30 (2000): 118; Trudo Lemmens, "Confronting the Conflict of Interest Crisis in Medical Research," *Monash Bioethics Review* 23 (2004): 19–40.

25. Benjamin Freedman, "Where Are the Heroes in Bioethics?," *Journal of Clinical Ethics* 7, no. 4 (1996): 297–99.

26. US Government Accountability Office, "Human Subjects Research: Undercover Tests Show the Institutional Review Board System Is Vulnerable to Unethical Manipulation" (Washington, DC: US Government Accountability Office, 2009), www.gao.gov/new.items/d09448t.pdf.

27. Alicia Mundy, "Sting Operation Exposes Gaps in Oversight of Human Experiments," *Wall Street Journal,* March 26, 2009; "Congress Vents Outrage at FDA, OHRP and IRBs," *Clinical Trials Advisor* l4 (2009).

28. The commercial has been posted on YouTube, www.youtube.com/watch?v = 6FnpRihlUH8.

29. Barry Meier, "An Overseer of Trials Draws Fire," *New York Times,* March 26, 2009.

30. Dueber's testimony is posted on the Energy and Commerce Committee website, http://energycommerce.house.gov/Press_111/20090326/testimony_ dueber.pdf.

31. See page 77 of the transcript of the hearing.

32. Jill Fisher, *Medical Research for Hire: The Political Economy of Pharmaceutical Clinical Trials* (Piscataway, NJ: Rutgers University Press, 2008), 11.

33. Heidi Ledford, "Trial and Error," *Nature* 448 (2007): 530–52.

34. www.goodwynirb.com/Home.htm.

35. The transcripts of the presentation are online, www.bioethics.gov/transcripts /sep02/session2.html.

36. Alicia Mundy, "Was a Three-Legged Dog Head of a Review Board?," *Wall Street Journal Health Blog,* March 26, 2009, http://blogs.wsj.com/health/2009/03/26 /was-a-three-legged-dog-head-of-a-review-board/.

37. The video, called *The Viagra Case Study,* is online, www.washingtonspeakers .com/scripts/flvplayer/flvplayer.cfm?src = art_caplan-d_viagra_case_study.

38. Janet Moore, "U Medical School Plan: Ban All Gifts to Doctors," *Minneapolis Star Tribune,* October 21, 2008.

39. Reed Abelson, "Whistle-Blower Suit Says Device Maker Generously Rewards Doctors," *New York Times,* January 24, 2006.

40. www.harrisinteractive.com/harris poll/index.asp?PID = 611. See also Margaret Eaton, "Managing the Risks Associated with Using Biomedical Ethics Advice," *Journal of Business Ethics* 77 (2008): 99–109.

PART III

Emerging Biotechnologies, Extreme Ideologies

The Recent Past and Near Future

The late twentieth and early twenty-first centuries saw a dizzying pace and array of biotechnological developments. During the 1970s, the first baby was born after in vitro fertilization, and "recombinant DNA" experiments with bacteria brought together genetic material from different organisms for the first time. During the 1980s, the US Supreme Court ruled in favor of patents on genetically modified organisms; the first "transgenic" animal—a mouse—was created by changing the genes of an early-stage embryo so that the modifications would be transmitted to future generations; and the Human Genome Project was launched, with the goal of determining the sequence of our species' three billion DNA base pairs and identifying all its genes. The 1990s saw the first human gene therapy experiment (in a four-year-old girl) and the first derivation of human embryonic stem cells. Scientists in Scotland announced the birth of the first cloned mammal, a sheep known as Dolly; and headlines spread about the prospect of cloned and genetically modified humans.

Biotechnology developments accelerated further after the turn of the millennium. During the 2000s,

- A "working draft" of the human genome sequence was announced to great fanfare in the White House;
- Scientists in Oregon created the first transgenic primate, a rhesus monkey engineered with jellyfish genes;
- Reports of covert efforts to clone human beings appeared in the media;
- Korean researcher Hwang Woo Suk gained worldwide fame when he claimed to have created cloned human embryos and extracted stem cells from them (a claim found to be fraudulent the following year);

- Advertisements for direct-to-consumer genetic tests were launched; and

- Scientists created the first genetically modified human embryo, the first cloned human embryo, and the first animal-human hybrid embryo.

The 2010s saw more landmarks:

- The synthesis of a complete bacterial genome and its use to take over a cell;

- The commercial marketing of prenatal tests that sequence fetal DNA extracted from maternal blood in very early pregnancy;

- The creation of embryonic stem cells from cloned human embryos;

- The creation of mice using sperm and eggs grown from induced pluripotent stem cells;

- The birth of the five-millionth IVF baby and of the first baby with its whole genome sequenced in utero; and

- The use of powerful new "gene-editing" techniques to modify human embryos.

Some of these developments slipped by with little notice, but others—especially proposals to use genetic modification or cloning techniques in human reproduction—sparked widespread controversy and public debate. In the late 1990s and early 2000s, dozens of countries, but not the United States, decided to adopt laws prohibiting human reproductive cloning and heritable genetic modification.

However, since the late 1990s, a small but vocal group of technology enthusiasts in the United States and elsewhere has openly advocated the unfettered use of biotechnologies for extreme scenarios that include "designer babies" and "posthumans." They advance libertarian arguments for "enhancing" cognitive and physical traits beyond human limits and for "transcending" aspects of the human condition such as aging and dying.

The contributions in this section represent the skeptical and critical side of this debate. The first two articles were both written in 2002, a time when advocacy of inheritable genetic modification and "designer babies" had reached a loud pitch. The next three were written in 2015, after the development of gene editing once again made the prospect of creating genetically modified humans a headline-grabbing controversy.

In "The Genome as Commons," Tom Athanasiou and Marcy Darnovsky introduce the "biotech boosters" of the day and unpack their vision of a high-tech consumer eugenics. "The emerging human genetic and reproductive technologies are a turning point," they write. "Unless we harness our moral intelligence and political will to shape them, they will conform to the existing social divides and to the inadequacies of our democracy, and they will exacerbate both."

Ruth Hubbard and Stuart Newman characterize the era about which they are writing as one of "Yuppie Eugenics." Unlike twentieth-century eugenics, the eugenic potential of this new era is based on modern genetics and a hypertrophied notion of "choice." But "what it shares with the earlier doctrines," they argue, "is the goal of improving and perfecting human bloodlines and the human species as a whole." Hubbard and Newman conclude by recommending that we keep firmly in mind that "genetics will never tell us what it takes to make a worthy human being and that the major causes of human illness and death continue to be not enough healthful food and too much unhealthful work. Eugenic and other gene dreams will not cure what ails us."

"Brave New Genome" is authored by prominent geneticist Eric S. Lander, founding director of the Broad Institute of MIT and Harvard and cochair of President Obama's Council of Advisors on Science and Technology. Lander's article presents what he sees as the key issues in the coming debate about editing the human germ line, and concludes, "It has been only about a decade since we first read the human genome. We should exercise great caution before we begin to rewrite it."

In "Can We Cure Genetic Diseases without Slipping into Eugenics?" Nathaniel Comfort offers an historical examination of the US eugenics movement and of the social and political context in which groundbreaking gene-editing technology is being introduced. New gene-editing technologies such as CRISPR-Cas9 could correct genetic mutations for people affected by serious illness, he writes. But will it also create a new "eugenics of personal choice"?

The final contribution to part 3 is a wry account of the Singularity Summit, a "roving recruitment seminar" for Silicon Valley–based Singularity University. Writer Corey Pein—one of nine hundred registrants, seventy-five of them journalists—reflects on the visions and views of the Singularity University speakers, including one who opined that humanity is "headed for subspeciation—a proliferation of separate and unequal races creating something like a cyberpunk version of Tolkien's Middle Earth."

13. The Genome as Commons

Tom Athanasiou and Marcy Darnovsky

The atmosphere. The oceans and fresh waters. The land itself, and the fruits and grains our forebears bred and cultivated upon it. The broadcast spectrum. The attention spans of our children.

Does such a list adequately evoke "the commons," and the stakes we face in trying to save it—both for itself and as the foundation of our common future? Or must we add yet another, more shocking example? Perhaps we must put the human genome itself on this endangered commons list, and note that if this genetic commons too is lost to partition and privatization, if it too becomes the privilege of the affluent, then none of us on either side of the divide can be sure of retaining the "humanity" we like to think we've achieved.

The biotech boosters, of course, don't see things this way. Many of them insist that any conceivable application of human genetic engineering is essential to medical progress, and that the possibilities, no matter how speculative, trump all other considerations. Thus they shrug off the likely outcome of embryo cloning—that it will sooner or later lead to reproductive cloning, and then jump start both the technologies and justifications of inheritable genetic modification.

Some of them are even enthusiastically promoting "designer babies" and "post-humans" as the next new things.[1] Indeed, the techno-eugenic hard school is now promising that, within a generation, "enhanced" babies will be born with increased resistance to diseases, optimized height and

World Watch Magazine, July/August 2002, Volume 15, No. 4 © 2002 Worldwatch Institute. Authors Tom Athanasiou and Marcy Darnovsky. More information at http://www.worldwatch.org/bookstore/publication/worldwatch-paper-5-twenty-two-dimensions-population-problem. Reprinted by permission of Worldwatch.

weight, and increased intelligence. Farther off, but within the lifetimes of today's children, they foresee the ability to adjust personality, design new bodily forms, extend life expectancy, and endow hyperintelligence. Some actually predict splicing traits from other species into human children: In late 1999, for example, a Ted Koppel/*ABC Nightline* special on cloning speculated that genetic engineers will eventually design children with "night vision from an owl" and "supersensitive hearing cloned from a dog."

There are dark portents here in profusion, and many of them will seem familiar to environmentalists. But consider first the fundamental point: our patently inadequate ability to protect the resources of the global commons, to do them justice, to make them (in reality as well as in United Nations rhetoric) "the common heritage of humankind." Consider, through this lens, the likely fate of the human genome—the script which unites us as a biological species—as it too goes on the auction block.

And attend to this chilling bit of futurology from Lee Silver, a Princeton professor and self-appointed champion of the new techno-eugenics:

> [In a few hundred years] the GenRich—who account for 10 percent of the American population—[will] all carry synthetic genes. . . . All aspects of the economy, the media, the entertainment industry, and the knowledge industry [will be] controlled by members of the GenRich class. . . . Naturals [will] work as low-paid service providers or as laborers . . . [Eventually] the GenRich class and the Natural class will become . . . entirely separate species with no ability to cross-breed, and with as much romantic interest in each other as a current human would have for a chimpanzee.[2]

Silver's predictions, in case this isn't clear, are not voiced in opposition to a eugenically engineered future. Here and elsewhere, his tone alternates between frank advocacy of a new market-based eugenics and disengaged acceptance of its inevitability.

Is such a future likely? We hope not, and we take some comfort in the possibility that scenarios like these may long remain beyond technical reach. Notwithstanding the flesh-and-blood accomplishments of genetic scientists—glow-in-the-dark rabbits and goats that lactate spider silk—artificial genes and chromosomes may never work as reliably as advertised. Transgenic designer babies may be too ridden with unpredictability or malfunction to ever become a popular option.

Still, both the technological drift and the strength of ideological feeling among proponents compel us to take the prospect of a techno-eugenic future seriously. Some surprisingly influential figures—including controversial celebrities like Nobel laureate James Watson and philosopher-provocateur Peter Singer, as well as mainstream academicians like Daniel Koshland of U.C. Berkeley and John Robertson of the University of Texas— are publicly endorsing visions similar to Silver's.

These boosters frankly acknowledge that designer-baby techniques would be very expensive and that most cloned or genetically "enhanced" children would be born to the well-off. They concede that the technologies of human genetic redesign would therefore significantly exacerbate socio-economic inequality, and they speculate about a future in which a genetic elite acquires the attributes of a separate species. But they do not find in any of these possibilities reason to forego eugenic engineering. In *Children of Choice*, for example, John Robertson writes that genetic enhancements for the affluent are "simply another instance in which wealth gives advantages."[3]

So ask not if the techno-eugenic agenda will come true anytime soon. Ask instead why it's getting so much air time, and why Silver et. al. have not been taken even mildly to task, either by their scientific colleagues or by liberal and progressive intellectuals who might be expected to muster a bit of angst over such crass eugenic visions.

And they are crass. Note the coarse neoliberalism that underlies Silver's certainty about the eugenic future: "There is no doubt about it," he writes, "whether we like it or not, the global marketplace will reign supreme."[4] Moreover: "[I]f the cost of reprogenetic technology follows the downward path taken by other advanced technologies like computers and electronics, it could become affordable to the majority of members of the middle class in Western societies ... And the already wide gap between wealthy and poor nations could widen further and further with each generation until all common heritage is gone. A severed humanity could very well be the ultimate legacy of unfettered global capitalism."[5]

The techno-eugenic vision carries with it a deep ideological message. It urges us, in case we still harbor vague dreams of human equality and solidarity, to get over them. It tells us that science, once (and sometimes still) the instrument of enlightenment and emancipation, may bequeath us instead a world in which class divisions harden into genetic castes, and that there's not a damn thing we can do about it. The story of an "enhanced" humanity panders to some of the least attractive tendencies of our time: techno-scientific curiosity unbounded by care for social consequence, economic culture in which we cannot draw lines of any kind, hopes for our children wrought into consumerism, deep denial of our own mortality.

This last theme, the one that brings our life expectancies and bodily functions to center stage, is a powerful one. Its driver is medical biotech, and the market niche for it is clearly waiting: All those aging boomers now avidly dropping Viagra and DHEA and Human Growth Hormone are the natural constituency of the techno-eugenicists. Tell them that they'll live longer, and they'll follow you anywhere. As James Watson put it in a conversation about how to convince the public that eugenic manipulation of future children is acceptable, "We can talk principles forever, but what the public actually wants is not to be sick. And if we help them not be sick they'll be on our side."[6]

Watson, unfortunately, is tuned to the zeitgeist of the well-off and the well-funded. Those of us disinclined to embrace eugenic engineering will have to work harder to be heard above the din of wildly exaggerated biomedical claims. It won't be easy, but the bottom line is clear enough: we have to distinguish genetic techniques that are plausible and appropriate from those that are likely to be unsafe, ineffective, unjust, and pernicious.

The history of environmentalism is instructive here. Advocates of ecological sanity have for decades expended oceans of sweat and tears to show the need for caution in the face of powerful new technologies—nuclear power plants, large dams, Green Revolutions. To be sure, the precautionary principle is generally swatted aside by powerful political and economic interests, but many people, and a few courageous policy makers, have accepted its key assumption: that technologies shape lives and societies and thus are appropriate matters for both careful forethought and democratic oversight.

This elementary precautionary lesson, however, is seldom applied to medical technologies. Even those desensitized to the Sirens' song of triumphant technical progress may find themselves dreaming of new therapies, fountains of youth, and genetically enhanced memories. We may nurse, if only in the backs of our minds, the comforting assurance that this is all moving too quickly to be stopped.

The near exemption of biomedical technologies from the principles of precaution may help explain the sudden emergence of embryo cloning as a national issue and the Alice-in-Wonderland quality of the debate about it: the out-on-a-limb promises of near-term cures (would that Christopher Reeve, a spokesman for therapeutic cloning, could be Superman again); the overblown claims of research breakthroughs (those cloned human embryos? Actually, they stopped dividing at six cells); the loose talk of treating millions of sufferers with "therapeutic" cloning (after, of course, finding the women to "donate" millions of eggs).

Biomedicine's dispensation from the precautionary principle may also shed light on another oddity. The nation's pundits, noting that both pro-choice liberals and conservatives are now voicing caution about embryo cloning, are suddenly fixated on the "strange bedfellows" that make up the anticloning lobby. Yet they've entirely overlooked the more disturbing lapses that still characterize so much of the liberal/progressive reaction to the prospect of unrestricted human biotechnology.

What, for example, are we to make of a recent comment (made in an off-the-record meeting of a national progressive organization) that "we don't ban things—bad guys ban things"? What about ozone-depleting chemicals, above-ground nuclear testing, and medical experimentation on inadequately informed women in the global South? And what of a new eugenics based on high-tech reproduction, consumer preferences, and market dynamics? If we don't ban these things, who will?

And what are we to think when a columnist in an intelligent liberal journal like *The American Prospect* opines that "humans are part of the natural world and all their activities, science, cloning, and otherwise, are therefore hardly unnatural, even if they may be unprecedented."[7] Surely environmentalists have been adequately warned against the naturalistic fallacy and are well aware that appeals to "Nature" can be made to justify anything. So aren't we entitled to a similar level of sophistication from those inclined to see "Luddites" behind every bioengineered bush? Surely even liberals who staunchly maintain their faith in the onward march of science can see the political dangers of conflating categories, of erasing the difference between the products of millions of years of evolution and the products of commerce and fashion.

When liberals throw in their lot with libertarians, there is danger near. The tension between personal liberty and social justice is a necessary one, and should not be collapsed into uncritical support for individual (or corporate!) rights. Commitments to solidarity and fairness must not be allowed to wither and die. The right to terminate an unwanted pregnancy is very different than the "right" to modify the genetic makeup of future children. Biomedical researchers and fertility doctors have no "right" to develop species-altering technologies in their petri dishes. And despite the eagerness of venture capitalists and the willingness of the patent office, they certainly have no "right" to send them out into the world.

Which brings us back to the rich and the poor, and their respective claims on the various global commons. Any serious vision of the future must address this issue, and clearly. Remember Aldous Huxley's *Brave New World*? It was, first of all, a world of caste. All the rest—the meaningless drug-optimized sex, the soma, the feelies, even the bottled babies—was secondary, just more bricks in the wall.

The emerging human genetic and reproductive technologies are a turning point. Unless we harness our moral intelligence and political will to shape them, they will conform to the existing social divides and to the inadequacies of our democracy, and they will exacerbate both. Until the designer babies and "posthumans" begin to populate the planet, until we allow inequality to be inscribed in the human genome, we're all in this together.

NOTES

1. See, for example, Gregory Stock and John Campbell, eds., *Engineering the Human Germline: An Exploration of the Science and Ethics of Altering the Genes We Pass to Our Children* (New York: Oxford University Press, 2000).

2. Lee M. Silver, *Remaking Eden: Cloning and Beyond in a Brave New World* (New York: Avon Books, 1997), 4, 6, 7.

3. John A. Robertson, *Children of Choice: Freedom and the New Reproductive Technologies* (Princeton, NJ: Princeton University Press, 1994), 166.

4. Silver, *Remaking Eden*, 11.

5. Silver, "Reprogenetics: How Do a Scientist's Own Ethical Deliberations Enter into the Process?," *Humans and Genetic Engineering in the New Millennium* (Copenhagen: Danish Council of Ethics, 2000), www.etiskraad.dk/publikationer /genethics/ren. htm.

6. Stock and Campbell, *Engineering the Human Germline*, 86. See also http:// research.mednet.ucla.edu/pmts/Germline/panel.htm.

7. Chris Mooney, "Idea Log: Oh no! Bill McKibben's Said too Much. He's Said It All." *American Prospect Online*, March 28, 2002, www.prospect.org /webfeatures/2002/03/mooney-c-03–28.html. See also Bill McKibben, "Unlikely Allies against Cloning," *New York Times*, March 27, 2002, www.nytimes .com/2002/03/27/opinion/27 MCKI.html.

14. Yuppie Eugenics

Ruth Hubbard and Stuart Newman

We have entered the era of Yuppie Eugenics. A contemporary, ostensibly voluntary form of older ideas and practices, Yuppie Eugenics is based in modern molecular genetics and concepts of "choice," and has begun to raise the high-tech prospect of employing prenatal genetic engineering. What it shares with the earlier doctrines is the goal of improving and perfecting human bloodlines and the human species as a whole.

The eugenics movement arose in the late nineteenth century in the wake of new scientific thinking about animal and plant breeding that culminated in evolutionary biology and genetics. While thus part of the scientifically influenced "progressive" thinking of the time, it was based on the fallacious argument that since nature had yielded "fit" species by "natural selection" and humans had "improved" domesticated animal breeds by "artificial selection," there was now a scientific warrant to use social and political means to discourage propagation of biologically "inferior" sorts of people. It soon gave rise to a set of state-directed programs in the United States and Europe.

The legal assault against people of "bad heredity" began in the United States with compulsory sterilization bills. The first of these was introduced into the Michigan legislature in 1897 but was defeated. A second bill, aimed at "idiots and imbecile children," passed the Pennsylvania legislature in 1905 but was vetoed by the governor. Indiana was the first state to actually enact a compulsory sterilization law in 1907, and it was followed by some

Original publication by Ruth Hubbard and Stuart Newman as "Yuppie Eugenics." *Z Magazine: The Spirit of Resistance Lives* (March 2002): 36–39. Reprinted by permission of Stuart Newman, and Elijah Wald, son and legal representative of the Estate of Ruth Hubbard.

thirty others. California did not repeal its law until 1979 and, in 1985, around twenty states still had laws on their books that permitted the involuntary sterilization of "mentally retarded" persons. The United States was by no means alone. In that liberal paragon Sweden, compulsory sterilizations of "unfit" persons were performed into the 1970s. All these laws were meant to improve the genetic makeup of the population, and especially of poor people, by preventing those judged to be "defective" from passing on their "defects" to future generations.

State eugenics reached an abhorrent extreme in the Nazi extermination programs of the 1930s and 1940s. Initially directed at people with similar health or social problems as were targeted by the US and Swedish sterilization laws, these were eventually expanded to cover entire populations—Jews, Gypsies, Poles—judged by the Nazi regime to represent "worthless lives" ("lebensunwerte Leben"). While certain overt state policies such as the use of gas chambers are now avoided, "ethnic cleansing," practiced on three continents in recent times, shows that eugenic cruelties have far from disappeared.

Technologies developed in the past three decades, however, have permitted a change in focus in the implementation of eugenics, at least in more affluent countries, from the state to the individual. Increasing numbers of diagnostic tests have been developed that enable physicians to assess some aspects of a fetus's future health status early enough to permit termination of a pregnancy during the second or even the first trimester. Though all such predictions have pitfalls and problems, they have made it possible for prospective parents not to bear a child they expect to be too ill or disabled to knowingly make part of their family. Though the intent of these methods is to widen choice in matters of procreation, they are eugenic in that they are meant to prevent the birth of people who are expected to perpetuate certain types of inborn conditions, such as cystic fibrosis, Huntington's disease, sickle cell disease, or phenylketonuria (PKU). Some scientists and physicians, indeed, have explicitly argued that it is wrong to permit ill or disabled people to procreate unless society is prepared to provide them with the "choice" to abort any fetus likely to manifest a condition like their own.

The new profession of genetic counseling has arisen to meet the need for information about the availability and significance of appropriate preconceptive and prenatal tests and about the decisions with which such tests confront prospective parents. Now a central factor in the shift of "scientific" selection from coercive state programs to socially sanctioned personal initiatives, advice about Choice Eugenics has become a routine part of prenatal care. In fact, practitioners have been sued by parents of children born disabled for not offering such information. Meanwhile, other prospective parents complain that their obstetrician or genetic counselor was excessively insistent that they accept a prenatal test and terminate a pregnancy that was predicted to produce a disabled child. So, though Choice Eugenics is not comparable to the earlier, compulsory state practices, some people experience it as coercive. Indeed, anxiety about social disapproval can sometimes be a more compelling dictator of choice than the law.

Extending the range of such "choices" since the 1970s and increasingly in the 1990s, hospital-based, nonprofit fertility clinics, as well as a growing for-profit fertility industry, have been devising new technologies and social practices and expanding the use of the traditional ones, such as artificial insemination. The basis for most of the newer reproductive practices is in vitro fertilization (IVF). Initially, IVF was intended to help women whose ovaries and uterus were intact, so that they could produce eggs and gestate an embryo, but whose fallopian tubes were missing or blocked. It involves hyperstimulation of the ovaries to induce several ova to mature simultaneously and then extracting a few of these and incubating them with fertile sperm outside the body ("in vitro"). Once an egg and sperm have fused and the first few cell divisions have occurred, several embryos are inserted into the woman's uterus in the hope that at least one of them will become implanted and develop into a baby.

Since the first successful attempt, in 1977, resulted in the birth of Louise Brown, IVF has become a widely offered and virtually routine part of reproductive medicine. It is covered by some state and private insurance programs in the United States and by the national health insurance programs of other countries. In addition to its procreative potential, it also enables prospective parents to have predictive tests performed on the embryos before they are implanted. It has therefore become an option for Choice Eugenics, especially for couples who strongly object to aborting an initiated pregnancy.

The access to human embryos offered by IVF has also made the technology a point of departure for a range of previously unavailable manipulations that raise complex questions, not just for the individuals they affect directly, but for society at large. The technically least challenging of these involves the participation of two women, instead of just one, in producing a child—one who produces the egg, the other who gestates the embryo. This arrangement raises the novel question of which of them is the child's biological mother. In cases of disagreement, judges have often come down on the side of genes—hence, of the egg donor—but that simply acts out our current genomania, because it ignores the biological major role of the woman who gestates and gives birth.

The explosive proliferation of preconceptive and prenatal tests has provided more and more reasons to terminate pregnancies in the hope of a better roll of the dice. Given certain features of modern prosperous societies, there is an increasing tendency to exercise such options on the basis of notions of biological perfectibility. This tendency is transforming Choice Eugenics into Yuppie Eugenics. What was once a preventative choice has become a proactive entitlement, exacerbated by the sense prevailing among current elites that one has the right to control all aspects of one's life and shape them by buying and periodically upgrading the best that technology has to offer, be it a computer, a car, or a child. Because this trend enjoys broadly based, mainstream sanction in the United States, what may begin as elite yuppie-ism is poised to become more widely disseminated as the technologies become cheaper and their use becomes more routine.

Fair-minded people differ on the point at which Choice Eugenics grades into Yuppie Eugenics. For some people with congenital disabilities, much Choice Eugenics, directed to preventing the birth of people like themselves, is going too far. People with congenital disabilities typically feel whole, and consider themselves victimized not by their genes but by dis-accommodating social arrangements. At the same time, a lack of extended family and health support systems can, for many, shift the balance against "preventable diversity." All eugenics defines some people as biologically unacceptable. Each turn of the screw en route to Yuppie Eugenics potentially excludes more and more people.

The false hope that scientists can alter an embryo genetically so as to "enhance" its potential and to make it conform to the future parents' image of a desirable child is also part of Yuppie Eugenics. Such unrealistic expectations, built on intrinsically unreliable genetic foreknowledge as well as on unscientific notions of the correspondence of specific genes to complex traits, can tempt prospective parents to agree to novel biological manipulations that are at least as likely to introduce problems as to remedy them. Also called germ line genetic engineering or germ line "gene therapy," this possibility has aroused widespread opposition. There are religious but also secular philosophical reasons for resisting a technology that plays into the idea of the developing human as a perfectible item. The goal of perfection encourages a view of existing life as imperfect. It transforms life into an ahistorical object, without context and eventually artifactual.

Objections also center on the fact that a genetic alteration introduced into an embryo is likely to become a permanent part of the genetic endowment of the person into whom that embryo develops and thus also of all of her or his progeny. Considering that the procedures themselves are experimental and the results are unpredictable (laboratory mice on which such procedures are performed often produce progeny with malformations, behavioral abnormalities, or increased cancer rates), germ-line genetic engineering poses unacceptable risks for "persons" who have just barely been conceived. There is no justification for undertaking such manipulations. If the prospective parents of the child into whom the embryo would develop are concerned that it may not meet their expectations, they need not gestate it.

Unfortunately this insidious prospect of germ line engineering has found advocates among scientists such as James D. Watson, the Nobel Prize winning codiscoverer of the structure of DNA and the first director of the Human Genome Project, and Princeton University biologist Lee Silver. In his book *Remaking Eden* and in numerous appearances on television and the college speaking circuit, Silver has been trying to persuade the public to get used to the prospect of a world with genetic haves and have nots that would eventually lead to separate, and intentionally unequal, human species.

Fortunately, other more responsible scientists, writers, and activists have warned about the ominous safety and social implications of following this path, and have called for a ban on producing genetically engineered humans. Indeed many European countries already prohibit such procedures.

This makes it particularly discreditable that the American Association for the Advancement of Science (AAAS), the largest professional organization of scientists in the United States, has gone on record with a statement that dances around the hazards of germ line modification without even raising the possibility of a ban. Although the panel that prepared the AAAS report considers that the time is not yet ripe for implementing inheritable genetic modifications (IGM), it leaves the door open for future uses: "Although there are major technical obstacles to developing human IGM in the responsible ways that we have recommended," they write, "it is possible that at some time in the future scientific advances will make it feasible to undertake IGM." The AAAS panel failed to explain that only an unethical line of research, in contravention of the internationally endorsed Nuremberg Code on human experimentation, would have to be undertaken before any assurances can be made as to the risks of these procedures.

Cloning, another experimental genetic technology touted as a new "reproductive option," is being advocated by maverick physicians and scientists despite wide evidence of pathology in animals produced in this fashion. Some of these proponents, including a representative of the Raelians, a Canadian religious cult that claims to have received an extraterrestrial directive to clone its adherents, were given a respectful hearing at a forum at the prestigious National Academy of Sciences in August 2001.

Yuppie Eugenics also has its boosters among journalists and professional bioethicists. Michael Kinsley, writing in *Slate* (April 2000), suggested that genetic tests should eventually be used as qualifications for employment. He was seconded by Andrew Sullivan in the *New York Times Magazine* (July 2000), where he argued that genetic testing for future capacities is less objectionable than using SAT scores or letters of recommendation, since genetic tests are "more reliable." Arthur Caplan, Glenn McGee, and David Magnus of the University of Pennsylvania Institute of Bioethics follow the Kinsley-Sullivan thesis to its logical conclusion when they state, in a 1999 *British Medical Journal* article, "It is not clear that it is any less ethical to allow parents to pick the eye color of their child or to try to create a fetus with a propensity for mathematics than it is to permit them to teach their children the values of a particular religion or require them to play the piano." Here the intersubjective nature of the parent-child relationship is conflated with the one-way imposition of a chancy, irreversible genetic alteration during the earliest stages of embryonic development.

What is generally ignored in such prescriptions is that each gene contributes to numerous traits, and that any trait of significance depends on the functions of many different genes. Genes and other features involved in growth and development constitute integral wholes that genetic alterations are more likely to disturb than enhance.

Yuppie Eugenics builds on the mirage that applications of genetics and biotechnology will be able to make us more perfect as well as to counter all forms of pain, illness, and death. The best way to restore us to sanity is to remember that genetics will never tell us what it takes to make a worthy human being and that the major causes of human illness and death continue to be not enough healthful food and too much unhealthful work. Eugenic and other gene dreams will not cure what ails us.

15. Brave New Genome

Eric S. Lander

Fifty years ago, microbiologists sparked the recombinant-DNA revolution with the discovery that bacteria have innate immune systems based on restriction enzymes. These enzymes bind and cut invading viral genomes at specific short sequences, and scientists rapidly repurposed them to cut and paste DNA in vitro—transforming biologic science and giving rise to the biotechnology industry.

Ten years ago, microbiologists discovered that bacteria also harbor adaptive immune systems, and subsequent progress has been breathtakingly rapid.[1] Between 2005 and 2009, microbial genetic studies conducted by the laboratories of Mojica, Jansen, Koonin, Horvath, van der Oost, Sontheimer, Marraffini, and others revealed that bacteria have a programmable mechanism that directs nucleases, such as Cas9, to bind and cut invading DNA that matches "guide RNAs" encoded in specific bacterial genome regions containing clustered regularly interspaced short palindromic repeats (CRISPR). In 2010 and 2011, Moineau and Charpentier defined the critical components of the CRISPR-Cas9 system, and Siksnys showed that it could be reconstituted in new bacterial species. Biochemical studies in 2012, by Charpentier and Doudna and by Siksnys, confirmed these results in vitro. In 2013, Zhang and Church each described how to repurpose the CRISPR-Cas9 system to work in mammalian cells, creating a general-purpose tool for editing the genome in living human cells. Over the past two years, thousands of laboratories around the world have begun to use CRISPR-Cas9 in research.

From *The New England Journal of Medicine*, Eric S. Lander, "Brave New Genome," 373 no. 1, 5–8. doi: 10.1056/NEJMp1506446. Copyright © July 2015 Massachusetts Medical Society. Reprinted with permission from Massachusetts Medical Society.

Genome editing also holds great therapeutic promise. To treat human immunodeficiency virus (HIV) infection, physicians might edit a patient's immune cells to delete the $CCR5$ gene, conferring the resistance to HIV carried by the 1 percent of the US population lacking functional copies of this gene. To treat progressive blindness caused by dominant forms of retinitis pigmentosa, they might inactivate the mutant allele in retinal cells. To prevent myocardial infarctions that kill patients with homozygous familial hypercholesterolemia, they might edit liver cells to restore a functional copy of the gene encoding low-density lipoprotein receptors. Editing of blood stem cells might cure sickle cell anemia and hemophilia.

These goals will require overcoming serious technical challenges (such as avoiding "off-target" edits elsewhere in the genome, which might give rise to cancer), but they pose no unique ethical issues because they affect only a patient's own somatic cells.

However, the technology also raises a more troubling possibility: creating children carrying permanent, heritable changes to the human germline DNA. The press has dubbed such brave new progeny "designer babies" or "genetically modified humans."

When scientists realized in the mid-1970s that recombinant DNA posed potential hazards, they called for a voluntary moratorium on experiments and organized a now-famous gathering in Asilomar, California, to develop biosafety principles for handling recombinant organisms, setting the field on its successful course. Now, several groups have urged a moratorium on human germline editing,[2] and the National Academy of Sciences has announced a fall 2015 meeting, which it plans to coordinate with academies from other countries, to begin an international conversation on the topic.

The task now is to develop a clear framework for evaluating human germline editing. Here, I offer a starting point, focusing on four key issues. (When considering these issues, readers should note that the Broad Institute, which I head, has filed patents on some of this technology, as detailed in my disclosure statement.)

The first is technical: whether genome editing can be performed with sufficient precision to permit scientists to responsibly contemplate creating genetically modified babies. Currently, the technology is far from ready: Liang and colleagues recently applied genome editing to human tripronuclear zygotes (abnormal products of in vitro fertilization [IVF] that are incapable of developing in vivo) and documented problems including incomplete editing, inaccurate editing, and off-target mutations.[3] Even with improved accuracy, the process is unlikely to be risk-free.

The second issue is whether there are compelling medical needs that outweigh the risks—both from inaccurate editing and from unanticipated effects of the intended edits. Various potential applications must be considered.

The most common argument for germline editing concerns preventing devastating monogenic diseases, such as Huntington's disease. Though avoiding the roughly 3600 rare monogenic disorders caused by known disease genes is a compelling goal, the rationale for embryo editing largely evaporates under careful scrutiny. Genome editing would require making IVF embryos, using preimplantation genetic diagnosis (PGD) to identify those that would have the disease, repairing the gene, and implanting the embryo. Yet it would be easier and safer simply to use PGD to identify and implant the embryos that aren't at risk: the proportion is high in the typical cases of a parent heterozygous for a dominant disease (50 percent) or two parents who are carriers for a recessive disease (75 percent). To reduce the incidence of monogenic disease, what's needed most is not embryo editing but routine genetic testing so that the many couples who don't know they are at risk can avail themselves of PGD.

Genome editing would add substantial value only when all embryos would be affected—for example, when one parent is homozygous for a dominant disorder or both parents are homozygous for a recessive disorder. But such situations are vanishingly rare for most monogenic diseases. For dominant Huntington's disease, for example, the total number of homozygous patients in the medical literature is measured in dozens. For most recessive disorders, cases are so infrequent (1 per 10,000 to 1 per million) that marriages between two affected persons will hardly ever occur unless the two are brought together by the disorder itself. The most common situation would probably be two parents with recessive deafness due to the same gene (among the many that can cause inherited deafness) who wish to have a hearing child.

Another potential application is reducing the risk of common diseases, such as heart disease, cancer, diabetes, and multiple sclerosis. The heritable influence on disease risk is polygenic, shaped by variants in dozens to hundreds of genes. Common variants tend to make only modest contributions (for example, reducing risk from 10 percent to 9.5 percent); rare variants sometimes have larger effects, including a few for which heterozygosity provides significant protection against disease.

Some observers might propose reshaping the human gene pool by endowing all children with many naturally occurring "protective" variants. However, genetic variants that decrease risk for some diseases can increase risk for others. (For example, the *CCR5* mutations that protect against HIV also elevate the risk for West Nile virus, and multiple genes have variants with opposing effects on risk for type 1 diabetes and Crohn's disease.) The full medical effect of most variants is poorly characterized, let alone the combined effects of many variants. Safety studies would be needed to assess effects across various genetic backgrounds and environmental exposures. The situation is particularly dicey for rare protective heterozygous variants: most have never been seen in the homozygous state in humans and might have deleterious effects. Yet heterozygous parents would routinely produce homozygous children (one-quarter of the total)—unless humans forswore natural reproduction in favor of IVF.

Currently, the best arguments might be for eliminating the ε4 variant at the *APOE* gene (which increases risk for Alzheimer's disease and cardiovascular disease) and bestowing null alleles at the *PCSK9* gene (which reduces the risk of myocardial infarction). Still, our knowledge is incomplete. For example, APOE ε4 has also been reported to be associated with better episodic and working memory in young adults.

Some scientists might ask: Why limit ourselves to naturally occurring genetic variants? Why not use synthetic biology to write new cellular circuits that, for example, cause cells to commit suicide if they start down the road toward cancer? But such efforts would be reckless, at least for now. We remain terrible at predicting the consequences of even simple genetic modifications in mice. One cautionary tale among many is a genetic modification of the *tp53* gene that protected mice against cancer while unexpectedly causing premature aging.[4] We would also need to anticipate the potential interactions among the diverse genetic circuits that creative scientists will cast into the gene pool. Mistakes would be inevitable, and there would be no way to recall novel genes from the human population.

A more distant frontier would be to reshape nonmedical traits. Height may prove challenging (the hundreds of natural variants have tiny effects), but hair and eye color may be pliable. Disruption of the *MC1R* gene is associated with bright red hair, although it also heightens the risk of melanoma. Sports-minded parents might want to introduce the overactive erythropoietin gene that conferred high oxygen-carrying ability on a seven-time Olympic medalist in cross-country skiing. Nonnatural genetic modifications hold even bolder prospects—and risks.

The third key issue is who has the right to decide. Some people will argue that parents should have unfettered autonomy—that modifying one's progeny is akin to using PGD to avoid genetic diseases or choosing sperm donors on the basis of intellectual or athletic prowess. Yet parental autonomy must be weighed against the interests of future generations who cannot consent to the genetic modifications their flesh will be heir to.

The final issue concerns morality—what's right and wrong and how we ought to live as a society. Although scientists may be reluctant to debate ethics, we have a responsibility to do so and insights to offer. How would routine genome editing change our world? Would we come to regard our children as manufactured products? Would marketers shape genetic fashions? Would the "best" genomes go to the most privileged? If we cross this threshold, it's hard to see how we could ever return.

The recombinant-DNA moratorium of the 1970s was a temporary pause to establish safety rules for laboratory research. Today's debate concerns not research (which should proceed) but clinical applications to human beings that result in permanent changes to the human gene pool.

Genetic modification of human embryos is not a new idea. At least among Western governments, there has been a long-standing consensus that manipulating the human germ line is a line that should not be crossed. Some European countries have outlawed genetic modification of embryos. The United States lacks a legislative ban, but the Food and Drug Administration—whose approval is needed for introducing substances, including DNA, into embryos—has said it will not permit genetic modification, and the National Institutes of Health (NIH) Recombinant DNA Advisory Committee will not currently approve such work at institutions receiving NIH funding. In many other countries, the situation remains unresolved.

The discussions that will begin in the fall may solidify a broad international consensus that germ line editing should be banned—with the possible exception of correcting severe monogenic disease genes, in the few cases in which there is no alternative. For my own part, I see much wisdom in such a position, at least for the foreseeable future. A ban could always be reversed if we become technically proficient, scientifically knowledgeable, and morally wise enough and if we can make a compelling case. But authorizing scientists to make permanent changes to the DNA of our species is a decision that should require broad societal understanding and consent.

It has been only about a decade since we first read the human genome. We should exercise great caution before we begin to rewrite it.

NOTES

1. P.D. Hsu, E.S. Lander, and F. Zhang, "Development and Applications of CRISPR-Cas9 for Genome Engineering," *Cell* 157 (2014): 1262–78.

2. D. Baltimore, P. Berg, and M. Botchan, et al., "Biotechnology: A Prudent Path Forward for Genomic Engineering and Germline Gene Modification," *Science* 348

(2015): 36–38; E. Lanphier, F. Urnov, S.E. Haecker, M. Werner, and J. Smolenski, "Don't Edit the Human Germline," *Nature* 519 (2015): 410–11.

3. P. Liang, Y. Xu, X. Zhang, et al. "CRISPR/Cas9-Mediated Gene Editing in Human Tripronuclear Zygotes." *Protein Cell* 6 (2015): 363–72.

4. S.D. Tyner, S. Venkatachalam, and J. Choi, et al. "Mutant Mice That Display Early Ageing-Associated Phenotypes," *Nature* 415 (2002): 45–53.

16. Can We Cure Genetic Diseases without Slipping into Eugenics?

Nathaniel Comfort

On April 18, 2015, scientists at Sun Yat-sen University in Guangdong, China, published an article in the obscure open-access journal *Protein & Cell* documenting their attempt at using an experimental new method of gene therapy on human embryos. Although the scientific significance of the results remains open to question, culturally the article is a landmark, for it has reanimated the age-old debate over human genetic improvement.

The Chinese scientists attempted to correct a mutation in the beta-globin gene, which encodes a crucial blood protein. Mutations in this gene lead to a variety of serious blood diseases. But the experiments failed. Although theoretically the new method, known as CRISPR (short for "clustered regularly spaced short palindromic repeats") is extremely precise, in practice it often produces "off-target" mutations. In plain English, it makes a lot of changes in unintended locations, like what often happens when you hit "search/replace all" in a word-processing document. The principal conclusion from the paper is that the technique is still a long way from being reliable enough for the clinic. Nevertheless, the science media and pundits pounced on the story, and for a while "#CRISPR" was trending on Twitter.

CRISPR is the fastest, easiest, and most promising of several new methods known collectively as "gene editing." Using them, scientists can edit the individual letters of the DNA code, almost as easily as a copy editor would delete, a stray comma or correct a spelling error. Advocates wax enthusiastic about its promise for correcting mutations for serious genetic diseases like cystic fibrosis and sickle-cell anemia. Other applications might include editing HIV

This article originally appeared in the Aug. 3/10, 2015 issue of *The Nation* Magazine as "Can We Cure Genetic Diseases without Slipping into Eugenics?" by Nathaniel Comfort. Reprinted with minor updates by permission of author.

out of someone's genome or lowering genetic risks of heart disease or cancer. Indeed, every week brings new applications: CRISPR is turning out to be an extraordinarily versatile technique, applicable to many fields of biomedical research. I'm pretty immune to biomedical hype, but gene editing has the marks of a genuine watershed moment in biotechnology. Once the kinks are worked out, CRISPR seems likely to change the way biologists do experiments, much as the circular saw changed how carpenters built houses.

The timing of the paper was provocative. It was submitted on March 30 and accepted on April 1; formal peer review was cursory at best. Two weeks before, scientists in the United States and Europe had called for a moratorium on experiments using CRISPR on human "germ line" tissue (eggs, sperm, and embryos), which pass alterations on to one's descendants, in contrast to the "somatic" cells that compose the rest of the body. The embryos in the Chinese experiments were not implanted and in fact could not have become humans: they were the unviable, discarded products of in vitro fertilization. Still, the paper was a sensational flouting of the Westerners' call for restraint. It was hard not to read its publication as an East Asian Bronx cheer.

The circumstances of the paper's publication underline the fact that the core of the CRISPR debate is not about the technological challenge but the ethical one: that gene editing could enable a new eugenics, a eugenics of personal choice, in which humans guide their own evolution individually and in families. Commentators are lining up as conservatives and liberals on the issue. Conservatives, such as Jennifer Doudna (one of CRISPR's inventors) and the Nobel laureates David Baltimore and Paul Berg, have called for cautious deliberation. They were among those who proposed the moratorium on using CRISPR on human embryos. "You could exert control over human heredity with this technique," said Baltimore. George Q. Daley, of Boston Children's Hospital, said that CRISPR raises the fundamental issue of whether we are willing to "take control of our genetic destiny." Are we ready to edit our children's genomes to perfection, as in the movie *Gattaca?* Could the government someday pass laws banning certain genetic constitutions or requiring others?

The CRISPR liberals are optimists. They insist that we should proceed as rapidly as possible, once safety can be assured—for example, that an "edit" wouldn't inadvertently cause cancer while treating thalassemia. Some, such as the Oxford philosopher Julian Savulescu, insist that we have a "moral imperative" to proceed with engineering our genomes as fast as our sequencers can carry us. Savulescu believes it would be unethical to have the technology to produce better children and not use it. (For once, I'm with the conservatives.)

This debate is very familiar to a historian. Thus far, CRISPR is following the classic arc of breakthrough methods in genetics and biotech. First come millennialist debates over the new eugenics; then, calls for caution. A few cowboys may attempt rash experiments, which often fail, sometimes tragically. Finally, the technology settles into a more humdrum life as another useful tool in the biologist's kit.

Each instance of this pattern, however, occurs in a different context, both scientifically and culturally. And while scientists, philosophers, and other commentators have been discussing the scientific risks and merits of CRISPR ad nauseam, no one seems to be placing the debate itself in this broader historical setting. Over the last 150 years of efforts to control human evolution, the focus on the object of control has tightened, from the population, to the individual, to the gene—and now, with CRISPR, to the single letters of our DNA code. Culturally, during this period, the pendulum has swung from cooperative collectivism to neoliberalism. The larger question, then, is, with the emergence of gene editing during an era of self-interested free-market individualism, will eugenics become acceptable and widespread again?

Until relatively recently, the only way to create genetically better humans was to breed them. In 1865, Charles Darwin's half-cousin Francis Galton sought both to inspire society's richest, wisest, and healthiest to breed like rabbits and to persuade the sick, stupid, and poor to take one for the empire and remain childless. In 1883, he named the plan "eugenics," from the Greek *eugenes*, meaning "well-born" or "well-bred." In Galton's mind, eugenics was a much kinder approach to population management than ruthless Malthusian efforts to eliminate charity and public services. However misguided eugenics may seem today, Galton saw it as a humane alternative to simply letting the disadvantaged freeze, starve, and die.

In early twentieth-century America, Galton's plan suddenly seemed far too passive and slow. A new generation of eugenicists, spurred by novel experimental methods in genetics and other sciences, sought to take a firmer hand in controlling the reproduction of the lower classes, people of color, and the insane or infirm. Can-do Americans passed laws restricting marriage and immigration to prevent the degradation of an imagined American "stock." Some, such as the psychologist Henry Goddard and the biologist Charles Davenport, sought to round up the so-called feebleminded (those with a mental age below twelve) and institutionalize them as a sort of reproductive quarantine—adult swim in the gene pool. But others pushed for laws to simply sterilize those seen as unfit. That way, they could then marry or have sex with whomever they wanted without endangering the national germ plasm. Altering the body seemed more humane than confining it.

Involuntary sterilization soon lost any veneer of benevolence. In the United States, thousands of people were sterilized against their will, under eugenic laws passed in more than thirty states. For the most part, educated middle- and upper-class white Protestant men decided who was fit to reproduce, and naturally they judged fitness in their own image. In Germany, a decades-old program of *Rassenhygiene* or "race hygiene" took a cue from the vigorous American eugenics movement. The fingerprints of Davenport, his superintendent at the Eugenics Record Office, Harry Laughlin, and other American eugenicists are on the infamous 1933 Nazi sterilization law. Controlling bodies was not so humane after all.

Around midcentury, many American scholars and scientists turned to environmental and cultural solutions for social problems, including poverty, mental illness, and poor education. However, some thinkers—biologists and others—advocated for more and better biotechnology. How much cleaner and more rational it would be, they argued, to separate sex from reproduction and make babies in the laboratory, using only the highest-quality sperm and eggs. In his 1935 book *Out of the Night*, the geneticist Hermann Joseph Muller called this "eutelegenesis." He and others painted sunny pictures of free love and sperm banks. But three years earlier, in *Brave New World*, the English novelist Aldous Huxley had taken a much darker view of the scientific control of evolution. "Bokanovsky's Process"—test-tube human cloning—was a "major tool of social stability!" said his director of hatcheries and conditioning. It was the biotechnical core of "community, identity, stability," the motto of the One World State.

Since then, each step in the development of biotechnology has seemed to bring Bokanovsky's Process closer to realization. In 1969, the Harvard biologist Jonathan Beckwith and colleagues discovered how to isolate, or "clone," a gene. At about the same time, Dan Nathans and Hamilton Smith at Johns Hopkins discovered how to use a type of molecular scissors called restriction enzymes to snip, insert, and reattach DNA strands in the lab. Each enzyme cuts the DNA at a specific site. (CRISPR, too, is based on naturally occurring bacterial enzymes.) In the 1970s, researchers discovered more than a hundred different restriction enzymes, forming a battery of tools to cut DNA almost anywhere one wished. The new research enabled genes to be recombined—cut and pasted at will, even between species. To techno-optimists, genetic engineering would make the old, inhumane eugenics unnecessary. There would be no need to prevent people with bad genes from reproducing if one could simply repair those genes.

Public outcry. Eugenic angst. Predictions of enzymatic Armageddon. The city of Cambridge, Massachusetts—home to Harvard and MIT—banned recombinant DNA research outright. (Some of the schools' top scientists promptly decamped for New York, Maryland, and California.) In 1974, fearing a massive clampdown from on high, scientists self-imposed a moratorium on recombinant DNA research. Ten months later, at a meeting at the Asilomar Conference Center near Monterey, California, David Baltimore, Paul Berg, James Watson, and other scientific luminaries agreed on a set of guidelines for laboratory safety: how to prevent, for example, a lethal bacterium from escaping the lab and causing epidemics or massive agricultural or ecological disaster. Within five years, fears had subsided and recombinant DNA had become a standard laboratory technique—forming the basis of a burgeoning biotech industry, whose early triumphs included synthetic insulin, the cancer drug interferon, and exogenous erythropoietin, a hormone that regulates the production of red blood cells.

But Asilomar is not the only—or even the best—historical comparison for CRISPR. Since the early 1960s, visionary scientists had imagined an era of "genetic surgery," in which defective genes could simply be repaired or replaced. Rather than curing diseased patients, or segregating them from the "healthy" population, researchers said they would cure diseased molecules. In 1980, the UCLA researcher Martin Cline made the first primitive attempt at using engineered molecules therapeutically. Like the CRISPR researchers, he targeted the beta-globin gene. Cline, however, ignored more than just his colleagues' own recommendations: Flouting National Institutes of Health regulations, he went overseas and injected a live virus containing the beta-globin gene into the bone marrow of two young women. Fortunately, the dosage was too small to have any effect; the girls were not helped, but neither were they harmed. Cline, on the other hand, suffered: He was publicly censured and had his federal funding restricted.

By the late '80s, gene therapy seemed poised for a breakthrough. Led by NIH researcher W. French Anderson, starry-eyed biologists anticipated cutting and pasting their way to the end of genetic disease. Hundreds of grant applications were filed for gene-therapy research. In 1990, Anderson and colleagues conducted the first approved trial, on an exceedingly rare disease called adenosine deaminase deficiency, in which the loss of a single enzyme wipes out the entire immune system. The trial appeared to be a success. But the gene-therapy cowboys were humbled in 1999, when Jesse Gelsinger, a teenager suffering from a rare liver disorder, died of massive organ failure from the engineered virus used to ferry a gene into his cells. Then, in 2002, a French gene-therapy trial to correct immune-system failure was a success—at least until the subjects of the experiment developed leukemia, because the virus used as a delivery vehicle disrupted a gene required for normal cell growth. The FDA then suspended retroviral gene-therapy trials on bone-marrow cells until regulatory measures could be implemented. Unintended consequences killed the gene-therapy hype.

In the succeeding years, gene therapy has quietly returned. Old methods have been improved, new methods have been developed, and researchers have had limited success with treatments for a variety of cancers, AIDS, and several eye diseases. Hope remains high among the optimists, but even they acknowledge that the promise remains greater than the results.

The gene-therapy craze of the 1990s yielded two fundamental ethical distinctions. First, researchers distinguished engineering the germ line from engineering somatic cells. Germ-line modifications are not used to treat disease in an individual but to prevent it (or lower the risk) in future individuals. Unlike preventive public-health measures such as the quarantine, however, meddling with the genome has a high risk of unintended consequences. The genome is like an ecosystem, with every element ultimately connected to every other. Inadvertently damaging alterations could thus be seen as harming the genomes of the others without their consent. Yet Anderson was willing to consider germ-line modifications should somatic gene therapy eventually prove safe. (Scientists like Harvard's George Church make similar arguments about CRISPR today.) The second distinction was that gene therapy should only be used to treat disease—not to enhance or alter normal traits. In short, gene therapists considered therapeutic applications ethical but enhancement not—and creating a master race was right out. (Anderson was more principled about some things than others; he is currently serving time for child molestation.)

Parallel to the development of genetic engineering, advances in reproductive technology made Muller's and Huxley's vision of test-tube babies a reality. On July 25, 1978, Louise Brown was born through in vitro fertilization, a technique developed by Patrick Steptoe and Robert Edwards. Combining IVF with new genetic-screening technologies made it technically possible to reject embryos with undesirable traits—or select those with desirable ones. "You do not need the still distant prospect of human cloning to begin to get worried," wrote Anthony Tucker in *The Guardian*. James Watson, who had recently recanted his conservative position on recombinant DNA research, nevertheless predicted, "All hell will break loose."

An even braver new world dawned in 1996, when the Roslin Institute in Scotland announced the birth of Dolly the sheep—the first "cloned" large mammal. The technique, formally known as somatic-cell nuclear transfer, revived the debate over designer babies. The US National Bioethics Advisory Commission launched an investigation and, in 1997, published a report that led to the unusual step of restricting a procedure that did not exist. The NIH prophylactically prohibited cloning human beings with federal funds.

Researchers promptly announced plans to attempt it with private money. One was Brigitte Boisselier, who was supported by Clonaid, the research arm of the transhumanist group the Raëlians. Its leader, Raël (né Claude Vorilhon), claimed to have been contacted by extraterrestrials. On December 27, 2002, Boisselier announced that a cloned baby, called Eve, had been born, although Clonaid wouldn't reveal any data or produce Eve for inspection. The mainstream scientific community rolled its collective eyes. Once again, though, the dust eventually settled, and somatic-cell nuclear transfer remains a legitimate laboratory technique. Clonaid claims to be cloning away still. But no armies of Hitlers have stormed across Europe, and to date, no genetically optimized Superman has communed with the groovy dudes from the next galaxy.

One important result from the cloning debate was that the kibosh on genetic enhancement began to relax. In 2001, Julian Savulescu started to argue for "procreative beneficence," a principle that holds that people are morally obligated to have the best children possible—including through genetic-enhancement technologies. (Savulescu's *Enhancing Human Capacities*, published in 2011, continues the campaign.) The eugenically named, self-proclaimed visionary Gregory Stock published *Redesigning Humans* in 2002; it rosily envisioned writing "a new page in the history of life, allowing us to seize control of our evolutionary future." What could go wrong?

CRISPR, then, is the latest chapter in a long, darkly comic history of human genetic improvement. Like whole-gene engineering in the 1970s, gene editing is proving remarkably versatile in basic science research: new applications appear weekly. But conservative researchers such as Doudna, Baltimore, and Berg insist that the taboos against germ-line engineering and enhancement remain in place—for the time being. However, notwithstanding Baltimore's and Berg's reassurance that eugenics is "generally considered abhorrent," some commentators are actively and publicly advocating what they consider a new kind of eugenics, and, as she writes in her new book, *A Crack in Creation*, even Doudna is now willing to contemplate germ-line enhancement with CRISPR. Their argument is couched in technology, but it rests on politics.

The eugenics movement of the early twentieth century was rooted in a spirit of collectivism. Ideals of progress and perfection dominated American culture. Across the political spectrum, Americans sought social improvement through a variety of reforms, ranging from public health to food production to workplace environments and education. Such a project required collective effort. Government legislation was broadly accepted as a tool of positive change. Cooperation for the good of society was a sign of good citizenship. And science, epitomizing rationality, efficiency, and mastery over nature, was society's most potent tool of progress.

Eugenics, often referred to as "racial hygiene," was associated with the Progressive hygiene movement in public health. In 1912, Harvey Ernest Jordan, who later became dean of the University of Virginia's medical school, addressed a conference of eugenicists on the importance of their field for medicine. He asserted, with the buoyancy of the era, that the country was emerging from a benighted period of selfish individualism—which Mark Twain had dubbed the "Gilded Age"—into an enlightened phase of concern for one's fellow man. Eugenics was of vital interest to medicine, he wrote, because it sought to prevent disease and disability before it occurred.

Progressives had faith in government as an instrument of social—and biological—change. By the 1960s, that faith had eroded. The Cold War had sparked an antiauthoritarian New Left that criticized state control as a corruption of the collectivist spirit. Left-wing biologists sometimes found themselves in an awkward position. When Beckwith, a staunch leftist, cloned the first gene, he held a press conference warning against the dangers of his own research. "The work we have done may have bad consequences over which we have no control," he said. His graduate student James Shapiro commented, "The use by the Government is the thing that frightens us."

During the 1970s, New Deal liberalism began to give way to neoliberalism. At the turn of the twenty-first century, biotech and infotech had grown as dominant as Big Oil and Big Steel had been in 1900. The Internet has become our railroad system. The last thirty years have seen Jordan's "general change from individualism to collectivism" reversed: elites justify increasing inequality with a libertarian rhetoric of individual freedom.

Individualism, say the biotech cheerleaders, immunizes us against the abuse of reproductive genetics. In their free-market utopia, control over who gets to be born would be a matter of personal choice, not state order. Couples should have the freedom to undergo in vitro fertilization and to select the healthiest embryos—even the best ones, because "best" is no longer a matter of official mandate. For those who define eugenics as state control over reproduction, this is not eugenics.

Others adopt a definition closer to that of the 1921 International Eugenics Congress: Eugenics is "the self-direction of human evolution." Many critics over the years have argued that eugenics wasn't wrong; rather, it was done badly and for the wrong reasons. So it goes today. In 2004, Nicholas Agar published *Liberal Eugenics,* a philosophical defense of genetic enhancement. In a nutshell, he argues that genetic enhancements ought not to be treated differently from environmental enhancements: if we are allowed to provide good schools, we must be allowed to provide good genes. Like Savulescu, Agar insists that it is immoral to prohibit parents from producing the best children they can, by whatever means. In 2008, in the foreword to a reissue of Charles Davenport's 1912 *Heredity and Eugenics,* Matt Ridley, a viscount, zoologist, science writer, and Conservative member of the House of Lords, argued that the problem with eugenics was its underlying collectivist ideology. Selfishness would save the human race. "There is every difference between individual eugenics and Davenport's goal," he wrote. "One aims for individual happiness with no thought to the future of the human race; the other aims to improve the race at the expense of individual happiness." Similarly, Gregory Stock wrote that a "free-market environment with real individual choice" was the best way to protect us from eugenic abuse. Liberal eugenics is really neoliberal eugenics.

In which case, it's hard to see how individual choice and the invisible hand will defang the dangers that eugenics still poses. In a 2011 *Hastings Center Report,* the Australian bioethicist Robert Sparrow showed how libertarian, individualist eugenics would lead to the same ends as good old-fashioned Progressive eugenics. Savulescu's "best possible children" must naturally have the most opportunities to flourish and the fewest impediments to a happy, fulfilling life. Accordingly, parents should select the traits that society privileges. But in our current society, who has the most opportunities? Of course: a tall, white, straight, handsome man. If neoliberal genetic enhancement were to proceed unregulated, then social convention, cultural ideals, and market forces would drive us toward producing the same tired old Aryan master race.

Further, the free market commodifies all. Neoliberal eugenics creates a disturbing tendency to regard ourselves, one another, and especially our children as specimens to be improved. The view that "the genome is not perfect," as John Harris, another proenhancement philosopher, puts it, perpetuates the notion of genetic hygiene. Even cautious reports, like one that appeared recently in the *International Business Times,* propagate this idea: "any hope that [CRISPR] will help physicians ensure spotless genomes," they write, remains distant. Not to put too fine a point on it, but whatever the time line, the goal of a spotless genome implies genetic cleansing.

Scholars of disability have mounted a vigorous critique of the pursuit of genetic perfection. Call it the *Gattaca* defense: by granting individuals the power and permission to select against difference, we will be selecting for intolerance of difference. But Sparrow notes that enforcing diversity is itself morally problematic. Should gene editing become a safe and viable option, it would be unethical to prohibit parents from using it to correct a lethal genetic disease such as Tay-Sachs, or one that causes great suffering, such as cystic fibrosis or myotonic dystrophy.

Where, then, does one draw the line—and how easy would it be to enforce? Harris, Savulescu, Agar, and others say that once you let in any modifications, you have to allow them all.

What gets all too easily lost in this debate is that it takes place in a geno-centric universe. Even those opposed to genetic enhancement presume DNA to be the ultimate determiner of all that is human and biotechnology the most effective tool for solving social problems. Such genetic determinism is inherently politically conservative—whatever one's personal politics.

Here's why: sci-fi genetic fantasies, whether hand-waving or hand-wringing, divert our attention from other, more important determinants of health. Studies by the World Health Organization, the federal Office of Disease Prevention and Health Promotion, the Centers for Disease Control and Prevention, and academic researchers leave no doubt that the biggest factors in determining health and quality of life are overwhelmingly social. Genetics plays a role in disease, to be sure, but decent, affordable housing; access to real food, education, and transportation; and reducing exposure to crime and violence are far more important. In short, if we really wanted to engineer better, happier, healthier humans, we would focus much more on nurture than on nature.

The reason we don't is obvious: the very selfishness that neoliberals proclaim as the panacea for eugenic abuse. Genetic engineering primarily benefits industry and the upper classes. In vitro fertilization and genetic diagnosis are expensive; genetic therapy would be even more so. Genetic medicine is touted as the key to ending "one-size-fits-all" medicine, instead tailoring care to the idiosyncrasies of each individual. President Obama's Precision Medicine Initiative, announced in his last State of the Union address, extolled a vision of individualized care for all. But historians of medicine have shown that the rhetoric of individualized medicine has been with us at least since the days of Hippocrates. The reality is that for twenty-five centuries, individualized treatment has been accessible to the rich and powerful, while lower-status people in every era—be they foreigners, slaves, women, or the poor—have received one-size-fits-all care, or no care at all.

The idealists and visionaries insist that costs will drop and that technologies now accessible only to the rich will become more widely available. And that does happen—but new technologies continually stream in at the top, leading to a stable hierarchy of care that follows socioeconomic lines. Absent universal health care, ultra-high-tech biomedicine depends on a trickle-down ideology that would have made Ronald Reagan proud.

Further, molecular problems have molecular solutions. The eternal eugenic targets—disease, IQ, social deviance—are overdetermined; one can explain them equally as social or as biomedical problems. When they're defined as social problems, their solutions require reforming society. But when we cast them in molecular terms, the answers tend to be pharmaceutical or genetic. The source of the problem becomes the individual; the biomedical-industrial complex, along with social inequities, escape blame.

In short, neoliberal eugenics is the same old eugenics we've always known. When it comes to controlling our evolution, individualism and choice point toward the same outcomes as authoritarian collectivism: a genetically stratified society resistant to social change—one that places the blame for society's ills on individuals rather than corporations or the government.

I'll be excited to watch the workaday applications of techniques like CRISPR unfold, in medicine and, especially, basic science. But sexy debates over whether reproductive biotechnology will permit us to control our genetic evolution merely divert us from the cultural evolution that we must undertake in order to see meaningful improvement in human lives.

17. Cyborg Soothsayers of the High-Tech Hogwash Emporia

In Amsterdam with the Singularity

Corey Pein

One cool morning last fall I followed a canal from my hostel in Amsterdam to the chic DeLaMar Theater. The theater was once a Big Data archive, a custodian of the records of a Nazi labor program that sent Dutch civilians to toil in German fields and factories and was firebombed by underground resistors in 1945. Five suspects were executed for that crime against quantification, now long forgotten.

A prim doorman waved me inside the lobby, where I gave my name and claimed my lanyard. The other arrivals, mostly middle aged and business suited, chatted cheerfully. By all appearances, no one was in the mood for grim reminders of the barbarity beneath the surface of civilization. No, today's production at the DeLaMar was about the bold, boundless future soon to spread, eagle-like, across the horizon of the starry sky—a future in which every square inch of the planet will be recorded, measured, and analyzed; in which a race of genetically enhanced cyborgs will rule as omniscient philosopher-kings, mining asteroids, hoarding bitcoins, and going for years without bathing, thanks to an odor-eating microbiotic spray.

Those who will have to hit the proverbial dusty highway include skeptical journalists, "linear-thinking" academics who deny the new, "exponential" reality, organized labor, government regulators, conservationists, Luddites, and other obstacles to the reign of positive-thinking, entrepreneurial *doers*. In their future, after all, there will be no want, only abundance—a corporate cornucopia. Work will be voluntary. Thanks to ever-greater extensions of technology, the doorman will be a friendly robot, and I won't need a lanyard

Original publication by Corey Pein as "Cyborg Soothsayers of the High-Tech Hogwash Emporia: In Amsterdam with Singularity." *The Baffler*, 28 (2015) 130–147. Reprinted by permission of *The Baffler*.

because a computer chip embedded in my flesh will continually record my identity and my emotional state. Sabotage will be impossible; resistance, futile.

Such heady visions made my caffeine panic acute. It was quite the ambitious agenda for a mere two-day conference, but unrealistic ambition was the coin of the realm at the Singularity Summit, which sprawled throughout six rooms of 7,330 square feet at the DeLaMar, not counting the 950-seat grand hall that was decorated like a *Tron* set for the occasion. The summit, which takes place once or twice a year, is a roving recruitment seminar for the California-based Singularity University (SU).

SU was founded in 2008 by two friends in futurism: Peter Diamandis, founder of the X Prize Foundation and the would-be Commodore Vanderbilt of private space colonization, and Ray Kurzweil, a noted inventor, big-selling author, and prophet of a technological end times. Kurzweil's 672-page fever dream, *The Singularity Is Near* (2005), lent the institution both a name and a doctrine. By page twenty-one, the book predicts the imminent physical and metaphysical merger of humanity and computers, culminating in "epoch six" of the universe, in which the disembodied hive mind of what was once *Homo sapiens* transcends the laws of physics to unify the cosmos in eternal ecstasy.

Well, what can I say? Had I purchased tuition for the full package, a ten-week SU summer program at a corporate park outside Mountain View, California, I would have been looking at a $30,000 price tag. The $2,500 entry fee for the Amsterdam summit was a steal—and I even got comped. My press badge also granted access to the VIP area upstairs, where I found coffee. A weary Dutch publicist hired for the summit told me nine hundred people had registered, including seventy-five journalists. It was the largest SU event ever. As someone who hopes to live in the future, I wasn't the only one feeling called to, you know, climb the summit of omniscience with those who claim to see the path forward so clearly.

As I had contemplated my trip, I had pictured a modest gathering of earnest hacker types, too New Agey for Comic-Con and yet too dweeby for South by Southwest. My naïveté was instantly punctured by the crystal chandeliers and red velvet and custom tailoring inside the venue. I should've known the future would be copacetic with corporate luxury. Other attendees, I learned, shared my fascination with the looming Singularity, but some seemed drawn by the prospect of scouting for fresh intelligence on new technologies. At least that was the rationale for their employers to pick up the tab for a midweek conference in Europe's adult Disneyland.

The summit's opening video had all the technomystical bombast of a Syfy original series, and quickly gave way to a banal, buzzword-packed, thanks-to-our-sponsors speech in the customary style of the expense-account set. The massive multinational auditing firm Deloitte sponsored this summit because, in a tall glass tower somewhere, accounting majors were debating the actuarial implications of indefinite lifespans and running cost-benefit analyses on extraplanetary mining expeditions. Things got more interesting when SU chief executive Rob Nail took the stage to perform an initiation of sorts. He described himself as "one of those geeky Silicon Valley guys" and dressed the part in black denim jeans, a plaid blazer, and a pair of Google Glass that pinned back his shoulder-length hair. A Stanford-educated robotics designer, Nail sold his company, then "gave up . . . to go surfing." Attending an SU executive program ended this idle period and changed his life, like, forever.

"One of the caveats of getting involved with Singularity University is it's very difficult to get back out," Nail said. "You'll find out why in a moment."

The screen behind him displayed a black-and-white pattern like a melting checkerboard—an optical illusion. "In this photo is an image of a dog. Anybody see the dog? I'll give you a quick hint. There's the dog," Nail said, showing little patience for stragglers. "Everybody see the dog now?"

"Research actually shows that in thirty years if you go back and see this image, you'll still see the dog. In fact, you can't *not* see the dog," he went on. "So, without your permission, I literally just physically rewired your brain." He paused to let that sink in. A few people chuckled. "Over this next two days, we're going to do that on a totally different level. We're going to totally rewire your brain."

"You have a choice now," Nail continued. "I know most of you have paid a lot of money, so you're not going to walk out the door now—but this is the red pill, blue pill question, okay?" It was taken for granted that everyone had seen Laurence Fishburne dose Keanu Reeves in *The Matrix.* "Everybody signed up for the blue pill? Or the red pill, I guess," Nail said, fumbling the reference. "If you take the blue pill, you can leave now. Thank you very much. We can send you the notes. Okay. Everybody signed in. That was my disclaimer, so you can't complain to me later."

RED PILLED

"I figure, since you've taken the red pill, why don't we just jump straight into the rabbit hole," Rob Nail went on in his metaphor-mixing exec-speak. "What if we could just disrupt reality completely?" Nail's profound hypothetical might've inspired an interesting discussion about the limits of perception, or at least some fun stoner talk, like, *What if* The Matrix *is real, bro?* Instead, it segued into an overview of virtual reality gaming headsets—a product pitch. What do you know, a product from Google, an SU sponsor, came out the winner.

"I think it's probably the most exciting time that all of humanity has ever lived," Nail said. He held up his smartphone. "This is not a phone," he said solemnly. "This is a teacher, this is a doctor, this is so many other things. And the fact that there's tens of millions of apps now means we're in a totally different realm altogether . . . which creates a lot of opportunities for all of us."

"Ultimately," Nail concluded, "we're here not just to help you forecast where the future is going, but to engage in steering towards one that we all want to live in."

For what it's worth, I think they mean well. That is, I wouldn't attribute exceptionally nefarious motives to anyone I met from SU. I merely suggest that their program pursues undesirable objectives in ways likely to produce disastrous outcomes.

Nail's high-blown lunacy was unexceptional among the SU speakers. What set his enthused disrupto-babble apart was its relative benignity. Later that first morning, however, I felt the tenor of the proceedings change from silly to scary. The turning point came with the onstage arrival of SU past president Neil Jacobstein, a wry veteran of the revolving-door establishment who boasts consulting stints at NASA, the Pentagon, and large weapons contractors like Boeing. Jacobstein's matter-of-fact delivery belied the startling contents of his speech on artificial intelligence. He had a way of normalizing the outlandish and of creating the feeling that it's crazy to cling to romantic biological defects like mortality.

"Let's talk about augmentation. Do we really need to augment our brains? The answer is *yes*," Jacobstein said, brooking no incredulity. "The human brain hasn't had a major upgrade in over fifty thousand years. And if your cellphone or your laptop hadn't had an upgrade in, say, five years, you'd be appropriately concerned about that."

On the other hand, I thought, could you trust an Apple Store Genius with your mind?

But Jacobstein was already galloping off to address planned upgrades to the organization of work. He cited a recent study claiming that within the next twenty years, 47 percent of US jobs would be subject to some kind of automation. Certain professions, he noted, have obstinately resisted the trend—lawyers, teachers, and doctors. "I would say that assuming zero new technology breakthroughs, professional work—white-collar work—is ripe for disruption," Jacobstein said. "White-collar workers often have the reaction, 'Well, all jobs can be automated—except ours, of course.' But they're not immune either." Doctors, for instance, might be put out by the X Prize to develop *Star Trek*–style medical "tricorders" that can diagnose diseases with a wave of the hand. At first, Jacobstein said, "we'll probably deploy them in places with low guild protection, like Africa, rather than places with high guild protection, like the U.S."

Then Jacobstein shared with us an illustration of a ginormous silicon brain floating through a cloudy blue sky. Perplexingly, a bright red line crossed the brain. There was no caption. But Jacobstein helpfully explained:

> If you unwrap my neocortex or yours, it would be about the surface area of a large dinner napkin. One could imagine, without having the confines of the human skull, that we could build an artificial neocortex with the surface area of this auditorium. Or Amsterdam. Or Europe. Or the planet. And you might think, "Wow, that's a little excessive." But it turns out it's not excessive, and the reason is . . . the accelerating wave of human knowledge.

See, in order to contain the unforeseeable side effects of this nigh-infinite computational superpower, Jacobstein went on, society must adapt, and soon. We must have the courage to think up antivirus-style surveillance programs that could, say, be deployed to monitor the world for "anomalies and misbehavior," preserving order and preventing sabotage.

Notwithstanding the dubious record of desktop antivirus software—and putting aside the contentious matter of what constitutes misbehavior—this struck me as a bad idea. To be sure, I could see the appeal of law enforcement via algorithm, at least from the perspective of the rulers of an inescapable totalitarian superstate. For them, it'd be snazzy.

When I interviewed Jacobstein later, we had this exchange:

> ME: One of your slides was a superintelligent artificial neocortex the size of a planet.
>
> JACOBSTEIN: Mm-hmm.
>
> ME: Maybe it's my feeble human brain, but I just can't imagine what sort of governing structure could contain such a thing.
>
> JACOBSTEIN: I don't think that the issues are [that] simple. . . . People, because we're not familiar with that kind of computing power, tend to be afraid of it. . . . That doesn't mean that there aren't real risks associated with AI. It does mean that before we decide we need to be afraid of something, what we should first do is do our best to understand it, to focus

on the upside potential and what it can do in implementing our own concepts of morality. When I look at the world today, I see a lot of problems that offend me morally. . . . All of those problems are failures of our ability to implement our own moral code. Exponential technologies like AI are going to give us an opportunity to improve our response.

Maybe I should have taken the blue pill and skedaddled.

NOT NEARLY NEAR

As the discourse burbled on and slideshows ticked ahead, one thing became clear: Singularity University was no such thing. As an unaccredited, for-profit enterprise with a pedagogy of salesmanship, SU could pass for a university only in a society that sees the institutions of knowledge as fountains of cash. Hence SU founders flaunt their elite credentials to the point of exaggeration. For instance, one SU brochure lists its chancellor as "Ray Kurzweil, PhD." Kurzweil does not have a PhD, though he has used the appellation elsewhere. He does have (at last count) eighteen honorary doctorates. These are awarded by administrators, not faculty, often to secure a speaking engagement. (Mike Tyson, Jon Bon Jovi, and Kermit the Frog also have honorary doctorates.)

What is SU, then, if not a university? In keeping with the style of the symposium, I shall present my findings in bullet points:

- An investment hustle run by mad scientists for credulous suits.

- A religious revival for confirmed atheists and progressive capitalist utopians.

- A blue-sky strategy session for cold-blooded technocrats who see futurism as a game not of predictions, but of power.

Fine, bone-deep American traditions, all.

Founded as a nonprofit, SU was later restructured as a California "public benefit" corporation. This meant seeking funders beyond the Silicon Valley faithful and hitting up hidebound corporations like Deloitte. Based on ticket prices and attendance, I figured SU grossed two million dollars on the Amsterdam summit, hardly enough to witness the future being born. According to Nail, SU counts three thousand alumni and six hundred-plus faculty speakers (none tenured, obviously). The university runs a start-up incubator, and its corporate friends enjoy generous in-class promotion.

Not a university, not destined to be a great business, SU was not imminently dangerous, either, in the manner of a Heaven's Gate or an Aum Shinrikyo. I spent a lot of time looking around during my visit but found no body parts in the summit's freezer rooms. I simply came to think that the Singularity holds more interest as a cultural byproduct of our epoch of stagnation than as a scientific theory or business proposition. Taken as pure entertainment, it's even enjoyable. Imagine! A perpetual revelry of wishes fulfilled; a never-ending party with all of your favorite people, engrossed in far-out, hyperintellectual discussions and enjoying previously impossible sexual combinations. What's more, this amazing future definitely makes room for *you*. Just like heaven, everyone can get in (except maybe Luddites who deny the gospel). Scottish author Ken MacLeod was onto something when he mocked the Singularity as "the Rapture for nerds."

Kurzweil's Singularity is an overheated white paper by a zealot for the American dream of luxury and convenience. Certainly, his success owes something to the American fondness for homespun prophets. His specific focus—artificial life—springs from a much older and apparently undead tradition. Ilia Stambler, an Israeli academic who has written a book-length history of "life-extensionism," names as precedents the Greek myth of Prometheus, the golem of Jewish folklore, the occultism of alchemy and early Freemasonry, crude androids "allegedly constructed by Albertus Magnus and Descartes," and "Wolfgang von Kempelen's mechanical chess-player (proved to conceal a man)."

Stambler is himself a transhumanist who belongs to a number of the same clubs as Kurzweil and other SU faculty, so it's noteworthy that he concludes that the scientific pursuit of immortality is a "fundamentally conservative" enterprise. Kurzweil's forerunners in this area were a truly mixed bag of nuts. Consider Nikolai Fedorovich Fedorov, a scary Russian Orthodox librarian whose 1913 treatise, *Philosophy of the Common Task*, made a Christian case for physical immortality and the resurrection of the dead—all of them. Fedorov wrote that humanity would achieve victory over its enemy, death, "only with absolute, patriarchal monarchy, with a King, standing in place of the [Heavenly] Fathers."

Yet Kurzweil's closest precursor may be Charles Asbury Stephens, a doctor and writer of young adult fiction who, with philanthropic backing, founded a clinic in Maine in 1888 to pursue his death-defying gerontological research. Stephens also published a number of books on his cellular theories of aging, including *Salvation by Science* and *Immortal Life*. He died in 1931.

To assess their record in scientific terms, the immortalists have a failure rate of 100 percent.

Sources aside, Kurzweil's vision inspires strong feelings of wonder or terror, depending upon your persuasion. I imagine these are the same feelings that inspired phrases like "fear of god" and "old-time religion." In the consumer culture that nurtured Kurzweil's fame and regards his ersatz university as an improvement on the real thing, awe is a new iPhone; heaven is a holodeck.

And hell? Hell is not having the best new toy.

THE CREEP SHOW TO COME

The show-stealer in Amsterdam was a big, Texas-born biotech scientist and engineer named Raymond McCauley. Prior to joining SU, McCauley cofounded BioCurious, a Sunnyvale, California, nonprofit with a mission to normalize technologies like genetic engineering by providing a low-cost "community biology lab for amateurs, inventors, entrepreneurs, and anyone who wants to experiment with friends."

In his opinion, humanity was headed for subspeciation—a proliferation of separate and unequal races creating something like a cyberpunk version of Tolkien's Middle Earth. He predicted that the prevention of childhood disease would open the backdoor to widespread human genetic modification. "I don't know how long before we stretch that word *therapeutics* to mean *enhancement*," he said. "That thrills and concerns me."

McCauley himself inspired some thrills and concerns when he theatrically summoned two young transhuman activists to surgically implant an RFID chip—a short-range tracking device capable of storing information—into his hand. I think I speak for many in the audience when I say it was the most titillating act of cyber-exhibitionism I had ever seen. Rising from the operating chair, a still-bloodied McCauley fielded audience queries on the implications of the procedure, including one from some spoilsport who asked pithily, "No more natural selection?" McCauley replied:

> We're seeing a form of natural selection to select for success,
> economically, to select for people who are high-performance
> individuals. . . . We're having an arms race with natural selection.
> Genetics is pretty slow, and individuals don't advance, races do. But
> we're now more in control.

I had a chance to sit down with McCauley later. I found him to be the most genuine and engaging person I met from SU. He had recovered quickly. He showed me the stout bandage on his hand and the smartphone app he used to control the embedded chip. The app bore an all-caps warning: "DO NOT FORGET YOUR PASSWORD!"

McCauley was still figuring out what to do with his new implant. "I could have some spot on the wall that looks like a blank spot, and I can use this to unlock a hidden door or passage," he said. I asked him about the vibe at the SU campus. "In some ways, it's exactly like being a sophomore in college with the three smartest guys and girls in the dorm, and it's four o'clock in the morning and no one wants to go to bed," he said. "It's eighty people who are the smartest people in their countries, and they're all riffing on each other, coming up with new ideas and trying to solve the world's problems—and make a buck while doing it.

"Man—it's so powerfully addictive, it's almost like a drug," he said. "It's almost like being in a cult. If we wore silvery jumpsuits, I would be really worried."

"Really?" I asked, scribbling furiously in my notebook. "You don't go without food and then do a bunch of exercise, or do repetitive chanting—right?"

"We have plenty of protein," McCauley replied. "Groupthink tends to be *very* discouraged. . . . But there is this real belief in technological positivism that sometimes approaches an almost religious fervor."

There have been signs that the Singularitarians want to up their game. Last year, an SU staffer filed papers with the Federal Election Commission to form three separate fundraising committees, each of the kind known as super-PACs. However, the applications were withdrawn after a Center for Public Integrity reporter started asking questions. SU then claimed its employee had gone rogue and filed the papers without permission, contradicting the employee's earlier statements that SU leadership was on board with the plan.

Campaign donations are for buying access to government. They are largely unnecessary when your people are already in government. Last year, Obama appointed former Google executive Megan Smith, who contributed a chapter to a recent SU book, as the White House chief technology officer. Google chairman Eric Schmidt was deeply involved in the 2012 Obama reelection campaign. "On election night he was in our boiler room in Chicago," David Plouffe, a former White House adviser now working as a political fixer for Uber, told *Bloomberg Businessweek*. In December 2014, the Democratic National Committee invited Schmidt to a "victory task force" that will set strategy for coming elections. This is all relevant because Googlethink is, for all practical purposes, SUthink. The company remains a key university sponsor, and SU cofounder Kurzweil leads Google's AI efforts. The company's expansion from online advertising toward more ostentatiously futuristic projects, such as driverless cars and wearable computers, reflects its founders' enthusiasm for Kurzweil's vision.

SUPERHAVING IT ALL

Kurzweil's partner and SU cofounder, Peter Diamandis, who has laid out his own vision of a world of *Abundance* (his first book), a world without haves and have-nots—only "haves and superhaves"—closed the first day of the Amsterdam summit. Already, the relentless optimist shtick had grown a touch predictable, and paying customers sitting in the audience had some questions, like, what about killer robots in the future? "I'm not worried about AI," Diamandis said confidently. "Intelligence is a good thing." What about wars, terrorism, civil unrest? "I'm not talking about happy, la, la, la, whatever. There will be turbulence on the road to abundance."

"In the past we had revolutions," one man asked. "What are the new solutions?"

Did you guess *technology?* Congratulations. "Governments don't get disrupted gracefully," Diamandis said. "The technologies you've been hearing about today are hard to regulate and they're impossible to stop." Although privacy was long gone, in his view, the new world of "perfect knowledge" and universal surveillance would—somehow, magically, counterintuitively—end tyranny.

> We're going to enter a period of time where little will be done in secret. We have three private orbital satellite constellations today; that will grow to at least five. We'll have ubiquitous drones flying, whether or not they're legal. We'll have . . . cameras woven into your clothing. . . . The flip side of [the loss of privacy] is a more peaceful world and a world with less oppression.

The Orwellian dystopia is actually a utopia. Got it? The final question—and I'm sorry it wasn't mine—was, "What if Google goes crazy?" Or edits the "don't" out of its motto?

Diamandis said that he trusted Google to guard itself against "terrorists." Anyway, he added, "There are other search engines, if Google went down." I was dumbfounded by his ability to miss the point. Diamandis then threw out a few lines about the democratizing power of the Internet, before musing once more on the obsolescence of the current order:

> All of our governmental systems were designed for governments of one hundred, two hundred years ago. . . . We had representative democracy, we didn't have an actual democracy. It could exist right now, but it

would basically destroy the political-legal system and the people in power today to do that—and so they will not let go.

Who did he think "the people in power" were, I wondered, if they were not his billionaire pals?

"We've run out of time," the emcee cut in. "A huge round of applause for Peter!"

Diamandis spoke again at the summit's keynote—which I missed, because I was conducting an interview in a side room. Later, I learned that he, too, had gotten an RFID implant onstage. I wondered if he had felt upstaged. Maybe this is how the future gets done, chip by chip. No problems. Everyone is doing it.

As I left the DeLaMar, I considered what to do with my final day in Amsterdam. I thought I might visit the Anne Frank museum, to cheer myself up.

PART IV

Markets, Property, and the Body

The introduction to *Beyond Bioethics* identifies concern about the role of commerce and markets in the development, use, and governance of human biotechnologies as one of the themes that distinguishes the new biopolitics from mainstream bioethical approaches. This section looks at the increasingly blurred lines between private biotechnology companies and university researchers, between basic research and commercialized technology, and between the profit imperatives of private enterprise and the perception that scientists are committed to serving the public interest. It discusses many of the dynamics through which federal, state, and institutional policies, including judicial decisions, have fueled the rapid commercialization of biology and medicine over the past several decades and examines how conflicts of interest have become increasingly manifest. Contributors consider how these powerful forces affect researchers, doctors, patients, consumers, universities, institutional review boards, and the field of professional bioethics itself.

The section begins with Harriet Washington's "Flacking for Big Pharma," which documents in distressing detail the ways in which the $310 billion pharmaceutical industry undermines the independence and accuracy of top medical journals and skews the trajectory of medical research itself. How big of a problem is this? Editors of medical journals estimate that "95 percent of the academic-medicine specialists who assess patented treatments have financial relationships with pharmaceutical companies." Washington calls this situation "an open secret," and notes that the only control exerted by medical journals—asking that authors disclose their financial arrangements—is a poor substitute for objectivity.

Osagie Obasogie considers some of the legal and ethical issues surrounding property interest in human tissues from the vantage of several court decisions in "Your Body, Their Property." As he points out, most of us routinely sign forms at the doctor's office that give up many rights over our own bodily materials, "with the same inattention we reserve for the latest end user license agreement from iTunes or Microsoft Office." In a number of cases, the courts have seemed to affirm that "discovering, isolating, or excising parts of human bodies" does "grant scientists property interests that they can exploit for financial benefit." But in 2013, the Supreme Court issued a different finding. In a case challenging patents on cancer-related genes held by the biotechnology company Myriad Genetics, the Court ruled that naturally occurring DNA sequences, even when isolated from the body, cannot be patented.

In "Where Babies Come From: Supply and Demand in an Infant Marketplace," Debora Spar argues that while many don't like to recognize that there is a market in babies, it does and should exist—and that in the United States, public policy to regulate it is urgently needed.

In "Dear Facebook, Please Don't Tell Women to Lean In to Egg Freezing," Jessica Cussins explains the problem with offers by Facebook and Apple to cover twenty thousand dollars toward elective egg freezing for their female employees. "What we need," she writes, "are family-friendly workplace policies, not giveaways that will encourage women to undergo invasive procedures in order to squeeze out more work for their beloved company under the guise of 'empowerment.'"

To end the section, we have "The Miracle Woman," an excerpt from Rebecca Skloot's best-selling and widely acclaimed book *The Immortal Life of Henrietta Lacks.* Henrietta Lacks was a poor black woman who had cells from her cervical cancer taken by doctors in the "colored ward" of Johns Hopkins Hospital in 1951. The cells turned out to have extraordinary technical (and commercial) value for research and medicine, but Lacks's contribution and personal story were not widely known until the publication of Skloot's book. Her story spans issues of medical experimentation on African Americans and on other vulnerable subjects, ownership of one's body, and continuing racial disparities in health in the United States.

18. Flacking for Big Pharma

Harriet A. Washington

"Drug Makers Cut Out Goodies for Doctors" and "Drugmakers Pulling Plug on Free Pens, Mugs & Pads" read headlines in *The New York Times* and *The Wall Street Journal* Health Blog at the end of 2008 after, in a very public act of contrition, thirty-eight members of the pharmaceutical industry vowed to cease bestowing on prescribing physicians goodies such as pens, mugs, and other tchotchkes branded with their names. Some physicians and ethicists had long expressed concern about the "relationship of reciprocity" that even a pizza or cheap mug can establish between doctors and drugmakers, and branded trinkets also send a message to the patient, who might reason that Gardasil must be a good drug if her doctor wields a reflex hammer inscribed with its name. But while the popular press celebrated this sudden attack of nanoconscience and while we still gravely debate whether physicians' loyalties can really be bought for a disposable pen or a free lunch, the $310 billion pharmaceutical industry quietly buys something far more influential: the contents of medical journals and, all too often, the trajectory of medical research itself.

How can this be? Flimsy plastic pens that scream the virtues of Vioxx and articles published in the pages of *The New England Journal of Medicine* would seem to mark the two poles of medical influence. Scarcely any doctor admits to being influenced by the former; every doctor boasts of being guided by the latter. In fact, medical-journal articles are widely embraced as irreproachable bastions of disinterested scientific evaluation and as antidotes to the long fiscal arm of pharmaceutical-industry influence.

Original publication by Harriet A. Washington as "Flacking for Big Pharma" in *The American Scholar*, June 3, 2011. Reprinted by permission of Harriet A. Washington. Due to space limitations, portions of this chapter have been deleted or edited from the original.

And yet, "All journals are bought—or at least cleverly used—by the pharmaceutical industry," says Richard Smith, former editor of the *British Medical Journal*, who now sits on the board of Public Library of Science (PLoS), a nonprofit open-access group publishing scientific journals that eschew corporate financing and are freely available online to the public.

Big Pharma, as the top tier of the industry is known, starts modestly, inserting the thin edge of its wedge by advertising copiously—and often inaccurately—in medical journals. In 1981, concerned officials at the Food and Drug Administration recognized the educational nature of pharmaceutical advertising by establishing explicit standards for medical-journal ads that mandate "true statements relating to side effects, contraindications, and effectiveness," and a "fair balance" of statements about medication risks and benefits.

In 1992, the editors of the esteemed *Annals of Internal Medicine* decided to gauge how well their own advertisements met that standard. They tested 109 advertisements along with the references cited by those ads, sending each ad to three expert reviewers who evaluated them in light of the FDA standards. Fifty-seven percent of the ads were judged to have no educational value, 40 percent failed the fair-balance test, and 44 percent, the reviewers believed, would result in improper prescribing. Overall, reviewers would have recommended against publication of 28 percent of the advertisements, as the *Annals* revealed in its published report.

The FDA subsequently issued eighty-eight letters accusing drug companies of advertising violations between August 1997 and August 2002. But the *Annals* editors were in no position to bask in this validation: the journal was fighting for its life after large pharmaceutical companies withdrew $1.5 million in advertising. "Finally, the editors felt that to save the journal, they must resign," recalls Smith. The coeditor of the *Annals*, Robert Fletcher, remarked as he departed his job: "The pharmaceutical industry showed us that the advertising dollar could be a two-edged sword, a carrot or a stick. If you ever wondered whether they play hardball, that was a pretty good demonstration that they do."

A decade later, with a different editor at the helm and a restored pharmaceutical advertising base, the *Annals* planned an editorial on high drug prices. But this time, it took care to first invite commentary from the premier drugmakers' organization, the Pharmaceutical Research and Manufacturers of America (PhRMA). PhRMA in turn funded a piece by John E. Calfee of the American Enterprise Institute, whose essay began with the statement "Price controls could have a substantial negative effect on pharmaceutical research and development."

Pharmaceutical advertising impinges heavily on the editorial sphere of medical journals, sometimes with surprising brazenness. The drug epoetin is widely accepted for its role in prolonging survival in people with end-stage renal disease: Medicare alone spent $7.5 billion on the drug in the decade preceding 2002. Dennis Cotter is president of a nonprofit institute that scrutinizes conventional medical wisdom, and his group's analysis suggested that epoetin's benefits for people with end-stage renal disease were largely chimerical, based on flawed logic.

In 2003, Cotter submitted an editorial that detailed his questioning of epoetin's role to *Transplantation and Dialysis*, whose editor and peer reviewers agreed that it should be published. However, as the *British Medical Journal* reported in January 2004, Joseph Herman, *Transplantation and Dialysis*'s editor, rejected the piece because "unfortunately, I have been overruled by our marketing department with regard to publishing your editorial. The publication of your editorial would, in fact, not be accepted in some quarters . . . and apparently went beyond what our marketing department was willing to accommodate."

After a hue and cry was raised in the medical press, the journal reversed itself and offered to publish Cotter's work, but he demurred, preferring a less commercial venue.

Medical journals are utterly dependent upon pharmaceutical advertising, which can provide between 97 and 99 percent of their advertising revenue. By 2005, some major journals, including *Consultant, Geriatrics,* and *American Family Physician,* carried more advertising than editorial pages and glossy, full-color inserts that were longer than the journal's longest article. . . .

Moreover, drugmakers sometimes agree to buy journal advertising only if it is accompanied by favorable editorial mentions of their products. Or their in-house stables of writers or hired pens generate "advertorials," a Frankensteinian mix of medical content and marketing messages that can be indistinguishable from editorial material. . . .

Pharma's journal ads tout not only products but also its hundreds of thousands of subsidized "educational opportunities." Drug and medical-device makers spend two billion dollars annually for more than three hundred thousand seminars and training opportunities, often held in the Bahamas or the Caribbean. The wolfed-on-the-run free pizza for harried medical residents that the industry has so sanctimoniously forsworn bears little resemblance to the sumptuous feasts, flowing wines, chartered flights, cruises, luxurious lodgings, golfing, snorkeling, and remarkably attractive sales reps that characterize these island educational junkets.

"There's a lot of bribery involved—the kids get pizza, the grownups get trips to Hawaii," observed Marcia Angell, MD, professor of social medicine at the Harvard Medical School, former editor-in-chief of the *New England Journal of Medicine* (*NEJM*), and the author in 2004 of *The Truth About the Drug Companies: How They Deceive Us and What to Do About It*.

These pedagogic playdates familiarize doctors with pharmaceutical companies' patented products to the exclusion of cheaper and sometimes safer and more effective alternatives. By 2000, drugmakers were paying physicians a total of six billion dollars a year for trinkets, island "educational opportunities," and financial grants for their pet projects, from golfing jaunts to clinics; this doesn't include the speaking and consulting fees that the pharmaceutical industry pays influential and "high-prescribing" clinicians to discuss its products. "Drug companies have moved their gift-giving from drug reps to hiring 'thought leaders'—the best drug reps of all," says Angell. "They send experienced physicians out to give talks and ensconce them on well-paid speakers' bureaus. Then they claim that this is education, not marketing."

However, the industry's seduction doesn't end with the advertisements, junkets, and overpaid speaking engagements. Drugmakers have enticed or ensnared the very font of evidence-based medical knowledge—the peer-reviewed medical journal. Not content to turn these journals out to ply the streets for cash, the industry finds many ways to pervert the editorial content itself.

· · ·

This perversion is such an open secret that in 2003 the *British Medical Journal* published a tongue-in-cheek essay instructing researchers in the fine art of "HARLOT—How to Achieve positive Results without actually Lying to Overcome the Truth." David L. Sackett, director of Ontario's Trout Research and Education Center, and Andrew D. Oxman, director of the Department of Health Services Research at Norway's Directorate for Health and Social Welfare, wittily summarized strategies by which drugmakers use clinical trials to tart up drugs that are poorly performing, dangerous, or both.

The proper conduct of a research study requires that it pose an important medical question in a clear, unambiguous manner and that it is carefully planned and randomized to ensure that the results are accurate and broadly applicable. Large numbers of subjects are typically recruited to help ensure that the results do not arise by chance. Control groups are given placebos or the standard of care in order to allow a meaningful comparison with the study group. Statistical expertise helps the study designers minimize and tease out any sources of error or bias.

But this expertise can also be used to introduce intentional bias in order to attain the desired result: for the determined adept, there exist many ways to subvert the clinical-trial process for marketing purposes, and the pharmaceutical industry seems to have found them all.

HARLOT's advice to those who would serve Pharma includes, "test against placebo, test against minimal dose, test against maximal dose, and test in very small groups." This means that companies sometimes seek to make bad drugs look good by:

- *Comparing their drug to a placebo.* A placebo, such as a sham or "sugar" pill, has no active ingredient, and, although placebos may evoke some poorly understood medical benefits, called the "placebo effect," they are weak: medications tend to outperform placebos. Placebo studies are not ethical when a treatment already exists for a disorder, because it means that some in the study go untreated. However, if you care only that your new drug shines in print, testing against placebo is the way to go.

- *Comparing their drug to a competitor's medication in the wrong strength.* Too low a dose makes the rival drug look ineffective. Too high a dose tends to elicit worrisome side effects.

- *Pairing their drug with one that is known to work well.* This can hide the fact that a tested medication is weak or ineffective.

- *Truncating a trial.* Drugmakers sometimes end a clinical trial when they have reason to believe that it is about to reveal widespread side effects or a lack of effectiveness—or when they see other clues that the trial is going south.

- *Testing in very small groups.* Drug-funded researchers also conduct trials that are too small to show differences between competitor drugs. Or they use multiple endpoints, then selectively publish only those that give favorable results, or they "cherry-pick" positive-sounding results from multicenter trials.

An increasingly popular variant is the much-abused technique of "data mining," wherein small subgroups of an unsuccessful trial are relentlessly scrutinized in search of groups for whom a benefit emerges, or seems to. . . .

In 2005, BiDil, a congestive heart-failure medication, became the first FDA-approved drug for African Americans only. BiDil was not tailored for African Americans, as its proponents often claim, but began life as the only patented drug of the Lexington, Massachusetts, biotech firm NitroMed. In 1987, the FDA had rejected NitroMed's application based on feeble results in its clinical trials, but the company scrutinized the drug's data in search of some group where it might show efficacy. Peering into BiDil's efficacy in women and in other subgroups yielded no fruit, but before NitroMed had to resort to astrology, the NIH passed the FDA Modernization Act, an initiative for the inclusion of racial minorities in clinical trials. NitroMed suddenly detected evidence in the FDA-rejected 1980s data that its drug might work better for blacks than it had for whites, and in 1997 BiDil was reborn as a "black" drug.

BiDil proponents published studies that supported their claim of a racially mediated genetic anomaly that was addressed by BiDil, making it an ideal drug for blacks but not for whites. At the company's invitation, other physicians published papers arguing for this genetic racial difference, but they could do so only by giving short shrift to critically important environmental and behavioral differences between black and white patients, such as disparate diets, smoking rates, environmental exposures, and exercise levels. NitroMed won FDA approval of a new trial that included only 1,050 black subjects, with no white subjects to provide comparison data. Furthermore, BiDil was not tested alone, but only in concert with heart medications that are already known to work, such as diuretics, beta-blockers, and angiotensin-converting enzyme (or ACE) inhibitors. The published results of the trial were heralded as a success when subjects taking the drug combinations that included BiDil enjoyed 43 percent fewer heart-failure deaths. The zealous data mining and the pairing of BiDil with drugs that are known to work well are recognizable tenets of HARLOT. Moreover, excluding whites was a medically illogical but financially strategic move because it eliminated the possibility that the drug would test well in whites, thereby robbing NitroMed of its already thin rationale for calling BiDil a black drug. The "black" label was crucial, because BiDil's patent covering use in all ethnic groups expired in 2007, but the patent for blacks only allows NitroMed to profit from it until 2020. BiDil is a case study in research methodology "flaws" that mask strategies calculated to make a dodgy drug look good on paper, for profit.

The medical record is also effectively distorted through what is *not* said, suggests Marcia Angell. "Any reputable journal is at the mercy of what is submitted to it," she says, "and must choose from whatever comes over the transom. Many studies never see the light of day because their findings are negative. There is a heavy bias toward positive studies, and this negative bias is a real problem. A company may conduct 1,000 trials; if two are positive, they get FDA approval and are published. The other 998 never see the light of day." In fact, half of all study data is never published.

But aren't physicians, with their scientific training and medical expertise, able to see through the negative bias and data manipulation? Not according to the editor of the *Journal of the National Medical Association.* "A busy pediatrician who is seeing patients until eight at night doesn't have time to figure out whether an article has been vetted," explains Eddie L. Hoover, MD. "He depends upon the journal editors to make sure he is not reading trash."

"When you are published in a medical journal, especially one of the top ones, this gives the article a certain imprimatur that makes people less critical," adds Joel Lexchin, MD, a bioethicist at York University in Toronto. "'If it's in *The New England Journal of Medicine* it's got to be good': This mentality diminishes the critical reading of the study." Moreover, many inaccuracies cannot be detected because neither the journal nor the reader has access to all of the original trial data. In the end, explains Angell, "Journals get a heavily winnowed-out selection of trial findings, and so doctors come to believe that medications in trials are more effective than they are. Many psychiatric medications are little more than placebos, yet many clinicians have come to believe that SSRI [selective serotonin reuptake inhibitors, a newer class of antidepressants] drugs are magic, all through the suppression of negative studies." . . .

· · ·

When John Abramson, MD, author of *Overdosed America: The Broken Promise of American Medicine,* lectured at Harvard's 2008 Ethical Issues in Global Health Research course, he dismissed much of the content of contemporary US medical journals as "little better than infomercials." What prompted this harsh assessment?

Despite the ubiquitous mantra of "evidence-based medicine," a curious lack of skepticism pervades journals about experts who accept money from the makers of the products they evaluate. A medical reviewer who writes a comprehensive assessment for a medical journal is supposed to be an expert in the field who evaluates medications, devices, and practices, distilling her expertise and her informed, disinterested opinion for the journal's readership. The need for objectivity is clear, and journals do not pay the authors of such articles. But the makers of the drugs and products in question often do pay them.

Once, conscientious journals did not permit reviewers to take money from drugmakers. But so many physicians began to take Pharma money to subsidize their research, to give speeches on behalf of favored new products, and to switch their patients to newer, more profitable medications, that working as a medical reviewer in the pay of drugmakers has become normalized. Today, medical-journal editors estimate that 95 percent of the academic-medicine specialists who assess patented treatments have financial relationships with pharmaceutical companies, and even the prestigious *NEJM* gave up its search for objective reviewers in June 1992, announcing that it could find no reviewers that did not accept industry funds.

Instead, financial disclosure has been pressed into service as a substitute for objectivity. These notices inform the reader which company paid the author, but neither how much nor what nonmonetary relationship the author may be enjoying with the subject of his assessment. Medical journals usually set a ceiling on the payments that evaluating doctors are permitted to accept from drugmakers, but these ceilings are high and vaulted, with the top-tier journals tending to publish the work of the reviewers who receive the fattest Pharma paychecks. Under the *NEJM* policy, for example, doctors writing medical reviews can accept up to ten thousand dollars a year in speaking fees and consulting fees from each drug company. "So if a doctor is doing . . . business with four or five companies, he or she can get as much as $40,000 to $50,000 a year and not violate the *New England Journal* policy," says Dr. Sidney Wolfe, director of Public Citizen's Health Research Group.

When physician-researchers are paid by the pharmaceutical industry, their medical-journal findings exhibit clear bias in line with the interests of the sponsoring company. Drs. Paul M. Ridker and Jose Torres at Harvard Medical School found that 67 percent of the results of industry-sponsored trials published between 2000 and 2005 in the three most influential medical publications favored the sponsoring company's experimental heart drugs and often its devices. Trials funded by nonprofits, however, were as likely to support the drugs or devices as to oppose them, and studies that combined industry funding with nonprofit support fell between the two on the spectrum, with 57 percent offering favorable results. The findings, published in *JAMA*, indicate that large pharmaceutical and device makers pay for studies on new medical treatments in hopes of replacing the current standard of care with their new therapy.

Not all clinical trials are performed to evaluate treatments: some are marketing tools. Drugmakers conduct seeding trials that induce many physicians to prescribe the drug, and then its effects are reported selectively, so that many articles extol the drug's positive results while any troubling findings are ignored. In switching trials physicians are induced by drug reps to switch their patients from an older medication to a newer one, and again, positive results are selectively published. The positive data from such trials are submitted for publication by different investigators in different guises so that journal editors have no way of knowing whether they are publishing new data or a retread. . . .

· · ·

Many biased medical-journal articles are the work not of physicians or scientists, but of ghostwriters who script them in accordance with the drug-makers' marketing messages. A medical expert is found who, for a few thousand dollars, is willing to append his or her signature, and then the piece is published without any disclosure of the ghostwriter's role. . . . Some argue that ghostwriting is not problematic because it is based on research data. But Fugh-Berman's article "The Haunting of Medical Journals: How Ghostwriting Sold 'HRT'" from *PLoS Medicine* explained that when research data conflicted with the marketing message, the former had to give way. . . .

Ghostwriting has been used to promote many drugs, including the anti-depressant Paxil (paroxetine); the recalled weight-loss drug "Fen-Phen" (fenfluramine and phentermine); the antiepilepsy drug Neurontin (gabap-entin); the antidepressant Zoloft (sertraline); as well as the painkiller Vioxx (rofecoxib)—to name just a few.

In 2003, the medical-publishing industry seems to have hit a ghostwriting nadir from which its reputation has not recovered. That year, Elsevier, the Dutch publisher of both *The Lancet* and *Gray's Anatomy*, sullied its pristine reputation by publishing an entire sham medical journal devoted solely to promoting Merck products. Elsevier publishes two thousand scientific journals and twenty thousand book-length works, but its *Australasian Journal of Bone and Joint Medicine*, which looks just like a medical journal, and was described as such, was not a peer-reviewed medical journal but rather a collection of reprinted articles that Merck paid Elsevier to publish. At least some of the articles were ghostwritten, and all lavished unalloyed praise on Merck drugs, such as its troubled painkiller Vioxx. There was no disclosure of Merck's sponsorship. Librarian and analyst Jonathan Rochkind found five similar mock journals, also paid for by Merck and touted as genuine. The ersatz journals are still being printed and circulated, according to Rochkind, and fifty more Elsevier journals appear to be Big Pharma advertisements passed off as medical publications. Rochkind's forensic librarianship has exposed the all-but-inaccessible queen of medical publishing as a high-priced call girl.

· · ·

Not content to skew reports of clinical trials on the back end, pharmaceutical companies also manipulate medical studies to generate the desired data for those reports. Studies are constructed in a manner that presents drugmakers' products in the most positive light or throws doubts on the seemingly clear hazards of taking their drugs.

Around 1999, pharmaceutical firms instructed their sales representatives to heavily promote expensive new COX-2 inhibitors such as Pfizer's Celebrex (celecoxib) and Merck's Vioxx (rofecoxib) for common conditions like arthritis and painful menstruation. But what ultimately closed the deal for physicians was the publication of two major clinical trials, the Celecoxib Long-term Arthritis Safety Study in *JAMA* and the Vioxx Gastrointestinal Outcomes Research study in the *NEJM*. Both journal articles reassured physicians that COX-2 drugs triggered far fewer intestinal problems than did aspirin and the older, cheaper, off-patent over-the-counter painkillers. Celebrex became a blockbuster drug, and by 2000, 60 percent of Americans with arthritis were taking it. Worldwide, Vioxx was prescribed to over eighty million people.

What the advertisements did not mention and the journal articles tried at length to hide was the fact that Celebrex, Vioxx, and other COX-2 drugs were triggering heart attacks and strokes: the data that revealed the increased risks had been withheld from the submitted studies. When it discovered this, the *NEJM* published not one but two "expressions of concern" and an assailed Merck pulled Vioxx from the shelves in 2004. . . . [The Pharmaceutical Research and Manufacturers of America (PhRMA) staff refused three requests for an interview or a specific written response to this article.]

· · ·

How can journals best mitigate the flow of misinformation resulting from purchased bias, shrouded data, and statistical mischief? Some critics suggest that medical journals start by dispensing with pharmaceutical advertising altogether and instead accept other lucrative advertisers, like the makers of luxury goods. All the editors cited here think that ghostwriting should be banned outright, or they are, at least, like Bauchner, "very uncomfortable" with the practice.

"How to avoid corporate manipulation? That's an easy question," says Abramson. "Journals have to see the primary data. You cannot be irresponsible enough to publish an article and say that article has been through peer review when all the primary data and protocols haven't been made available to you. Journals are blessing something with a public statement of integrity when there's no way in hell they can know if it has any, so they're playing a role in that deception."

But what help exists for doctors, who need to know the potential sources of bias in peer-reviewed articles? Several books offer clear, impeccably researched guides to sniffing out manipulation. Angell's *The Truth About the Drug Companies* and Abramson's *Overdosed America* are likely to be most helpful to a busy clinician. The Pharmed-Out website (www .pharmedout.org) offers tools for detecting undue influence in medical research and publishing.

Leadership has also come from open-access journals, including the Public Library of Science publications. Their business models vary, but because they don't accept pharmaceutical advertising or funding and are usually freely accessible to all online—in contrast to journals that must maintain income to answer to stockholders—open-access publishers can keep their hands in their pockets and avoid the rest of the profession's rampant conflicts of interest.

19. Your Body, Their Property

Osagie K. Obasogie

Polio ravaged much of the United States during the twentieth century, leaving thousands sick, paralyzed, and dead. Those who were not afflicted with the virus were constantly haunted by the terror that their loved ones— particularly children, who were most vulnerable—would awaken one morning unable to walk and destined to a life of leg braces and iron lungs. That is until 1953, when Jonas Salk created a vaccine. There were more than 45,000 total cases of polio in the United States in each of the two years before the vaccine became broadly available. By 1962 there were only 910. Salk's invention was one of the greatest successes in the history of American public health.

Amidst the adulation and fame that came with saving untold numbers of lives, Salk did something that seems curious if not unwise by today's standards: he refused to patent the vaccine. During a 1955 interview, Edward R. Murrow asked Salk who owned the patent, leading a seemingly bewildered Salk to respond, "The people, I would say. There is no patent. Could you patent the sun?"

These days, amid a patent-driven biotech boom, it is difficult to imagine a researcher making a similar appeal to the commons. But this sensibility received a crucial endorsement in the Supreme Court's [2013] decision in *Association for Molecular Pathology v. Myriad Genetics*. The court held that Myriad, a biotech firm in Utah, could not patent naturally occurring objects such as the two cancer-related human genes in question.

The decision upended many aspects of American intellectual property law that emerged in the wake of *Diamond v. Chakrabarty* (1980), when the

Original publication by Osagie K. Obasogie. "Your Body, Their Property" in the *Boston Review*, September 30, 2013. Reprinted by permission of Osagie K. Obasogie.

court held that living organisms—specifically, man-made crude-oil bacteria—are patentable subject matter. *Chakrabarty* inspired a rush to patent not just living things but also a growing array of biological materials, including human genes.

Fast-forward a few decades and almost one-fourth of all human genes had been patented and controlled by private hands. This made it expensive for scientists to do research implicating these genes or to develop tests that examine how certain mutations might affect health outcomes. Until *Myriad* the sensibility evoked by Salk—that entities beneficial to all humankind should not be patented and privatized—had largely been treated as a distant memory of a bygone era, like the jukebox or rotary telephone.

However, Justice Clarence Thomas eloquently set the record straight in his unanimous opinion for the Court in *Myriad* by contrasting the company's patents to those upheld in *Chakrabarty:* "Myriad did not create anything. . . . It found an important and useful gene, but separating that gene from its surrounding genetic material is not an act of invention." Now, many existing human gene patents are in question.

It is rare that a legal dispute of this importance, technical complexity, and jurisprudential nuance is resolved by the court with such clarity, conviction, and common sense. Yet even after *Myriad,* the dispute over who can claim property interests in human biological materials, and in what circumstances, is far from over. Human gene patents are not the only means by which corporations and researchers assert rights to parts of human bodies, and many more legal reforms are needed to ensure that your body remains entirely yours.

• • •

An important example of this can be seen in the litigation surrounding John Moore's spleen. Moore was a Seattle businessman who suffered from hairy cell leukemia, a rare cancer that caused his spleen to grow to fourteen times its normal size. Moore first traveled to UCLA Medical Center in 1976 for treatment, where Dr. David Golde told him that he should have his spleen removed. Moore complied and returned to UCLA for follow-up examinations with Golde for several years after the surgery. During the visits he routinely gave blood, skin, and other biological materials. Moore was told that these return visits and sample withdrawals were a necessary part of his ongoing treatment. What he was not told, however, was that Golde and the university were cashing in.

Researchers quickly realized that Moore's cells were unique. The scientists took portions of Moore's spleen to distill a specialized cell line—

affectionately called "Mo"—and found that the cells could be useful in treating various diseases. Golde, researcher Shirley Quan, and UCLA were assigned a patent for the cell line in 1984. At the time, analysts estimated that the market for treatments stemming from Moore's spleen was worth roughly three billion dollars. Golde worked with a private company and received stock options worth millions, and UCLA also received hundreds of thousands of dollars in outside funding. Moore, whose spleen made all of this possible, received no compensation.

Moore sued the researchers and UCLA, claiming not only that they deceived him for their own financial benefit, but also that he was entitled to a portion of the revenues stemming from the Mo cell line because his property—his spleen and other biological materials—was taken from him and commercialized without his consent. In 1990 the California Supreme Court found that Golde and UCLA did not fulfill their disclosure obligations. Yet Moore was not owed a penny since the court found that he no longer had a property interest in his own spleen once it was removed and used for research.

Moore v. Regents of the University of California enshrined a principle in property law that haunts us to this day: patients have virtually no property interest in most of the nonreproductive cells or tissues taken from them, even when these materials turn out to be profitable to researchers and institutions. This conclusion by the California Supreme Court has been followed by almost every jurisdiction.

Although the US Supreme Court denied an appeal to review *Moore* in 1991, it may be time to revisit this holding in light of the underlying sentiment embraced in *Myriad*, [in] that discovering, isolating, or excising parts of human bodies should not grant scientists property interests that they can exploit for financial benefit. Each time we visit the doctor and undergo a medical treatment that involves a tissue biopsy, we contribute to the more than 270 million human tissue samples held in research biobanks. Patients—the donors of this raw material—not only lack any say over how their tissues are used, but they also are not compensated if their materials lead to profitable innovations that would not have been possible but for their contribution.

Most of us unknowingly turn over this property interest in our own bodily materials as part of the terms and conditions of receiving medical treatment. The release is in black and white on one of the many boilerplate forms that we often read and sign in a doctor's office with the same inattention we reserve for the latest end user license agreement from iTunes or Microsoft Office. But *Moore v. Regents of the University of California* is what legitimizes this default transaction. It is thus worthy of reconsideration if the Supreme Court's rejection of privatizing human genes in *Myriad* is to take hold as a broader principle for how we should think about third-party property interests in human body parts. Rather than pretend that patients seeking medical attention are in a position to negotiate the transference of property interest in their excised tissues, we ought to have a series of default rules and practices that treat patients equitably and as true partners in research endeavors whenever possible.

Some have questioned, with less than convincing evidence, the sincerity of Salk's statement against patents by claiming that his vaccine was not patentable to begin with. These claims misunderstand Salk's enduring legacy and miss the bigger point that he was trying to make. At its core, his response to Murrow was about much more than patents. It was about respecting our shared humanity by not reducing the profession of healing to a series of biomedical land grabs in which doctors and patients are pit against each other. Thankfully, the US Supreme Court has taken one aspect of this issue off of the table by striking down human gene patents. Let's hope that the holding in *Moore v. Regents of the University of California* is next in their crosshairs.

20. Where Babies Come From

Supply and Demand in an Infant Marketplace

Debora Spar

Say the word "market," and what comes to mind? Financial markets, maybe, or supermarkets. There are markets in real estate, markets in used cars, markets filled with farmers selling green beans and cheese. But a market in human fertility—sperm, eggs, hormones, surrogate mothers, embryos? Babies, or the means to make them, aren't supposed to be sold. They aren't supposed to be bought. They aren't supposed to have prices fixed upon them.

But there *is* a market for babies, one that stretches across the globe and encompasses hundreds of thousands of people. This market doesn't work like the markets for green beans or mortgages. Its high prices are more stubborn than the usual adjustments in supply and demand normally produce, it can never fully provide all the goods that are desired, and the role of property rights—the underpinning of most modern markets—remains either ambiguous or contested.

What has created this market is a deep and persistent demand from people who have been denied the blessings of reproduction, along with a wide and steadily increasing supply of ways to produce babies when nature proves inadequate. It includes businesses such as for-profit fertility clinics and drug companies that sell their wares in this market, charging often hefty sums along the way. In 2001, nearly forty-one thousand children in the United States were born via in vitro fertilization (IVF). Roughly six thousand sprang from donated eggs; almost six hundred were carried in surrogate, or rented, wombs.

Some people lament the very existence of this baby trade, insisting that reproduction—like love or honor—should never be sold. Some argue that the cutting edge of reproductive science violates the rules of nature and degrades all of the participants. Yet the baby business is alive and well and growing. And the demand for it is so widespread and powerful that any attempts to stamp it out would almost certainly fail or do harm. If the baby trade were to move to a donor model like the one governing organs, for example, it would probably face comparable shortages: fewer women willing to donate eggs, fewer surrogate mothers, and a smaller supply of sperm. At the same time, a black market for these components would probably arise, much as it has for kidneys and other vital organs. Similarly, if governments were to outlaw the trade completely, people desperate for children would scramble to find illicit providers, subjecting themselves to legal and medical risk. Unlike outlawed trades such as drugs or prostitution, the baby business creates a product—children, for people who want them—that is inherently good. The market may make people uncomfortable, but it's more efficient than the alternatives, and it provides inestimable value to those who choose to purchase.

Even so, the United States is the only major country in the world whose national government has chosen not to address the complex issues of equity, access, and cost that are raised by the baby trade. The federal government has been exceedingly wary of imposing limits on high-tech baby-making, preferring to let the courts, the markets, or the state legislatures sort things out.

Part of this reluctance may be rooted in America's typical laissez-faire response to emerging markets. Unlike its European counterparts, the US government historically has been loath to constrain high-growth, high-technology markets and industries. The US mobile phone and Internet industries, for example, arose in largely deregulated markets.

In the baby business, legislators' reluctance to regulate is exacerbated by a profound fear of religious or ethical entanglement. Because the abortion debate in the United States has been so divisive, politicians have avoided pursuing any policy agenda that touches it even lightly. As a result, there are no national policies on IVF, which requires creating and often discarding embryos; none addressing genetic engineering; and none on the permissibility of preimplantation genetic diagnosis (PGD), which involves examining an embryo for signs of genetic disease before its implantation in the uterus.

Other countries are far more explicit. In Israel, for example, assisted reproduction is touted as a national good. Accordingly, Israel permits (and only lightly regulates) most forms of high-tech reproduction and pays for a couple's fertility treatments until the couple has two children. In Germany, a deep-seated wariness of genetic manipulation has manifested itself in more restrictive legislation: no egg transfers, no surrogacy, no PGD. And the United Kingdom has an agency, the Human Fertilization and Embryology Authority, that oversees all aspects of Britain's reproductive trade. It licenses and monitors all IVF clinics, sets price caps for egg donation, and assesses applications for PGD.

In the United States, this kind of authority would almost certainly come under attack. As an over-fifty mother of newborn twins said: "I had my babies. I paid for my babies. I could afford my babies. Why is it any more complicated than that?" Similar (though perhaps more subtle) sentiments are voiced by many practitioners in the fertility industry, who worry that regulation of their trade could become expensive, unwieldy, and unfair. One prominent specialist argues that any regulation would slow medical progress in his field: "We have been able to sail under regulatory visibility," he notes. "If we had been under scrutiny, many steps would have been forbidden."

Although US fertility practitioners generally seem delighted to remain in the gray area of self-regulation, the history of technological development in other sensitive trades suggests that some widening of availability and the introduction of property rights, rules, and institutional policies would make the baby trade more responsive to the social, medical, and ethical issues that are emerging from its science. Right now, there is little reason for any provider—drug company, fertility clinic, sperm bank—to wrestle with these concerns. Should there be age limits on fertility treatments, for example? (Fertility clinics' most profitable patients are those women least likely to conceive.) Should new procedures be subject to rigorous testing protocols? Should multiple births, which often result from the multiple implantations intended to increase the likelihood of at least one embryo's viability, be controlled or limited?

In the absence of outside pressure, the market will try to satisfy most client desires in the interest of creating new business. Yes, a sixty-three-year-old woman can choose to undergo IVF. Yes, a family with a daughter has the right to ensure that its next child is a son. Yes, a couple can proceed with the birth of IVF-induced quintuplets. And perhaps these individuals should indeed have these freedoms. But one could argue that these are not the kinds of questions that markets answer best. What happens when the sixty-three-year-old sues the fertility clinic for damages because she has given birth to a severely deformed child? And what if hospitals and insurance companies balk at covering the costs for quintuplets? In the absence of accepted guidelines, private firms and independent physicians are likely to pass along to the rest of society the costs of decisions they are not fully authorized to make. . . .

A BETTER MARKET

If the baby trade is to develop into a broader and more normal market, it will need to acquire at least some semblance of property rights. Defining these rights would not resolve the deep moral issues that this market raises. They would certainly not ease the concerns of those who view reproductive medicine as a step toward the commoditization of both children and families. But property rights could at least provide a framework for discussion and for clarifying, at a minimum, who has the right to create, dispose of, implant, and exchange embryos. Similar guidelines could easily cover the component side of the market, clarifying ownership rights involving eggs, sperm, and wombs. Supported by rules and policies, the fertility trade would have a better chance of producing happy, healthy children, matched more consistently and at a lower cost to the parents who so desperately want them. . . .

A system of contracts and property rights—even a rudimentary one—could . . . delineate not only who has rights to what forms of genetic or social offspring, but under what conditions those rights can be expanded. It could establish by law not necessarily who has ownership of a particular child but who has the right or responsibility to parent that child.

Establishing a property rights regime for sperm, for example, should be relatively easy. Laws could establish whether men have any lingering rights to the children created by their sperm and whether (or under what conditions) these children could uncover their genetic heritage. Similar rules could cover eggs and, with somewhat more difficulty, wombs. One could imagine a regime that worked as follows. Female donors (or surrogates) could agree at the outset to the kind of relationship that would exist between them and an eventual child. They would then know, explicitly, what rights they had with regard to this child and what kinds of decisions were theirs to make. Theoretically, such a system could incorporate provisions from the world of adoption, including a mandatory waiting period after the child's birth, during which the birth mother (or egg donor) would retain all rights to the child. Once she agreed to relinquish the baby, though, she would forfeit any subsequent rights to parent the child.

A system of property rights could also define the limits of the market, separating those elements—sperm, eggs—that can be sold or, in the case of wombs, leased, from those that cannot. In particular, it could draw a much clearer line between the components of reproduction and the babies themselves, ensuring that parents do not profit from relinquishing their offspring. Men could still sell their sperm, women could receive compensation for reasonable pregnancy expenses, and intermediaries could charge fees for managing these exchanges. But while mothers and fathers could renounce their right to parent a particular child, they could not sell their parenting rights to others. Essentially, this is the fine line that already exists—and generally works—in the area of adoption.

Note that in this type of system, potential offspring would not be treated as property per se. They wouldn't be bargained for or sold, and their humanity would never be called into question. Instead, market participants would be furnished with a sense of order and predictability, a set of norms that would prescribe behavior and establish the limits of acceptability. These norms would allow people to transact more securely and to know the rules beforehand. Creating property rights for the baby business would neither, as critics assert, turn children into commodities nor mothers into baby machines. It would not tarnish reproduction or turn intimate relations into financial ones. Instead, property rights would help codify transactions and procedures that already take place and thus resolve disputes that lead too often to tragedy.

BEYOND PROPERTY RIGHTS

A baby business governed by a system of property rights is a vital intermediate step that would bring clarity and predictability to the market. The introduction of these rights would not, however, tell us which pieces of this emerging technology are acceptable or for whom. Therefore, society also needs to decide how much control parents can exert over their child's conception and genetic makeup and what part of the conception society should pay for. These are exceedingly difficult, Solomonic choices. Yet at the moment, we are making them in a purely ad hoc way—depending on the particular state, the local court system, and the finances of the individuals involved. A far better approach would be for Americans to decide, as a society, just what we consider acceptable in the baby trade. Are we comfortable allowing commercial exchange in the pursuit of procreation? Are we willing to permit parents and their doctors to manipulate the embryos that will become their offspring? How will we determine which procedures push the trade too far? Any one person—this author included—is likely to have strong views on each of these questions. But the process here is far more important than any single set of conclusions. Americans need to debate these questions. For without such a process, we will never be able to arrive at a regulatory strategy that sticks.

Admittedly, the politics of this process will be tough. Americans may never agree on the moral issues that surround the baby trade or on the rules that should guide it. But if we break this debate into several manageable pieces, thinking of the baby business in terms of principles rather than problems or technologies, we may find a route to consensus and thus to effective policies. There are five areas that should be considered in the debate over making the baby business both efficient and decent.

Access to Information

Most Americans view information as a public good. We are happy to have the government provide it for free (or require others to do so), and we believe others should have access to information as well. This set of preferences is particularly strong in matters that relate to health and safety, which explains why the United States has long had warning labels on consumer products and dosage information on drugs. Applying this preference to the field of reproductive medicine would be relatively straightforward. It would simply suggest a light-handed regulatory regime in which providers of assisted reproductive services would be required to inform potential clients of the costs, benefits, and potential dangers of their services. The government could subsequently decide to aggregate some of these data or to commission additional studies of longer-term risks. In any case, the essential idea would be to determine what information is important to the health and safety of the American population and then to provide it.

Already, the outlines of such provisions are in place. In 1992, Congress passed the Fertility Clinic Success Rate and Certification Act, which requires fertility clinics to submit basic statistical information to the Centers for Disease Control. In 2004, the President's Council on Bioethics recommended stiffer penalties for clinics that don't report their data, and it recommended longitudinal studies of children born through assisted reproduction. Should Americans decide that they need more information about the effects of high-tech baby-making—about the impact of hormone treatments, for example, or the costs of labor and delivery for mothers over forty—the release of different kinds of data could similarly be required.

Equity

The United States guarantees equal education for children and gives all citizens equal protection under the law. Although there's no guarantee of equal access to health care, Americans do in many situations apply the notion of equity to the medical realm: Donor kidneys can be had even by the poorest patients, for instance, and free prenatal care is extended, by statute and regulation, to nearly all women. Various aspects of the baby trade could receive similar treatment. Legislators could, for example, decide to designate infertility (under certain conditions) a disease and require that treatment be distributed equitably among its sufferers. Or they could decide that having children is a basic right and that society therefore needs to find some way to provide at least one child to everyone who wants to be a parent.

Note that the equity principle does not dictate a particular policy outcome. Instead it provides a relatively neat way of framing an otherwise messy and complicated debate. What is it about reproduction that society might want to distribute fairly? Is it a pregnancy? A child genetically related to both parents? Simply a chance at parenthood? If it's the first of these options, then implementation would involve providing assisted reproduction services to all kinds of prospective parents and either subsidizing or reimbursing the cost. If it's the second, then taxpayers wouldn't need to cover forms of reproduction that involve third-party sperm or eggs. And if it's the third, citizens would probably want policies that favor adoption over fertility treatments. Yet the logic in these cases is precisely the same. As a society, we need to consider what, if anything, we want to distribute equitably. Then we need to decide how to accomplish this distribution and cover its inevitable costs.

Legality

While the baby business is full of parents who don't get the children they crave, certain clinics and middlemen arguably carry the pursuit of babies too far. A central question, then, is where to draw the line between legitimate and illegitimate practice. Currently, there are few laws in this field and few politicians willing to tackle a subject that touches both on the question of abortion and on the deeply held desires of those most likely to be affected by any restrictions or bans. Yet even in this intimate area, Americans could still consider where in the baby business they want to limit either technology or parental choice. Some of these lines exist already. . . . US laws prohibit birth mothers from selling their children and define the legitimate boundaries of reimbursement. Americans could, if they chose, draw similar lines in other areas of the baby business, explicitly limiting parental choice or the reach of technology.

Cost

In the baby business, even private transactions can impose costs on the rest of society. Consider, for instance, the babies born to twenty-five-year-old Teresa Anderson of Mesa, Arizona, in April 2005. Anderson was a gestational surrogate who, for $15,000, had agreed to carry a child for Enrique Moreno, a landscaper, and his thirty-two-year-old wife, Luisa Gonzalez. To increase the chance of pregnancy, doctors transplanted five embryos into Anderson's womb. They all survived, and Anderson subsequently bore quintuplets for the couple. When the babies arrived, the news media showed the smiling surrogate, the delighted couple, and the five relatively healthy babies. These babies, however, were extraordinarily expensive: The costs of delivery almost certainly ran to well over $400,000. Gonzalez and Moreno paid to conceive these children, but US consumers—through increased insurance fees and hospital costs—are paying, too. According to one recent study, the total cost of delivering a child born through IVF ranges between $69,000 and $85,000. If the child is born to an older woman, the cost rises to between $151,000 and $223,000. The prospective parents in these cases pay for part of these costs—the IVF, the hormones, the multiple medical visits—but their fellow citizens are paying as well.

Society also pays the costs that accumulate as these children grow up. Currently, about 35 percent of births resulting from IVF and intracytoplasmic sperm injection (ICSI), a relatively common procedure in high-tech pregnancies, are multiples. While most of these newborns are perfectly fine, a significant portion arrive prematurely or underweight, conditions that can burden them with problems later in life. Approximately 20 percent of low-birth-weight children suffer from severe disabilities, while 45 percent need to attend special-education programs. So individual choices about procreation generate costs for society at large, not to mention for the children themselves.

In these cases, Americans may choose to pay the steep costs of assisted reproduction and to embrace the technologies that impose those costs. Or they may not. The cost consideration merely helps frame the policy debate. In so doing, it forces society to address the question of how much it values the various outcomes of the baby business. If the cost of delivering quintuplets is exceedingly high, then perhaps there should be a limit on the number of embryos that can be transferred in a single cycle of IVF (most

European countries already have such limits). If the overall costs of IVF babies are deemed too great, then perhaps access to the technology should be limited to clients who can pay.

Parental Choice

In choosing to conceive a child, parents have to make decisions that run from the prosaic (Is this the right time?) to the profound: Should I create a second child in the hope that some of his or her bone marrow could save my first, who is suffering from leukemia? Am I too old? Too sick? Too single? Since the advent of assisted reproduction, Americans have shied away from interference in these choices, believing that their rights to privacy and procreation shield essentially all aspects of reproduction from government prying.

As the baby business expands, however, a zone of parental privacy may be increasingly difficult to maintain. Confronted with the costs of delivering high-tech babies, of educating disabled children, and perhaps of caring for youngsters orphaned by elderly parents, society may become more willing to set limits on who can make use of assisted reproduction and when.

Other questions are potentially even more vexing. Should society take it in stride if gender ratios are shifted by a generation of parents separately making the private decision to conceive either a boy or a girl? Should parents be allowed to manipulate their gene pool to produce offspring who are taller, smarter, or more athletic than they would have been otherwise? What if cloning were to become a realistic reproductive option? At that point, procreative choices would become more than personal. They would affect the very core of how people reproduce themselves and their society.

Thus, as the technology of procreation evolves, society may well want to revisit the bounds of privacy and parental choice. What kind of control should parents have over the fate of their offspring? And what controls should they be denied? Already, Americans draw these lines in more mundane realms. Under US law, for example, parents can choose to educate their children at home, have them tutored, or send them to any of a wide variety of schools. They cannot, however, choose to deprive their children of an education. Similarly, while parents can opt to serve their teenagers beer or give them guns, they cannot let their children purchase beer or attend school with guns. Parents' rights to administer or withhold medical care are also circumscribed. In all these cases, society sets clear limits on what families can do and where parents' desires for their own children must give way to the interests of others.

As changing demographics and social mores collide with exploding technological advances, more and more people will desire the goods and services that allow them control over conception. They will want to decide when they conceive and how they conceive and even, increasingly, the characteristics of the children they raise.

If this market isn't to balloon out of control, society has only four options. First, it could leave the baby business to the vagaries of market forces, allowing supply and demand alone to determine its shape. In that case, supply would increase, but only the rich would enjoy the benefits. Second, society could vainly attempt to ban the baby business altogether after deciding that its risks and inherent inequities are simply too great. Third, it could treat high-technology reproduction as it treats organ transplants, allowing the science to flourish but removing it completely from the market. Societal pressure to maintain an open market, however, would be all but insurmountable; unlike the organs of living or recently deceased people, a supply of eggs, wombs, and embryos is already available, and the players are in place.

Which leaves us, really, with the best and most feasible option: US society needs to decide how to regulate the baby trade and how to make the market work better and more equitably.

21. Dear Facebook, Please Don't Tell Women to Lean In to Egg Freezing

Jessica Cussins

The workforces of Facebook and Apple are 69 percent and 70 percent male, and the companies have been getting a lot of flack for those figures. In their latest bid to attract and retain more women, the tech giants have come up with a technical fix: offering female employees a twenty-thousand dollar benefit toward elective egg freezing.

According to a statement from Apple about the program, "We want to empower women at Apple to do the best work of their lives as they care for loved ones and raise their families."

Surely what they meant to say was, "We want women at Apple to spend more of their lives working for us without a family to distract them."

The Facebook version might be, "We don't want women leaning out to start families, so we're paying them not to!"

The move by Apple and Facebook is a boon for the companies marketing social egg freezing in Silicon Valley, New York City, and elsewhere. But despite what EggBanxx wants wealthy Manhattanites to believe, freezing your eggs is not a magic wand that will allow you to raise a family at your own pace, away from the pressures of your workplace and biological clock.

Unfortunately, when you work for a company that wants you to spend your entire life at the office, in a society that underprioritizes all occupations traditionally undertaken by women, there will never be an ideal time to start a family.

Moreover, the chance that a frozen egg will actually result in a child is still low—much lower than the smiling babies on the fertility clinic and

Original publication by Jessica Cussins. "Dear Facebook, Please Don't Tell Women to Lean in to Egg Freezing." *Huffington Post*. October 16, 2014. Reprinted by permission of Jessica Cussins.

egg-freezing websites would lead you to believe. But as Robin Marantz Henig put it after attending EggBanxx's infamous egg-freezing cocktail party, in "an evening of 'The Three F's: Fun, Fertility, and Freezing'—[there are] no F's left over for 'Failure Rates.'"

In fact, egg freezing is still explicitly discouraged for elective, nonmedical reasons by both the American College of Obstetricians and Gynecologists and the American Society for Reproductive Medicine. Not only does egg freezing fail to guarantee that you'll end up with a child, it also poses serious and understudied short- and long-term health risks to women and children.

The process of egg retrieval involves weeks of self-delivered hormone injections to hyperstimulate your ovaries, which can lead to nausea, bloating and discomfort, not to mention blood clots, organ failure, and hospitalization in rare cases. The surgery to remove your eggs involves a needle being inserted into your pelvis, with risk of internal bleeding and infection. Long-term impacts on women's health are understudied, but seem to include increased rates of breast, ovarian, and endometrial cancer.

Additionally, those frozen eggs can only become children if you use in vitro fertilization, which means greater risk of multiple gestation, preterm birth, and fetal anomalies. It is not yet known if freezing eggs for multiple years will further impact children's health outcomes.

What we need are family-friendly workplace policies, not giveaways that will encourage women to undergo invasive procedures in order to squeeze out more work for their beloved company under the guise of "empowerment."

The United States is the only developed country in the world without paid maternity leave. At the point that women do have children, no matter what age they are, they end up taking pay cuts whose effects can last for decades. Having one's employer pay for egg freezing doesn't push back against the status quo, but puts the onus on women to change themselves (Gee, why does that sound familiar?).

This policy could also send the problematic message that young women who don't choose this option are less serious about their careers. "You want time off when? Oh by the way, do you know about this new perk we offer?"

Facebook and Apple are right to want more women in their workforce. This latest move, however, is more likely to alienate than attract them.

22. The Miracle Woman

Rebecca Skloot

When Henrietta Lacks was diagnosed with cancer in 1951, doctors took her cells and grew them in test tubes. Those cells led to breakthroughs in everything from Parkinson's to polio. In an excerpt from her book, The Immortal Life of Henrietta Lacks, *Rebecca Skloot tells her story.*

In 1951, at the age of thirty, Henrietta Lacks, the descendant of freed slaves, was diagnosed with cervical cancer—a strangely aggressive type, unlike any her doctor had ever seen. He took a small tissue sample without her knowledge or consent. A scientist put that sample into a test tube, and, though Henrietta died eight months later, her cells—known worldwide as HeLa—are still alive today. They became the first immortal human cell line ever grown in culture and one of the most important tools in medicine: Research on HeLa was vital to the development of the polio vaccine, as well as drugs for treating herpes, leukemia, influenza, hemophilia, and Parkinson's disease; it helped uncover the secrets of cancer and the effects of the atom bomb, and led to important advances like cloning, in vitro fertilization, and gene mapping. Since 2001 alone, five Nobel Prizes have been awarded for research involving HeLa cells.

There's no way of knowing exactly how many of Henrietta's cells are alive today. One scientist estimates that if you could pile all the HeLa cells ever grown onto a scale, they'd weigh more than fifty million metric tons—the equivalent of at least a hundred Empire State Buildings.

Today, nearly sixty years after Henrietta's death, her body lies in an unmarked grave in Clover, Virginia. But her cells are still among the most widely used in labs worldwide—bought and sold by the billions. Though those cells have done wonders for science, Henrietta—whose legacy involves the birth of bioethics and the grim history of experimentation on African Americans—is all but forgotten.

On January 29, 1951, David Lacks sat behind the wheel of his old Buick, watching the rain fall. He was parked under a towering oak tree outside Johns Hopkins Hospital with three of his children—two still in diapers—waiting for their mother, Henrietta. A few minutes earlier she'd jumped out of the car, pulled her jacket over her head, and scurried into the hospital, past the "colored" bathroom, the only one she was allowed to use. In the next building, under an elegant domed copper roof, a ten-and-a-half-foot marble statue of Jesus stood, arms spread wide, holding court over what was once the main entrance of Hopkins. No one in Henrietta's family ever saw a Hopkins doctor without visiting the Jesus statue, laying flowers at his feet, saying a prayer, and rubbing his big toe for good luck. But that day Henrietta didn't stop.

She went straight to the waiting room of the gynecology clinic, a wide-open space, empty but for rows of long, straight-backed benches that looked like church pews.

"I got a knot on my womb," she told the receptionist. "The doctor need to have a look."

For more than a year Henrietta had been telling her closest girlfriends that something didn't feel right. One night after dinner, she sat on her bed with her cousins Margaret and Sadie and told them, "I got a knot inside me."

"A what?" Sadie asked.

"A knot," she said. "It hurt somethin' awful—when that man want to get with me, Sweet Jesus aren't them but some pains."

When sex first started hurting, she thought it had something to do with baby Deborah, whom she'd just given birth to a few weeks earlier, or the bad blood David sometimes brought home after nights with other women—the kind doctors treated with shots of penicillin and heavy metals.

About a week after telling her cousins she thought something was wrong, at the age of twenty-nine, Henrietta turned up pregnant with Joe, her fifth child. Sadie and Margaret told Henrietta that the pain probably had something to do with a baby after all. But Henrietta still said no.

"It was there before the baby," she told them. "It's somethin' else."

They all stopped talking about the knot, and no one told Henrietta's husband anything about it. Then, four and a half months after baby Joseph was born, Henrietta went to the bathroom and found blood spotting her underwear when it wasn't her time of the month.

She filled her bathtub, lowered herself into the warm water, and slowly spread her legs. With the door closed to her children, husband, and cousins, Henrietta slid a finger inside herself and rubbed it across her cervix until she found what she somehow knew she'd find: a hard lump, deep inside, as though someone had lodged a marble the size of her pinkie tip just to the left of the opening to her womb.

Henrietta climbed out of the bathtub, dried herself off, and dressed. Then she told her husband, "You better take me to the doctor. I'm bleeding and it ain't my time."

Her local doctor took one look inside her, saw the lump, and figured it was a sore from syphilis. But the lump tested negative for syphilis, so he told Henrietta she'd better go to the Johns Hopkins gynecology clinic.

The public wards at Hopkins were filled with patients, most of them black and unable to pay their medical bills. David drove Henrietta nearly twenty miles to get there, not because they preferred it, but because it was the only major hospital for miles that treated black patients. This was the era of Jim Crow—when black people showed up at white-only hospitals, the staff was likely to send them away, even if it meant they might die in the parking lot.

When the nurse called Henrietta from the waiting room, she led her through a single door to a colored-only exam room—one in a long row of rooms divided by clear glass walls that let nurses see from one to the next. Henrietta undressed, wrapped herself in a starched white hospital gown, and lay down on a wooden exam table, waiting for Howard Jones, the gynecologist on duty. When Jones walked into the room, Henrietta told him about the lump. Before examining her, he flipped through her chart:

> Breathing difficult since childhood due to recurrent throat infections and deviated septum in patient's nose. Physician recommended surgical repair. Patient declined. Patient had one toothache for nearly five years. Only anxiety is oldest daughter who is epileptic and can't talk. Happy household. Well nourished, cooperative. Unexplained vaginal bleeding and blood in urine during last two pregnancies; physician recommended sickle cell test. Patient declined. Been with husband since age 14 and has

no liking for sexual intercourse. Patient has asymptomatic neurosyphilis but canceled syphilis treatments, said she felt fine. Two months prior to current visit, after delivery of fifth child, patient had significant blood in urine. Tests showed areas of increased cellular activity in the cervix. Physician recommended diagnostics and referred to specialist for ruling out infection or cancer. Patient canceled appointment.

It was no surprise that she hadn't come back all those times for follow-up. For Henrietta, walking into Hopkins was like entering a foreign country where she didn't speak the language. She knew about harvesting tobacco and butchering a pig, but she'd never heard the words *cervix* or *biopsy.* She didn't read or write much, and she hadn't studied science in school. She, like most black patients, only went to Hopkins when she thought she had no choice.

Henrietta lay back on the table, feet pressed hard in stirrups as she stared at the ceiling. And sure enough, Jones found a lump exactly where she'd said he would. If her cervix was a clock's face, the lump was at 4 o'clock. He'd seen easily a thousand cervical cancer lesions, but never anything like this: shiny and purple (like "grape Jello," he wrote later), and so delicate it bled at the slightest touch. Jones cut a small sample and sent it to the pathology lab down the hall for a diagnosis. Then he told Henrietta to go home.

Soon after, Howard Jones dictated notes about Henrietta and her diagnosis: "Her history is interesting in that she had a term delivery here at this hospital, September 19, 1950," he said. "No note is made in the history at that time or at the six weeks' return visit that there is any abnormality of the cervix."

Yet here she was, three months later, with a full-fledged tumor. Either her doctors had missed it during her last exams—which seemed impossible—or it had grown at a terrifying rate.

· · ·

Henrietta Lacks was born Loretta Pleasant in Roanoke, Virginia, on August 1, 1920. No one knows how she became Henrietta. A midwife named Fannie delivered her in a small shack on a dead-end road overlooking a train depot, where hundreds of freight cars came and went each day. Henrietta shared that house with her parents and eight older siblings until 1924, when her mother, Eliza Lacks Pleasant, died giving birth to her tenth child.

Henrietta's father, Johnny Pleasant, was a squat man who hobbled around on a cane he often hit people with. Johnny didn't have the patience for raising children, so when Eliza died, he took them all back to Clover, Virginia, where his family still farmed the tobacco fields their ancestors had worked as slaves. No one in Clover could take all ten children, so relatives divided them up—one with this cousin, one with that aunt. Henrietta ended up with her grandfather, Tommy Lacks.

Tommy lived in what everyone called the home-house, a four-room wooden cabin that once served as slave quarters, with plank floors, gas lanterns, and water Henrietta hauled up a long hill from the creek. The home-house stood on a hillside where wind whipped through cracks in the walls. The air inside stayed so cool that when relatives died, the family kept their corpses in the front hallway for days so people could visit and pay respects. Then they buried them in the cemetery out back.

Henrietta's grandfather was already raising another grandchild that one of his daughters left behind after delivering him on the home-house floor. That child's name was David Lacks, but everyone called him Day, because in the Lacks country drawl, *house* sounds like *hyse*, and *David* sounds like *Day*. No one could have guessed Henrietta would spend the rest of her life with Day—first as a cousin growing up in their grandfather's home, then as his wife.

Like most young Lackses, Day didn't finish school: He stopped in the fourth grade because the family needed him to work the tobacco fields. But Henrietta stayed until the sixth grade. During the school year, after taking care of the garden and livestock each morning, she'd walk two miles—past the white school where children threw rocks and taunted her—to the colored school, a three-room wooden farmhouse hidden under tall shade trees.

At nightfall the Lacks cousins built fires with pieces of old shoes to keep the mosquitoes away, and watched the stars from beneath the big oak tree where they'd hung a rope to swing from. They played tag, ring-around-the-rosy, and hopscotch, and danced around the field singing until Grandpa Tommy yelled for everyone to go to bed.

Henrietta and Day had been sharing a bedroom since she was 4 and he was 9, so what happened next didn't surprise anyone: They started having children together. Their son Lawrence was born just months after Henrietta's 14th birthday; his sister, Lucile Elsie Pleasant, came along four years later. They were both born on the floor of the home-house like their father, grandmother, and grandfather before them. People wouldn't use words like *epilepsy, mental retardation,* or *neurosyphilis* to describe Elsie's condition until years later. To the folks in Clover, she was just simple. Touched.

Henrietta and Day married alone at their preacher's house on April 10, 1941. She was 20; he was 25. They didn't go on a honeymoon because there was too much work to do, and no money for travel. Henrietta and Day were lucky if they sold enough tobacco each season to feed the family and plant the next crop. So after their wedding, Day went back to gripping the splintered ends of his old wooden plow as Henrietta followed close behind, pushing a homemade wheelbarrow and dropping tobacco seedlings into holes in the freshly turned red dirt.

A few months later, Day moved north to Turner Station, a small black community outside Baltimore where he'd gotten a job working in a shipyard. Henrietta stayed behind to care for the children and the tobacco until Day made enough money for a house and three tickets north. Soon, with a child on each side, Henrietta boarded a coal-fueled train from the small wooden depot at the end of Clover's Main Street. She left the tobacco fields of her youth and the hundred-year-old oak tree that shaded her from the sun on so many hot afternoons. At the age of 21, she stared through the train window at rolling hills and wide-open bodies of water for the first time, heading toward a new life.

After her visit to Hopkins, Henrietta went back to her usual routine, cleaning and cooking for her husband, their children, and the many cousins she fed each day. Less than a week later, Jones got her biopsy results from the pathology lab: "epidermoid carcinoma of the cervix, Stage I." Translation: cervical cancer.

Cervical carcinomas are divided into two types: invasive carcinomas, which have penetrated the surface of the cervix, and noninvasive carcinomas, which haven't. The noninvasive type is sometimes called "sugar-icing carcinoma," because it grows in a smooth layered sheet across the surface of the cervix, but its official name is carcinoma in situ, which derives from the Latin for "cancer in its original place."

In 1951 most doctors in the field believed that invasive carcinoma was deadly, and carcinoma in situ wasn't. So they hardly treated it. But Richard Wesley TeLinde, head of gynecology at Hopkins and one of the top cervical cancer experts in the country, disagreed—he believed carcinoma in situ was simply an early stage of invasive carcinoma that, left untreated, eventually became deadly. So he treated it aggressively, often removing the cervix, uterus, and most of the vagina. He argued that this would drastically reduce cervical cancer deaths, but his critics called it extreme and unnecessary.

TeLinde thought that if he could find a way to grow living samples from normal cervical tissue and both types of cancerous tissue—something never done before—he could compare all three. If he could prove that carcinoma in situ and invasive carcinoma looked and behaved similarly in the laboratory, he could end the debate, showing that he'd been right all along, and doctors who ignored him were killing their patients. So he called George Gey (pronounced "guy"), head of tissue culture research at Hopkins.

Gey and his wife, Margaret, had spent the last three decades working to grow malignant cells outside the body, hoping to use them to find cancer's cause and cure. But most of the cells died quickly, and the few that survived hardly grew at all. The Geys were determined to grow the first immortal human cells: a continuously dividing line of cells all descended from one original sample, cells that would constantly replenish themselves and never die. They didn't care what kind of tissue they used, as long as it came from a person.

So when TeLinde offered Gey a supply of cervical cancer tissue in exchange for trying to grow some cells, Gey didn't hesitate. And TeLinde began collecting samples from any woman who happened to walk into Hopkins with cervical cancer. Including Henrietta.

Jones called Henrietta on February 5, 1951, after getting her biopsy report back from the lab, and told her the tumor was malignant. Henrietta didn't tell anyone what Jones said, and no one asked. She simply went on with her day as if nothing had happened, which was just like her—no sense upsetting anyone over something she could just deal with herself.

The next morning Henrietta climbed from the Buick outside Hopkins again, telling Day and the children not to worry.

"Ain't nothin' serious wrong," she said. "Doctor's gonna fix me right up."

Henrietta went straight to the admissions desk and told the receptionist she was there for her treatment. Then she signed a form with the words operation permit at the top of the page. It said:

> I hereby give consent to the staff of The Johns Hopkins Hospital to perform any operative procedures and under any anaesthetic either local or general that they may deem necessary in the proper surgical care and treatment of:
>
> _____.

Henrietta printed her name in the blank space. A witness with illegible handwriting signed a line at the bottom of the form, and Henrietta signed another.

Then she followed a nurse down a long hallway into the ward for colored women, where Howard Jones and several other white physicians ran more tests than she'd had in her entire life. They checked her urine, her blood, her lungs. They stuck tubes in her bladder and nose.

Henrietta's tumor was the invasive type, and like hospitals nationwide, Hopkins treated all invasive cervical carcinomas with radium, a white radioactive metal that glows an eerie blue. So the morning of Henrietta's first treatment, a taxi driver picked up a doctor's bag filled with thin glass tubes of radium from a clinic across town. The tubes were tucked into individual slots inside small canvas pouches hand-sewn by a local Baltimore woman. One nurse placed the pouches on a stainless steel tray. Another wheeled Henrietta into the small colored-only operating room, with stainless steel tables, huge glaring lights, and an all-white medical staff dressed in white gowns, hats, masks, and gloves.

With Henrietta unconscious on the operating table in the center of the room, her feet in stirrups, the surgeon on duty, Lawrence Wharton Jr., sat on a stool between her legs. He peered inside Henrietta, dilated her cervix, and prepared to treat her tumor. But first—though no one had told Henrietta that TeLinde was collecting samples or asked if she wanted to be a donor—Wharton picked up a sharp knife and shaved two dime-size pieces of tissue from Henrietta's cervix: one from her tumor, and one from the healthy cervical tissue nearby. Then he placed the samples in a glass dish.

Wharton slipped a tube filled with radium inside Henrietta's cervix, and sewed it in place. He then sewed a pouch filled with radium to the outer surface of her cervix and packed another against it. He slid several rolls of gauze inside her vagina to help keep the radium in place, then threaded a catheter into her bladder so she could urinate without disturbing the treatment.

When Wharton finished, a nurse wheeled Henrietta back into the ward, and a resident took the dish with the samples to Gey's lab, as he'd done many times before. Gey still got excited at moments like this, but everyone else in his lab saw Henrietta's sample as something tedious—the latest of what felt like countless samples that scientists and lab technicians had been trying and failing to grow for years.

Gey's 21-year-old assistant, Mary Kubicek, sat eating a tuna salad sandwich at a long stone culture bench that doubled as a break table. She and Margaret and the other women in the Gey lab spent many hours there, all in nearly identical cat's-eye glasses with fat dark frames and thick lenses, their hair pulled back in tight buns.

"I'm putting a new sample in your cubicle," Gey told Mary.

She pretended not to notice. "Not again," she thought, and kept eating her sandwich. Mary knew she shouldn't wait—every moment those cells sat in the dish made it more likely they'd die. But they always died anyway. "Why bother?" she thought.

At that point, there were many obstacles to growing cells successfully. For starters, no one knew exactly what nutrients they needed to survive or how best to supply them. But the biggest problem facing cell culture was contamination. Bacteria and a host of other microorganisms could find their way into cultures—from people's unwashed hands, their breath, and dust particles floating through the air—and destroy them. Margaret Gey had been trained as a surgical nurse, which meant sterility was her specialty—it was key to preventing deadly infections in patients in the operating room.

Margaret patrolled the lab, arms crossed, leaning over technicians' shoulders as they worked, inspecting glassware for spots or smudges. Mary followed Margaret's sterilizing rules meticulously to avoid her wrath. Only then did she pick up the pieces of Henrietta's cervix—forceps in one hand, scalpel in the other—and carefully slice them into one-millimeter squares. She sucked each square into a pipette, and dropped them one at a time onto chicken-blood clots she'd placed at the bottom of dozens of test tubes. She covered each clot with several drops of culture medium, plugged the tubes with rubber stoppers, and wrote "HeLa," for Henrietta and Lacks, in big black letters on the side of each tube. Then she put them in an incubator.

For the next few days, Mary started each morning with her usual sterilization drill. She'd peer into all the incubating tubes, laughing to herself and thinking, "Nothing's happening." "Big surprise." Then she saw what looked like little rings of fried egg white around the clots at the bottom of each tube. The cells were growing, but Mary didn't think much of it—other cells had survived for a while in the lab.

But Henrietta's cells weren't merely surviving—they were growing with mythological intensity. By the next morning, they'd doubled. Mary divided the contents of each tube in two, giving them room to grow, and soon she was dividing them into four tubes, then six. Henrietta's cells grew to fill as much space as Mary gave them.

Still, Gey wasn't ready to celebrate. "The cells could die any minute," he told Mary. But they didn't. The cells kept growing like nothing anyone had seen, doubling their numbers every twenty-four hours, accumulating by the millions. "Spreading like crabgrass!" Margaret said. As long as they had food and warmth, Henrietta's cancer cells seemed unstoppable.

Soon, George told a few of his closest colleagues that he thought his lab might have grown the first immortal human cells.

To which they replied, Can I have some? And George said yes.

George Gey sent Henrietta's cells to any scientist who wanted them for cancer research. HeLa cells rode into the mountains of Chile in the saddlebags of pack mules and flew around the country in the breast pockets of researchers until they were growing in laboratories in Texas, Amsterdam, India, and many places in between. The Tuskegee Institute set up facilities to mass-produce Henrietta's cells, and began shipping twenty thousand tubes of HeLa—about six trillion cells—every week. And soon, a multibillion-dollar industry selling human biological materials was born.

HeLa cells allowed researchers to perform experiments that would have been impossible with a living human. Scientists exposed them to toxins, radiation, and infections. They bombarded them with drugs, hoping to find one that would kill malignant cells without destroying normal ones. They studied immune suppression and cancer growth by injecting HeLa into rats with weak immune systems, who developed malignant tumors much like Henrietta's. And if the cells died in the process, it didn't matter—scientists could just go back to their eternally growing HeLa stock and start over again.

But those cells grew as powerfully in Henrietta's body as they did in the lab: Within months of her diagnosis, tumors had taken over almost every organ in her body. Henrietta died on October 4, 1951, leaving five children behind, knowing nothing about her cells growing in laboratories around the world.

Henrietta's husband and children wouldn't find out about those cells until twenty-five years later, when researchers from Johns Hopkins decided to track down Henrietta's family to do research on them to learn more about HeLa.

When Henrietta's children learned of HeLa, they were consumed with questions: Had scientists killed their mother to harvest her cells? Were clones of their mother walking the streets of cities around the world? And if Henrietta was so vital to medicine, why couldn't they afford health insurance? Today, in Baltimore, her family still wrestles with feelings of betrayal and fear, but also pride. As her daughter Deborah once whispered to a vial of her mother's cells: "You're famous, just nobody knows it."

PART V

Patients as Consumers in the Gene Age

In his January 2015 State of the Union address, President Obama announced the Precision Medicine Initiative, a $215 million project to collect genetic and health information from a million Americans. The plan accepts the ambitious and increasingly popular claim that precision medicine will harness advances in genetic research and information technology to revolutionize health care, creating a new era in which disease prediction and medical treatment will be tailored to each individual's genetic characteristics.

The precision medicine project depends on building a huge database of genetic sequences, medical records, lifestyle reports, and other information, and on developing the ability to meaningfully interpret such truly vast amounts of data. Different aspects of the effort are being undertaken by a range of research institutions and by private enterprises, from start-up direct-to-consumer genetic-testing companies to Big Data behemoths, including Google, Apple, and Amazon.

Even before the launch of the Precision Medicine Initiative, observers raised many questions about whether the promises of precision medicine were likely to be realized and about what its unexpected consequences might be. Will precision medicine in fact lead to more effective medical treatments? If so, for which diseases and which patients, under what circumstances, and at what cost? Will it improve public health outcomes, or divert our attention from more basic measures? What should we consider in terms of our genetic and medical privacy, our influence on how our data are used, and who profits from it? Who is developing the infrastructure of medicine in the gene age— the genetic tests, genetic databases, biobanks, information technology, interpretive procedures, and other components? What business models are involved? How do we weigh the promises of precision genetic medicine against its perils, the possibilities against the opportunity costs?

These are some of the questions taken up by contributors to this section.

In "What Is Your DNA Worth?" David Dobbs looks more closely at gene therapy, genome-wide association studies, and other aspects of "the big-data branch of human genetics." He recounts the series of unfulfilled promises made in the name of genetics and warns that medical genetics continues to be oversold. And he points out that the "hype about the hope, the silence about the disappointments" is not innocent: It "gobbles up funding that we might spend better elsewhere, warps the expectations of patients and the incentives of scientists, and has implications even for people who pay genetics scant attention."

Today, patients are not only consumers but also research subjects and providers of raw materials for medical research—though sometimes without knowing it. Faced with a "Terms and Conditions of Service" document at a medical appointment, sociologist Jenny Reardon reflects on this situation in her commentary, "Should Patients Understand That They Are Research Subjects?" Reardon, herself a scholar of research ethics, found that the document failed to answer key questions and considers whether today's biotech boom means that risks and ethical issues are being overlooked.

In "Direct-to-Consumer Genetic Tests Should Come with a Health Warning," Jessica Cussins looks at the clinical usefulness of direct-to-consumer gene tests sold by companies such as Google-backed 23andMe and explores the concerns about this model, which are widely shared by researchers and by government agencies, including the US Food and Drug Administration and the US Government Accountability Office. Despite the FDA's crackdown on its health claims, Cussins notes that 23andMe continues to market its product in dozens of countries.

Karuna Jaggar, executive director of Breast Cancer Action, takes issue with a specific kind of genetic testing: the recommendation that all women should routinely be tested for the "breast cancer genes" BRCA1 and BRCA2. "The answer to the breast cancer epidemic does not rest with mass genetic testing," she argues. "The more we look for answers in individual women's bodies, the less we focus on the societal, systemic factors driving the epidemic in the first place."

Troy Duster's contribution, "Welcome, Freshmen: DNA Swabs, Please.," describes the controversy surrounding a project at the University of California at Berkeley, which asked incoming freshmen in the fall of 2010 to voluntarily submit their DNA for genetic testing. The faculty members behind the program explained that they simply wanted to introduce students to the future of personalized medicine by giving them the chance to have a few of their own genes analyzed. Duster points to a host of problems, including the error rates and uncertainty in interpreting what particular variants of the genes might mean, violations of human subjects standards, and social coercion to participate. Duster concludes with a recommendation that such projects be redesigned to avoid the "substantial intellectual risk" that students will be "institutionally introduced into misunderstanding the precision, interpretation, and historically problematic execution of [genetic] research, and the subtle, unexamined undercurrent of coercion in their participation."

Donna Dickenson's "Me Medicine" asks whether we should believe the promises being bandied about for genetic personalized medicine and for related commercial developments, including private banks for umbilical cord blood and speculative enhancement technologies that supposedly enable us to become "transhuman." She argues that the huge advances in human health and longevity over the last two centuries have come from public health measures, sanitation programs, and vaccination against infectious diseases, and that this kind of "we" medicine will likely continue to improve our health more than the "me" focus of modern biomedicine.

Ronald Bayer and Sandro Galea address similar questions in "Public Health in the Precision-Medicine Era." Owing to the well-understood importance of social determinants of health, and of prioritizing investments toward advancing public health and reducing health inequities, they are skeptical about what the precision medicine agenda offers. "Without minimizing the possible gains to clinical care from greater realization of precision medicine's promise," they write, "we worry that an unstinting focus on precision medicine by trusted spokespeople for health is a mistake—and a distraction from the goal of producing a healthier population."

23. What Is Your DNA Worth?

David Dobbs

"Success in sight: The eyes have it!" Thus the scientific journal *Gene Therapy* greeted the news, in 2008, that an experimental treatment was restoring vision to twelve people born with a congenital disorder that slowly left them blind. Healthy genes were injected to replace the faulty mutations in the patients' retinas, allowing an eight-year-old to ride a bike for the first time. A mother finally saw her child play softball. Every patient, the researchers reported, showed "sustained improvement." Five years in, a book declared this "breakthrough"—a good-gene-for-bad-gene swap long pursued as a silver bullet for genetic conditions—as *The Forever Fix*.

In early 2015, two of the three research teams running these trials quietly reported that the therapy's benefit had peaked after three years and then begun to fade. The third trial says its patients continue to improve. But in the other two, all the patients tracked for five years or more were again losing their sight.

Not all gene therapy ends in Greek-caliber tragedy. But these trials serve as a sadly apt parable for the current state of human genetics. This goes especially for the big-data branch of human genetics called Big Genomics. In five years of talking to geneticists, biologists, and historians, I've found that the field is too often distinguished by the arc shown here: alluring hope, celebratory hype, dark disappointment.

We live in an age of hype. But the overselling of the Age of Genomics—the hype about the hope, the silence about the disappointments—gobbles up funding that we might spend better elsewhere, warps the expectations of patients and the incentives of scientists, and has implications even for

Originally published by David Dobbs. "What Is Your DNA Worth?" *BuzzFeed.* May 21, 2015. Reprinted by permission of David Dobbs.

people who pay genetics scant attention. Many hospitals, for instance, are now collecting genetic information from patients that they may market to "research partners" such as drug companies. Some take more care than others do to secure informed consent. (Had blood drawn lately? Read everything you signed that day?) It's not just that they're selling you this stuff. They may well be selling you. And the sale depends on an exaggerated picture of genetic power and destiny.

To be sure, medical genetics has chalked up some sweet victories. Our growing ability to spot rare mutations, for instance, is helping doctors diagnose and sometimes treat nasty rare diseases. Last fall, for instance, doctors in St. Louis sequenced an infant dying of liver failure, saw that he had inherited a rare mutation that both his parents happened to pass to him, devised a way to counter the mutation's disruption of his immune system, and saved his life.

But when it comes to how genes shape the traits and diseases that matter most to us—from intelligence and temperament to cancer and depression—genetic research overpromises and underdelivers on actionable knowledge. After 110 years of genetics, and 15 years after the $3.8 billion Human Genome Project promised fast cures, after more billions spent and endless hype about results just around the corner, we have few cures. And we basically know diddly-squat.

I know—*diddly-squat* is rough talk. Yet this is hardly a radical claim. Geneticists and doctors outside of Big Genomics—people studying genetics in songbirds, sea urchins, monkeys, microbes, fruit flies, and roundworms, for instance—often voice it privately. Others are eager to tell us what genes can't do or warn that "precision medicine" will let us down. One of the world's most respected geneticists, Britain's Steve Jones, gives quite an entertaining lecture on our humble state of knowledge. "The more we learn, the less we understand," he says. "We know almost nothing of genetics."

The press, of course, too often falls hard for ludicrous memes such as "the slut gene." But much of the time, the media is simply amplifying the signal sent by Big Genomics. Big Genomics outfits like the National Institutes of Health and the Broad Institute regularly assure us that their careful reading of the genome's text will find crucial misspellings that generate disease—and let us revise, delete, or write around those errors.

In doing so, they continue a tradition as old as genetics itself. Historian Nathaniel Comfort, in *The Science of Human Perfection*, calls the history of genetics "a history of promises." Cambridge geneticist William Bateson coined the term genetics in 1905; by 1927, biologists made the first of many assertions that genetics would cure cancer. In 1940, Canadian physician and embryologist Madge Macklin promised "a world in which doctors come to their patients and tell them what diseases they are about to have, and then begin treatments before the patient feels even the first symptoms." In 1967, Stanford geneticist Joshua Lederberg predicted gene replacement therapy—the kind that is now failing in the blindness trial—"within a few years."

In 2000 the leaders of the Human Genome Project doubled down. Standing next to President Bill Clinton, they announced that the project had sequenced the human genome, exposing the full genetic code to view. "Personalized genetic medicine," an accompanying White House Statement said, would soon "cure diseases like Alzheimer's, Parkinson's, diabetes and cancer by attacking their genetic roots." Francis Collins, the project's director (now head of the National Institutes of Health), said the genomic revolution could reduce cancer to zero and would make gene-tailored personalized medicine common by 2010.

A century of hype is a lot, but this is particularly inspirational ground. The gene, especially after Franklin, Watson, and Crick gave us a peek at DNA in 1953, looked promising as hell. For decades, the gene was seen as the key to all of biology—or as President Clinton would eventually put it, "the language in which God created life." In its code we would read the story of life, evolution, disease, and death.

But when the Genome Project finally revealed the links in Franklin, Watson, and Crick's deceptively simple structure, it found few of the strong gene-to-trait connections one might have hoped for. Instead, it found a mess. Our DNA held far fewer genes than expected, almost twenty thousand, which was confusing. Few held obvious function. Some seemed to do nothing. Some seemed to work fine one day but not the next, or to do one thing in one situation and another in another. And these genes were surrounded by vast stretches of DNA material that aren't really genes, and which some geneticists called junk, starting a big fight.

To clarify this mess—to figure out what did what, and to identify medically relevant genes—researchers started using sequencing machines to scan the genomes of tens or even hundreds of thousands of people for gene variants that appear more often in people with some condition, disease, or trait. These overrepresented genes are then presumed to contribute to the condition or trait in question.

Unfortunately, GWAs [genome-wide association studies] seldom revealed the sort of neat or consistent gene-to-trait relationships that allow decisive treatment. Instead, they usually found "many genes of small effect": handfuls and sometimes hundreds of gene variants carried by most (but not all) people with the condition in question, whose effects were seldom clear, and whose presence in a given person did little to predict risk.

"Many genes of small effect" became a sort of tepid curse. I myself prefer the stronger, more memorable phrase, "Many Assorted Genes of Tiny Significance," or MAGOTS—a mass of barely significant genes explaining little.

MAGOTS infest most GWA studies for a simple, brutal reason: If a gene variant reliably plays a large role in causing disease, both the variant and the disease it causes tend to be rare, because its carriers tend to die without leaving offspring. This is why the genetic contributions for common diseases and conditions usually come from MAGOTS—the effects of which, it bears repeating, are usually maddeningly obscure and unpredictable. This applies even to diseases and traits that run in families. Take height: Hundreds of genes of small effect, few clues to how they contribute, and no real target to tweak if, say, you want to make someone tall. The best way to engineer a tall person? Tell two tall people to tango.

Similarly, deep digs at cancer, schizophrenia, heart disease, hypertension, diabetes, intelligence, bipolar disorder, and height have found mostly MAGOTS. The biggest schizophrenia study so far, for instance, published [in July 2014] to great fanfare, found 128 gene variants that appeared to account for perhaps 7 percent of a given person's actual risk.

The genomic age's signature finding is not any great discovery. It is the yawning gap between the genetic contributions that geneticists assume exist and the genetic contributions they can spot. It is as if they cracked a safe they knew was packed with cash and found almost nothing. The money's got to be somewhere. But where?

Researchers in the field are quick to point to one of the handful of effective drugs to come from genomic insight, such as Gleevec, a leukemia drug developed in 2001. But Gleevec, however potent, falls far short of the medical miracles forecast fifteen years ago. As science writer Ed Yong pointed out in a recent Twitter conversation about this, "Treasure was promised. Gleevec's a coin."

At this point, the problem is not so much that genetics fell short of its early promises. The problem is that big genomics players keep making similar promises.

Take, for instance, that schizophrenia study rife with MAGOTS. When the study came out last July, John Williams, head of neuroscience and mental health at the Wellcome Trust, Britain's biggest biomedical funder, saw it as cause for humility. "What this research screams to me," he wrote, "is how little we know about schizophrenia, and how far we are from biological tests and treatments for mental health disorders compared to other major diseases."

Yet last July, the Broad Institute, a genomics powerhouse that played a big role in that schizophrenia study, triumphantly unveiled it as part of an announcement that a donor had given Broad $650 million to expand research at its Stanley Center for Psychiatric Genomics. Broad's director called the study part of "a revolution in psychiatric disease." Francis Collins, apparently deaf to how closely his promises echoed those he'd made fifteen years before, when the Human Genome Project was unveiled, said psychiatric genomics now stood "poised for rapid advances." The promises were a decade old, the rhetoric a century. The only things new were the event's over-the-top staging and production—it views like an awards ceremony—and how boldly, even after fifteen years of the "genomic age" with little to show, the Broad conjured big money from thin results.

Big Genomics is converting hype to cash at unsettling speed. After the FDA told consumer genomics company 23andMe it could no longer sell people health data, the company began selling that data to drug and biotech companies. An entire industry, potentially fed by almost anyone who draws blood, spit, or biopsies from you, is emerging to do likewise. Its growth, along with the increasingly routine collection of genetic data by hospitals, will feed the genomics bubble while putting private genetic and health information at increased risk.

Meanwhile, it's becoming routine for researchers and research centers to leverage genomic findings into industry jobs or startups.

None of this is to say we should pull the plug on Big Genomics. Some suggest—and I agree—that we'd do well to take some of the billions spent chasing genes for conditions like Type II diabetes, heart disease, or stroke and spend it instead on finding ways to change risk-elevating behaviors like smoking, overeating, overdrinking, and avoiding exercise.

It would be responsible, however, for researchers to temper their hype—though this seems unlikely, because hype pays.

So let me offer a hype filter. This one comes courtesy of the oceanographer Henry Bryant Bigelow, who helped found Woods Hole Oceanographic Institute. A century ago, Bigelow opened a letter his brother had written him from Cuba. His brother reported that while weathering a hurricane there, he had seen, flying by, what he was almost sure was a donkey.

With three words, Bigelow gently told his brother he didn't quite believe him—and stated a maxim for maintaining the ever-curious but ever-skeptical stance that marks the good scientist.

"Interesting if true," he wrote.

24. Should Patients Understand That They Are Research Subjects?

Jenny Reardon

While checking in last month for an appointment with my doctor, I was handed a clipboard and pen and asked to sign a Terms and Conditions of Service form.

With a bit of time on my hands, I decided to actually read the document, instead of just checking the box and signing it. I was half reading and half listening to the Cooking Channel tell me how to make the perfect polenta when I reached point 4: "I also understand that my medical information and tissues, fluids, cells and other specimens (collectively, "Specimens") that UCSF may collect during the course of my treatment and care may be used and shared with researchers."

Surely, I thought, this must be wrong. It wasn't the idea that researchers might use my tissues for research that startled me. It was the phrase "I understand."

For decades, the question of whether and under what conditions medical professionals should collect human tissue samples for research has raised complicated ethical questions that are far from settled and certainly are not easy to understand.

I have spent two decades studying this minefield, and even I had a hard time making sense of what it might mean for these researchers to have access to my samples. For example, UCSF would be required by law to make my samples "anonymous," yet research published in *Science* the day of my visit revealed that even anonymous samples can be reidentified. Does this mean that information gained from my samples might be linked back to me?

Originally published by Jenny Reardon. "Should Patients Understand That They Are Research Subjects?" *San Francisco Chronicle*, March 3, 2013. Reprinted by permission of *San Francisco Chronicle* / Polaris.

The form went on to state: "I understand that under California law, I do not have any rights to any commercially useful products that may be developed from such research."

Does this mean that UCSF could own or perhaps even patent my bodily material?

Questions about ownership and patenting in the biological realm have been debated intensely the past few years. Just last month the US Supreme Court set April 15 [2013] as the date for oral arguments on the breast cancer gene patents case. At issue is whether it is legal to patent DNA isolated from human tissues that corresponds to human genes. Given that these property questions are under debate by even the highest court in the land, how can I be expected to understand this issue?

UCSF and the city of San Francisco seek to establish the Bay Area as a hub of biomedical innovation. Already the Bay Area ranks first in the nation for the number of biotechnology patents per million citizens. Collecting samples from patients might seem an obvious step forward. However, are risks and ethical issues being overlooked or minimized in pursuit of this biotech boom?

I recently traveled to the University of Washington, and in the course of my research learned that this university is implementing a method of requesting a patient's permission for research use of samples collected in the course of clinical care. Patients will be able to refuse permission without it in any way jeopardizing their clinical care—their doctors will not know what answer they gave. While federal policy that governs human-subjects research allows the University of Washington, just as it would allow UCSF, to use samples as long as they de-identify them, this institution chooses to ask patients for permission. Although not the norm, this policy is in line with findings of University of Washington and Group Health investigators, who last year reported that their patients were on the whole supportive of genomic research, but they wanted to be informed about the purpose for which their samples would be used, as well as about data-sharing agreements.

It is this effort to build understanding—and seek permission—that was missing from my experience at UCSF. In the end, after a period of discomfort, I took the form and quietly stuffed it into my backpack, leaving the clipboard and pen to one side.

I hope that by the time of my annual visit next year, UCSF's terms will have changed. I would like our region to be known for our biotech innovation, but also for our commitment to addressing the challenging questions of ethics, justice, and democracy that today are emerging in biomedicine.

25. Direct-to-Consumer Genetic Tests Should Come with a Health Warning

Jessica Cussins

In late 2014, the Google-backed company 23andMe announced that it would start selling its direct-to-consumer (DTC) genetic test in Canada and the United Kingdom—despite being banned by the US Food and Drug Administration (FDA) from selling it in the United States following misleading marketing.

The genetic test provides information for around 108 health conditions for which some degree of evidence exists, including 44 inherited conditions, 12 drug responses, 12 genetic risk factors, and 41 traits. Some conditions on the list are both obvious and innocuous. For example, hair color, eye color, and height are among the 41 traits. But many of the results are about important health conditions and are clearly supposed to incite a change in behavior.

Genetic testing is appropriate—and can be life saving—when doctors and genetic counselors interpret complex results and map out the various courses of action. However, DTC genetic-testing companies, such as 23andMe, deliberately eschew the framework between clinician and patient. Under the banner of personal empowerment, DTC companies proclaim that their products confer a new level of control over one's health, and that to have your genome sequenced is a liberating act that is both exciting and personally responsible. In practice, however, results may lead to unnecessary anxiety or a false sense of security.

First published in *The Pharmaceutical Journal*, 17 January 2015, Vol 294, No 7845, online | URI: 20067564 (http://www.pharmaceuticaljournal.com/opinion/comment /direct-to-consumer-genetic-tests-should-come-with-a-health-warning/20067564 .article). Reprinted by permission of *The Pharmaceutical Journal*.

FALSE ALARM

Lukas Hartmann submitted a saliva sample to 23andMe in November 2010 to find out whether he was genetically predisposed to any conditions. He received a report reassuring him that he would probably die when the time came from a mix of heart attack and prostate cancer; nothing out of the ordinary there. But a year later, he received an email from 23andMe asking whether he wanted it to reveal some potentially disturbing, life-altering detail about his genome found in his original test. He clicked through, and was told that he carried two mutations linked to limb-girdle muscular dystrophy that could cause him to lose his ability to walk and potentially end in early death.[1]

A self-described nerd, Hartmann responded by downloading his raw genetic data from 23andMe and scouring the Internet for any further information. After several sleepless nights of research, he discovered that while he did have the two mutations of concern, they were actually located on different genes, unlike the mutations associated with the studies that prompted the email alert. Hartmann was able to catch this critical distinction, but 23andMe's algorithm was not.

NOT SO FAST

The FDA has expressed its concerns over misleading marketing of DTC genetic tests for some time. In November 2013, the agency sent a letter to 23andMe chief executive Anne Wojcicki noting that after "many interactions with 23andMe, we still do not have any assurance that the firm has analytically or clinically validated the [Personal Genome Service] for its intended uses." The letter informed 23andMe that it was prohibited from providing US consumers with health information until it provided evidence that its test works, noting the gravity of harm that can result from inaccuracies.[2]

The FDA is not alone in raising concerns about the accuracy of DTC gene test results. A 2010 report by the US Government Accountability Office (the investigative arm of Congress charged with examining matters relating to the receipt and payment of public funds) found that "identical DNA samples yield contradictory results" from four different DTC companies and the results were "misleading and of little or no practical use." Also that year, the Nuffield Council on Bioethics in the United Kingdom (an independent body advising policy makers and stimulating debate in bioethics) cautioned: "The powerful rhetoric used to promote these developments should be treated with caution, since it can downplay potential harms and exaggerate the usefulness of the technologies concerned."[3]

In 2013, a study published in *Nature* found that predictions of disease risk from different DTC genetic-testing companies still varied significantly.[4] The research team, led by epidemiologist Cecile Janssens at Erasmus University Medical Center in Rotterdam, the Netherlands, compared methods and results from 23andMe, deCODEme, and Navigenics by simulating genotype data for a hundred thousand people and predicting the risks of six diseases based on each company's method. They found substantial differences among predicted risks owing to diverging formulas and the way that each company determined average risk. The variation in outcome led each company to encourage radically different courses of action based on the same DNA. deCODEme, based in Reykjavík, Iceland, and Navigenics, based in Foster City, California, have both stopped selling their DTC tests.

GROWING ELSEWHERE

While deCODEme and Navigenics have gone out of business, and despite the FDA's limitations in the United States on 23andMe, it continues to market its product elsewhere. It has accumulated some 750,000 clients in more than fifty countries.

Unlike the FDA, the Department of Health in England and Health Canada have determined that 23andMe's test is not a medical device. And the company's UK website claims its reports are intended for informational purposes only and do not diagnose disease or illness. But when a test tells you that you carry a *BRCA* gene mutation linked to hereditary breast and ovarian cancer, you are unlikely to receive the news as nonmedical.

The *BRCA* variants account for only about 5 percent of breast cancers. Women who have the mutations on which 23andMe reports have a significantly elevated risk, yet the majority will not go on to develop cancer. Nonetheless, Angelina Jolie's announcement in May 2013 that her *BRCA1* mutation led her to obtain a preventative double mastectomy caused what is now known as the "Angelina Jolie effect," with the number of DNA tests for breast cancer mutations increasing by two-and-a-half times.[5]

One of the FDA's explicit concerns is that some of the women worried about their risk of these cancers will opt for DTC genetic tests and receive overly deterministic or inaccurate information, leading to unnecessary surgeries, treatments, or screenings—or, for those who find they do not have these particular variants, to a false lull of complacency.

SOCIAL IMPACTS

Unrealistic claims about gene tests also encourage a false sense of genetic determinism. Increasing numbers of "previvors" or "worried well" would cause huge burdens for the medical community and tie up resources for a predominately privileged citizenry.[6] Most diseases and traits do not reliably stem from the tiny snippets of DNA that current DTC genetic tests examine. In reality, most of us can learn more from stepping on the bathroom scale than from a DTC gene test.

NOTES

1. L. Hartmann, "Why 23andMe Has the FDA Worried: It Wrongly Told Me I Might Die Young, *Quartz; Obsessions,* November 27, 2013.

2. A. Gutierrez, 23andMe, Inc., 11/22/13, "US Food and Drug Administration: Inspections, Compliance, Enforcement, and Criminal Investigations," document number GEN1300666, November 22, 2013.

3. "Direct-to-Consumer Genetic Tests: Misleading Test Results Are Further Complicated by Deceptive Marketing and Other Questionable Practices," GAO-10–847T; 2010; Medical Profiling and Online Medicine: The Ethics of 'Personalised Healthcare' in a Consumer Age, *Nuffield Council on Bioethics* 2010.

4. R. Kalf, R. Mihaescu, and S. Kundu, et al., "Variations in Predicted Risks in Personal Genome Testing for Common Complex Diseases, *Genetics in Medicine* 16 (2013): 85–91.

5. H. Briggs, "Breast Cancer Test 'Angelina Jolie Effect' Found," *BBC News,* September 19, 2014.

6. D. Dickenson, "In Me We Trust: Public Health, Personalized Medicine, and the Common Good," *Hedgehog Review* 16, no. 1 (2014).

26. Genetic Testing for All Women?

Not a Solution to the Breast Cancer Epidemic

Karuna Jaggar

The perseverance and scientific innovation of Dr. Mary Claire King, the geneticist whose efforts to identify the BRCA genes [linked to breast cancer] were initially scoffed at by her peers, should not be underestimated or underappreciated. Her drive and desire to find better tools to address the breast cancer epidemic are shared by activists, researchers, geneticists, and patients alike. But her new recommendation, published in the *Journal of the American Medical Association*—that all women in the United States over age thirty be screened for BRCA mutations linked to breast and ovarian cancer—fails as a matter of public health to recognize the significant limitations, and harms, of mass genetic testing in the current health care environment.

In proposing that genetic testing for the BRCA1 and BRCA2 genes become routine for all women over age thirty, Dr. King focuses only on what she sees as the potential benefit, of identifying women with genetic mutations that increase their risk of cancer, and minimizes the harms that result from exposing all 105,000,000 women age thirty and over in the United States to genetic testing without access to genetic counselors who can explain tests or results.

Genetic testing is complex and brings with it a wide range of medical, ethical, and scientific issues—despite years of medical training, even many physicians don't fully understand all the complexity involved. And though I share Dr. King's concern that "only 19 percent of US primary care physicians accurately assessed family history for BRCA1/BRCA2 testing," I

Original publication by Karuna Jaggar. "Genetic Testing for All Women? Not a Solution to the Breast Cancer Epidemic." *Huffington Post.* September 24, 2014. Reprinted by permission of Karuna Jaggar.

disagree that we can solve that problem by removing doctors and genetic counselors from the testing process.

Failure to discuss and acknowledge the limitations, harms, and options of any medical test or procedure deprives patients of their right to make informed decisions about their bodies and their health care. Understanding genetic testing is particularly complex and requires trained genetic counselors, both before and after testing. Genes are not simply binary switches and knowledge of a mutation brings a complex set of choices that are challenging to navigate for patients. The very nature of inherited genes means that these issues extend to other generations—meaning this complex set of ethical, legal, medical, and life-course questions extend from individuals to their blood relatives, who may be effectively deprived of privacy and the right to self-determination. Unfortunately, there aren't even enough trained genetic counselors to attend to the number of people who are getting BRCA testing now, let alone with mass population-wide screening.

Furthermore, the current BRCA tests leave a significant number of people (between 3 and 10 percent of those who undergo testing) with the knowledge that they have a mutation on the BRCA gene that may or may not be harmful, what is called a "variant of unknown significance." But we all have any number of genetic mutations, many of which are benign and do not increase our risk of cancer. As Dr. King herself says, "The challenge is not identification, but interpretation, of making sense of [the BRCA mutations that are] identified."

Dr. King's solution to these mysterious mutations? Only tell patients about those mutations which we know to be linked to cancer and exclude all variants of unknown significance, which knowing about could, in Dr. King's words, "increase confusion and compromise clinical management." Rather than simplifying matters, this proposal underscores the deep problems of population-wide BRCA testing without genetic counseling; women whose test reports finding no mutation "with definitive effect on cancer risk" may be misled about their cancer risk when information about variants of unknown significance is omitted. To suggest that we simply not inform women about mutations we don't fully understand creates a whole different ethical dilemma and is a violation of the right to full information about the medical procedures to which we consent.

It is wrong to ask people to undergo genetic testing without receiving professional counseling to explain the limitations of the test and informing them of the test's results—even the ambiguous results. Limiting population-wide testing to only variants we currently understand is ultimately an inadequate way to address the gaps in Dr. King's proposal.

It is sobering that Dr. King's drastic proposal is based on one small study in Israel of Ashkenazi Jewish men, an ethnic group known to have a much higher prevalence of BRCA mutations. While BRCA mutations are rare in the general population (only about one in six-to-eight hundred women), they are common among Ashkenazi Jews (about one in forty women). And, despite common misperceptions about inherited breast cancer risk, only an estimated 5 to 10 percent of breast cancers are even linked to BRCA mutations.

Initiating mass genetic testing for a rare genetic mutation that is linked to only a fraction of all breast cancers is not going to stem the breast cancer epidemic. And we must question a public health strategy that will cost hundreds of millions of dollars—even after last year's Supreme Court ruling opened up BRCA testing, commercial tests still start at $1,000 a person and up—to find people who are at increased risk of cancer, but may never get cancer. Just an initial screening of the current population of women over thirty would cost $105 billion.

As the head of an organization dedicated to public health interventions to address and end the breast cancer epidemic, if Dr. King's proposal would make a dent in the breast cancer epidemic, even at a cost of $105 billion, I'd be among the first to advocate for it. But there are so many other parts of the breast cancer–care continuum that could benefit from even a fraction of $105 billion. With that kind of money, we could give all women who need it access to affordable, high-quality, culturally competent care; we could fund research to further our understanding of how to treat metastasis, and prevent it in the first place; we could address the appalling, vast racial disparities in breast cancer death rates; and we could even fund research to help understand why some women with BRCA mutations never get cancer, and thus learn more about how to prevent breast cancer.

I am all for access to genetic testing when medically indicated—in fact, Breast Cancer Action was the only breast cancer organization to join the lawsuit challenging Myriad Genetics for their patent stranglehold on the BRCA mutations, which we won at the US Supreme Court last summer. And I agree with Dr. King that women need better risk assessment tools for breast cancer, one piece of which includes genetic and inherited risk.

But the answer to the breast cancer epidemic does not rest with mass genetic testing—and the more we look for answers in individual women's bodies, the less we focus on the societal, systemic factors driving the epidemic in the first place.

27. Welcome, Freshmen: DNA Swabs, Please

Troy Duster

When the University of California at Berkeley announced a project that asks incoming freshmen for the fall of 2010 to voluntarily submit their DNA for genetic testing, the plan quickly garnered national attention and generated heated debate. *The New York Times* carried its account in the May 19 issue—and within twenty-four hours the Center for Genetics and Society (on the West Coast) and the Council for Responsible Genetics (on the East Coast) called for a halt to the project.

One of the plan's sponsors, Mark Schlissel, Berkeley's dean of biological sciences, said he has been surprised by the furor and negative response from some quarters. Schlissel noted that the project had been cleared by the campus institutional review board that assesses risks and benefits to human subjects. After having their DNA analyzed, the incoming students would be given information about three areas of their genetic code that would inform them of their tolerance for alcohol, lactose (in milk), and possible need for folic acid (in leafy greens). The website announcing the project proudly proclaimed that "the information Berkeley students will glean from their genetic analysis can only lead to positive outcomes." (According to a recent article at Genome Web, both Stanford and Duke Universities are considering even larger projects with their students.)

Why has this seemingly benign educational and health project become controversial? The answer lies in a better understanding of the social context in which human genetic research is carried out. Direct-to-consumer

Original publication by Troy Duster. "Welcome, Freshmen. DNA Swabs, Please." *The Chronicle of Higher Education.* May 28, 2010. Reprinted by permission of Troy Duster.

genetic testing, to which this project is tied, was begun on hotly contested terrain. In the early years, protocols required a genetic counselor or a physician to interpret test results for patients and consumers. In recent years, ancestry testing companies have proliferated, unregulated, often providing wildly conflicting answers about "origins." When it comes to informing subjects about their potential health risks, the terrain is even more controversial. The California Department of Public Health actually stepped in to caution companies to stop short of even seeming to provide implicit medical analysis and advice. Bear in mind that those cases involved consumers who themselves initiated and then voluntarily submitted their DNA samples.

The House of Representatives Energy and Commerce Committee is investigating the claims made by consumer-genetic-testing companies. And as the Council for Responsible Genetics put it in a letter to Berkeley administrators:

> The American Medical Association, the American Society for Human Genetics, and the American Clinical Laboratory Association have all issued strong statements against direct-to-consumer genetic testing and recommended that a genetics expert be involved in ordering and interpreting genetic tests, consumers be made fully aware of the capabilities of genetic tests, the scientific evidence on which tests are based be available and stated so that the consumer can understand it, the laboratories conducting the tests be accredited, and consumers be made aware of privacy issues associated with genetic testing.

The UC Berkeley DNA project is being promoted with the explicit goal of introducing students to a future of personalized medicine, via the analysis of an individual's genes. But one of the most important things we have learned in the last two decades of human molecular genetics is that there is high variability in the clinical expression of even single genes. In lay terms, this means that two people with the same genetic structure can and often do have very different manifestations of the phenomenon under investigation. Two subjects with the identical delta F508 gene for cystic fibrosis can have remarkably different outcomes—one with relatively mild symptoms, the other with debilitating lung problems requiring continuous medical treatment. The same is true for sickle-cell anemia and many other genetic conditions. That means that there are important environmental triggers that can cause the gene to express itself in different ways.

Two people with the same genetic variant can and do have different triggers that shape differential expressions as phenotype. Even a sophisticated and well-read human subject getting the results of a scientific project from a university-approved lab is likely to treat this information not as a complex interactive feedback loop but as a personalized finding of a level of susceptibility. The person could well be motivated to act upon this apparently precise information by reorienting his or her behavior in a certain direction. A planned public lecture by research scientists to explain the findings is hardly a substitute for personalized interpretation from a clinician or a genetic counselor.

Another problem with the project is that it violates the well-established principle of protecting human subjects, namely, by ignoring or trivializing the power/status relationship between the subject of the research and those in charge of the project. More than a half century of case law has drawn attention to extreme cases, such as research on imprisoned subjects and prisoners of war. But the continuum is long, and far more subtle cases are also covered, as when employers ask employees or lab assistants to participate as research subjects; when professors ask their students to participate; or when doctors ask their patients to enroll in studies with which the doctors are associated or from which they receive compensation. When refusal to participate can have the effect of an increased sense of vulnerability, institutional review boards have routinely flashed a red light.

The Berkeley project falls into this category. Not only are incoming students being asked by faculty members and administrators to participate, but a decision to decline also means that they will not be on the same proverbial page as their fellow students who do participate. In this situation, there is subtle social coercion in the underlying scaffolding of the decision. Moreover, just to underscore the links, the announcement informs students that, among "alternatives to participation," they are free to seek out similar information from private companies such as 23andMe.

Then there is yet another problem. The project announcement acknowledges a 1 percent error rate for such tests. Assuming that 3,000 of the 5,500 incoming students take the test (x 3 would be 9,000 tests), 90 tests would be incorrect, but because data will be anonymized (only the individual tested will have the matching bar code) no one will know whose results are erroneous, not even the lab technicians. One of the ethical dilemmas of research on human subjects is that while anonymization of samples prevents breaches of privacy or unauthorized use of the data, it also prevents follow-up counseling or notification.

There is a strong tendency for molecular geneticists to reduce attitudes toward the world of science into the binary of "for" or "against." That leaves little room for nuanced, engaged critiques.

In the late 1990s, I served as a member of the National Advisory Council to the Human Genome Project and later chaired its Ethical, Legal, and

Social Issues Advisory Group. In that capacity, I attended a large national conference in Washington of scientists and policy makers. At one point, a scientist brought the house down with laughter and prolonged applause when he said from the podium something like this: By 2010, when all the Ethical, Legal, and Social Issues people have gotten out of our way, we will have better understood and solved a host of human genetic disorders, provided therapies for them, and harmed no human subjects along the way.

In the early 1990s, articles filled the scientific literature and spilled into the mass media about the discovery of "the gay gene" and then "the obesity gene" and then "the violence gene." In all those cases, pundits predicted the application of this new knowledge to daily life. There was deafening silence from leading geneticists, who should have cautioned the public that these discoveries wouldn't simply and quickly revolutionize policy and practices regarding obesity or violent crimes. Other members of the academic community did counsel caution and suggested integrated approaches to better understand complex behavioral outcomes. It is now common knowledge that reductionist notions of "the gene for" this or that have been almost completely abandoned in favor of a more modest, nuanced understanding of emerging fields of epigenetics and proteomics.

A recent well-publicized ruling on the use of genetic tests highlights the minefields of ethical and legal issues in how genetic tests are used and reported. The Havasupai Indians of Arizona have just received a favorable settlement in their suit against Arizona State University. Scientists had given tribal members the impression that their genetic research centered on susceptibility to diabetes. While some subjects had signed a rather vague consent form that included language covering "other behavioral problems"—the overwhelming majority believed that the work on their stored blood samples would help in their struggles with widespread diabetes. The institutional review board of Arizona State University had approved the research project—even though the researchers were looking at genetic susceptibility for schizophrenia and working on genetic information about migration patterns.

I do not mean to imply that the current Berkeley project has any semblance of a relationship to the Havasupai story on matters of consent, save for where I began: The social context of research on people is delicate. As Schlissel, the Berkeley dean, has acknowledged, we need a full-scale exploration of the scientific and social issues surrounding such research, but we need it long before gene tests are offered to students.

Might a Berkeley freshman get a false understanding of his or her reaction to alcohol, or intolerance for milk, or need for kale? Well, possibly, but that is not the main point. Rather, the substantial intellectual risk is that they'll be institutionally introduced into misunderstanding the precision, interpretation, and historically problematic execution of such research, and the subtle, unexamined undercurrent of coercion in their participation. Until students have a firm comprehension of all those aspects, such projects shouldn't be planned.

Stanford, Duke, and other universities, take note: There are far more sensible and effective ways of introducing incoming students to their classmates and to their college.

28. Me Medicine

Donna Dickenson

[For some years, advocates of personalized genetic medicine—also known as "precision medicine—have been making lofty claims about imminent breakthroughs in the very nature of medical care.] Francis Collins, Director of the US's National Institutes of Health, guides us through the [anticipated] upheaval in his [2011] book *The Language of Life—DNA and the Revolution in Personalized Medicine*. As he puts it, "We are on the leading edge of a true revolution in medicine, one that promises to transform the traditional 'one size fits all' approach into a much more powerful strategy that considers each individual as unique and as having special characteristics that should guide an approach to staying healthy. But you have to be ready to embrace this new world."

This seismic shift toward genetic personalized medicine promises to give each of us insight into our deepest personal identity—our genetic selves—and let us sip the elixir of life in the form of individually tailored testing and drugs. But can we really believe these promises?

Genetic personalized medicine isn't the only important new development. Commercial ventures like private blood banks play up the uniqueness of your baby's umbilical-cord blood. Enhancement technologies like deep-brain stimulation—"Botox for the brain"—promote the idea that you have a duty to be the best "me" possible. In fact, modern biotechnology is increasingly about "me" medicine, the "brand" being individual patients' supposed distinctiveness.

But all these technologies remain more hype than reality—and sometimes dangerous hype. Personalized genetic testing [has been investigated] by the US Congress and the US Food and Drug Administration for misleading customers into thinking we know much more than we actually do about the link between particular genes and the probability of developing particular illnesses.

Likewise, privately banked cord blood has been shown to be clinically *less effective* than *publicly* banked and pooled blood, leading to two skeptical reports from leading obstetricians' associations warning against its routine collection at childbirth. And enhancement technologies, supposedly enabling us to become "transhuman," have attracted much publicity, but remain largely speculative.

Credit for the greatest advances in human health and longevity over the last two centuries should go to "we" medicine, not "me" medicine. Public health and sanitation programs, polio and smallpox vaccinations, and tuberculosis screening in schools and workplaces have contributed the most to improved health in the Western world and beyond.

But when parents buy into scares linking childhood vaccines to autism, when media pundits scoff at public health measures to prevent swine flu from spreading, or when a UK researcher claims that "the scourge of aging is worse than smallpox," vaccination, epidemic prevention, and screening fall by the wayside. Conversely, there is an unchallenged and unthinking preference for "me" medicine, partly because it pushes all the right buttons in our psyches, the ones marked "choice," "individuality," and "special."

The new biomedicine was originally funded and promoted as a *public* health initiative that would benefit all of us. Hopes for widespread cures were high when the Human Genome Project—financed by a private medical charity, a UK research council, and the US National Institutes of Health—was completed ten years ago.

Instead, one-fifth of the human genome is now subject to *private* patents, meaning that patients can't afford tests for genes that cause cancer, and researchers can't make progress if another team owns the patents on the genes that they want to study. What went wrong?

Part of the answer has to do with the genetic mystique: the notion that I simply *am* my genes, and that's why I'm unique. The genetic mystique plays on the individualism of Western culture and seems to give it a scientific basis. There are also powerful commercial interests at stake, meaning that research frequently concentrates on genetic links to diseases whose diagnosis and treatment will produce the greatest profit, rather than the greatest reduction in global mortality.

Some genetically personalized treatments may well be signs of progress, such as *pharmacogenetics*, which promises drug regimes tailored to the patient's own genome. If this new technology works, it could lessen the direst side-effects of chemotherapy for cancer care: oncologists would no longer have to prescribe one-size-fits-all regimes if patients who are genetically more receptive to the drug could be differentiated and given lighter regimes.

But the high cost of developing new drugs means that pharmaceutical companies need substantial patient markets to make their investments profitable. Will minority ethnic groups miss out?

For a while it looked as if the reverse would happen, with niche markets in pharmacogenetics targeting ethnic minorities. Race-based medicine hit the scene in 2005 when the FDA approved BiDil, the first drug to treat a specific racial group—African Americans suffering from heart failure. But there was no real clinical evidence that the drug worked better in African Americans, and it has been withdrawn from the market.

Nevertheless, major advances like the Human Genome Project have certainly geneticized medicine: there is a growing popular tendency to define all conditions as genetically determined. And that means that public health measures are likely to be neglected in favor of individual genetic scans or personalized genetic-testing services. Genetic tests, if properly administered, can save lives, but they also tend to create a feeling that the responsibility for your health rests with *you*, the individual patient.

The genetic mystique, the legally doubtful view that we own our bodies, and the widening reach of the market in our lives lead many to believe that "me" medicine is the only game in town. When health care is paid for by individuals or their employers, "me" medicine becomes a natural outlook.

But, like the drunk who looks for his lost keys only under the streetlight, biomedicine is in danger of concentrating only where the glare is brightest—not on the most effective health interventions, but on the most personalized and profitable, which nowadays go hand-in-hand.

29. Public Health in the Precision-Medicine Era

Ronald Bayer and Sandro Galea

That clinical medicine has contributed enormously to our ability to treat and cure sick people is beyond contention. But whether and to what extent medical care has transformed morbidity and mortality patterns at a population level and what contribution, if any, it has made to the well-being and life expectancy of the least-advantaged people have been matters of contention for more than a century. This debate has taken on renewed importance as the scientific leadership at the National Institutes of Health (NIH), National Academy of Medicine, and US universities have taken up the challenge of personalized or precision medicine. It is a challenge given all the more salience by President Barack Obama's announcement in his State of the Union address that his administration would seek to fund a major new initiative. Responding to the President's words, Harold Varmus, director of the National Cancer Institute, and Francis Collins, director of the NIH, have written that "What is needed now is a broad research program to build the evidence base needed to guide clinical practice."[1]

The enthusiasm for this initiative derives from the assumption that precision medicine will contribute to clinical practice and thereby advance the health of the public. We suggest, however, that this enthusiasm is premature. "What is needed now" is quite different if one views the world from the perspective of the broad pattern of morbidity and mortality, if one is concerned about why the United States has sunk to the bottom of the list of comparable countries in terms of disease experience and life expectancy,

From *New England Journal of Medicine,* Ronald Bayer and Sandro Galea, "Public Health in the Precision-Medicine Era," 373, no. 6, 499–501. Copyright © (August 6, 2015) Massachusetts Medical Society. Reprinted with permission from Massachusetts Medical Society.

or if one is troubled by the steep social gradient that characterizes who becomes sick and who dies. The burgeoning precision-medicine agenda is largely silent on these issues, focusing instead on detecting and curing disease at the individual level.

So is this approach indeed "What is needed now"? Our skepticism about what precision medicine has to offer is predicated on both a reading of the evidence regarding social determinants of population health and recognition of what that evidence means for the priorities that should guide our investments to advance public health and reduce health inequities.

There are many frameworks that recognize multiple contributors to the production of population health. In a 2008 report, the World Health Organization Commission on Social Determinants of Health articulated the importance to population health of improving daily living conditions and tackling the inequitable distribution of money, power, and resources. The centrality of social policy and public health approaches to improving population health and reducing health inequities is emphasized in several publications from the Centers for Disease Control and Prevention (CDC),[2] and these aims were recognized as critical twin goals in the Healthy People 2010 and 2020 plans.

The evidence that clinical intervention, however important, cannot remedy health inequalities emerges from a broad range of empirical studies. Perhaps most convincingly, the Whitehall Studies of the British Civil Service in the United Kingdom revealed that even when health care services were provided as a matter of right and the cost of care was no longer a barrier to treatment, a marked social gradient persisted, as a substantial proportion of the population fared poorly on health indicators. Moreover, the inequity did not manifest simply as a gap between the rich and privileged and the poor and disenfranchised: people at every income level did better than those at the level just below them. In 1980, the Black Report on health inequality concluded that "Thirty years of the welfare state and the National Health Service have achieved little in reducing social inequality in health." In sum, there is now broad consensus that health differences between groups and within groups are not driven by clinical care but by social-structural factors that shape our lives.

Yet seemingly willfully blind to this evidence, the United States continues to spend its health dollars overwhelmingly on clinical care. It is therefore not surprising that even as we far outpace all other countries in spending on health, we have poorer health indicators than many countries, some of them far less wealthy than ours. In 2013, the National Research Council (NRC) and the Institute of Medicine (IOM) issued a bleak report on life expectancy and well-being in the United States. Shorter Lives, Poorer Health documented the extent to which Americans were at a disadvantage at every stage of the life cycle as compared with their counterparts in peer countries. In terms of birth outcomes, heart disease, motor vehicle accidents and violence, sexually transmitted disease, and chronic lung disease, Americans fared worse than residents of all other high-income countries. Only for people over seventy-five years of age was the story better.

In attempting to explain this state of affairs, the NRC–IOM report considered the problematic question of access to health care services for the uninsured and poorly insured in the United States—but then dismissed those gross inequities as a possible explanation for the US disadvantage. "Even if health care plays some role," said the report, "decades of research have documented that health is determined by far more than health care." Amplifying this message, a recent analysis concluded that "In many ways the American health care system is the most advanced in the world. But whiz-bang technology just cannot fix what ails us."[3]

It is against this backdrop that the claims of the most ardent defenders of precision medicine must be read. Francis Collins has written that "the 21st century is the century of biology. The nation that invests in biomedical research will reap untold rewards in its economy and the health of its people."[4] Research undertaken in the name of precision medicine may well open new vistas of science, and precision medicine itself may ultimately make critical contributions to a narrow set of conditions that are primarily genetically determined. But the challenge we face to improve population health does not involve the frontiers of science and molecular biology. It entails development of the vision and willingness to address certain persistent social realities, and it requires an unstinting focus on the factors that matter most to the production of population health.

Unfortunately, all the evidence suggests that we, as a country, are far from recognizing that our collective health is shaped by factors well beyond clinical care or our genes. The NIH's most recent Estimates of Funding for Various Research, Condition, and Disease Categories report shows, for example, that total support in fiscal year 2014 for research areas including the words "gene," "genome," or "genetic" was about 50 percent greater than funding for areas including the word "prevention." Our investment in public health infrastructure, including local health departments, lags substantially behind that of other high-income countries, and the CDC's annual budget is dwarfed (by a factor of about five) by investment in the NIH, even as the latter pursues an approach increasingly focused on science and treatments that aim to promote individual health. The proportion of

NIH-funded projects with the words "public" or "population" in their title, for example, has dropped by 90 percent over the past ten years, according to the NIH Reporter.

Our public investments in broad, cross sectoral efforts to minimize the potential effect of such foundational drivers of poor health as poverty and racial residential segregation are pitifully few in comparison with those of other countries. Perhaps unsurprisingly, recent high-profile police shootings of minority men have triggered civil rights protests across the United States, reflecting widespread dissatisfaction with unequal distribution of resources within a deeply divided society.[5]

Without minimizing the possible gains to clinical care from greater realization of precision medicine's promise, we worry that an unstinting focus on precision medicine by trusted spokespeople for health is a mistake—and a distraction from the goal of producing a healthier population.

NOTES

1. F.S. Collins and H. Varmus, "A New Initiative on Precision Medicine," *New England Journal of Medicine* 372 (2015): 793–95.

2. Conclusion and Future Directions: CDC Health Disparities and Inequalities Report—United States, 2013, supplement 3, *MMWR Surveillance Summaries* 62 (2013): 184–86.

3. E. Porter, "Income Inequality Is Costing the U.S. on Social Issues," *New York Times*, April 28, 2015, www.nytimes.com/2015/04/29/business/economy/income-inequality-is-costing-the-us-on-social-issues.html.

4. F.S. Collins. "Exceptional Opportunities in Medical Science: A View from the National Institutes of Health, *JAMA* 313 (2015): 131–32.

5. M.T. Bassett, "#BlackLivesMatter—A Challenge to the Medical and Public Health Communities," *New England Journal of Medicine* 372 (2015): 1085–87.

PART VI

Seeking Humanity in Human Subjects Research

Human subjects research has been a key concern within the field of bioethics, from its earliest days to the present. Guiding principles for such research have been laid out in landmark documents, including the Nuremberg Code (1947), the Declaration of Helsinki (1964), the Belmont Report (1979), and the United Nations Educational, Scientific, and Cultural Organization's *Universal Declaration on Bioethics and Human Rights* (2005). Many countries, including the United States, have adopted detailed regulations and encouraged the establishment of institutional review boards to ensure that these guidelines are followed.

Despite all this, significant problems and outright scandals in human subjects research continue to surface, particularly among vulnerable populations such as prisoners, people living in the Global South, those living precariously in the Global North, and those serving in the military. The persistence of such abuses is often inadequately acknowledged, as in the *Wikipedia* entries for "Unethical human experimentation" and "Unethical human experimentation in the United States" (accessed November 11, 2016). Despite their length, these entries mention only a couple of incidents after 1980, when the commercialization of biotechnology took off.

Contributions in this section reveal some of these modern failures in human subjects research and examine the pressures and incentives in the research profession—for example, economic, career, status—that drive them.

Allen Hornblum's and Osagie Obasogie's commentary, "Medical Exploitation: Inmates Must Not Become Guinea Pigs Again," criticizes efforts by the Institute of Medicine to reintroduce medical research in US prisons. The authors note that prisoners, who are not permitted to make decisions about most aspects of their daily lives, "from when to eat to when

273

to sleep," are hardly in a position to give voluntary consent to taking part in a medical experiment. As with other vulnerable populations, the conditions in which they live make "a mockery of informed consent, medical research's foundational principle."

In "The Body Hunters," Marcia Angell, former editor-in-chief of the *New England Journal of Medicine*, reviews *The Constant Gardener*, a film based on the novel by John le Carré. It's the story of a global pharmaceutical company that, in testing a dangerous drug on unsuspecting poor children in Kenya, fails to report resulting deaths and murders Westerners who uncover evidence of the scandal and try to blow the whistle. Angell uses the review as a springboard for describing the "ordinary drug company practices" that "distort research to make their drugs look better and safer than they are," and the conditions under which clinical trials are conducted in poor regions of the real world.

Carl Elliott's "Guinea-Pigging" explores the world of the unemployed, college students, the formerly incarcerated, and others (mostly on the margins) who are trying to make a living by enrolling in clinical trials that test the safety of new drugs. In past years, most clinical trials were conducted in medical schools and teaching hospitals, but now more than 70 percent take place in the private sector, leading Elliot to the question, "What happens when both parties involved in a trial see the enterprise primarily as a way of making money?"

In "Human Enhancement and Experimental Research in the Military," Efthimios Parasidis looks at US military experiments through the eyes of their human subjects. He begins with the government's involvement during the mid-to-late twentieth century in unethical, if not illegal, experimental research on American civilians and service members; mustard gas, nuclear weapons, and psychotropic drugs were used. He then examines the military's recent involvement in two areas: its "efforts to mandate use of medical products for off-label or investigational purposes, and its emphasis on developing biotechnologies that seek to facilitate the cognitive and physical enhancement of service members."

The section concludes with "Non-Consenting Adults," a commentary by Harriet Washington about the numerous and ongoing instances of breaches of consent in a variety of research settings. Washington summarizes recent examples that show how marginalized groups in the United States are "being coerced into studies that violate their right to consent." This includes the waiver given to the military that allowed it to dose millions of troops with experimental anthrax vaccines between 1990 and 2005; a 1996 law that allows "nonconsensual research on trauma victims on the pretext that they are unconscious and unable to give their consent"; and Pfizer's non-consensual administration of an experimental meningitis drug on children in Nigeria. Invoking the Nuremberg Code, which was set up after the Nazi atrocities of World War II to protect the human subjects of research, Washington suggests that it is "time we live up to our stated ideals."

30. Medical Exploitation

Inmates Must Not Become Guinea Pigs Again

Allen M. Hornblum and Osagie K. Obasogie

Any day now the US Department of Health and Human Services may decide to turn back the clock to a time when doctors went unchallenged, medical investigators could do no wrong, and vulnerable people were grist for the research mill.

... [In 2006,] the Institute of Medicine (IOM) released a controversial report recommending the return of federally funded medical research to our nation's prisons. Propelled by historical amnesia and corporate greed, a resurgence of such research may do much harm.

Although Tuskegee (black sharecroppers), Fernald (orphans), and Willowbrook ([developmentally disabled] children) are infamous examples of how medical researchers exploited vulnerable populations, prisoners were scientists' guinea pigs of choice during the twentieth century.

Prisoners across the country were routinely incorporated into dangerous medical experiments that were unthinkable for other populations: testicular transplants and radiation studies, injections of live cancer cells, dioxin slatherings, and exposure to psychotropic chemicals and mind-control agents. University of Pennsylvania researchers set up labs inside Holmesburg Prison for easy access.

Commercial interests, the military, and the CIA were behind many of these dubious initiatives. It wasn't until the late 1970s that policy makers curbed these brutal practices.

Original publication by Allen M. Hornblum and Osagie K. Obasogie. "Medical Exploitation: Inmates Must Not Become Guinea Pigs Again." *The Philadelphia Inquirer.* September 13, 2007. Reprinted by permission of the authors.

The IOM now thinks that new "guidelines," institutional "transparency," and increased "monitoring" would safeguard today's prisoners from past transgressions. Such views are wildly optimistic.

Prisons are unusual institutions. Oppressive, paramilitary, and sequestered from society, they are the reason the Nuremberg Code's first principle precludes those in "constrained" and "coercive" environments from participating in medical research.

Perhaps even more troubling than the recommendation itself is how the IOM came to it.

First, and most shocking, the IOM admits to having visited only one prison during its two-year investigation. How is it possible to make sound policy decisions without taking a thorough look at the conditions faced by those most affected? By failing to acknowledge that nearly every aspect of prisoners' daily lives—from when to eat to when to sleep—is imposed at the barrel of a gun, the IOM committee makes a mockery of informed consent, medical research's foundational principle.

Second, the committee based its decision on a review of articles about trends in ethics since the late 1970s, when current restrictions on research with prisoners were put in place. But decisions of such consequence cannot be based solely on changes in the academic wind. What also needs to be considered is whether the appalling conditions giving rise to the current protections have been eliminated. And all evidence suggests that they have only gotten worse.

Last, the committee isolates its inquiries from other moral commitments relevant to prisoners' well being—namely, human rights. Vesting internationally agreed upon human rights in every person and creating ethical standards for medical research are two sides of the same coin. But with the wide-ranging human-rights violations in today's prisons—including sexual assault and decrepit living conditions—attempts to isolate medical research from human-rights standards can lead one ethical norm to undermine the other, exposing prisoners to even greater abuse.

Bioethical dilemmas involving prisoners will be with us for some time. South Carolina, for example, is considering a proposal to relieve its shortage of kidneys for transplant by shaving 180 days off inmates' sentences if they agree to become donors.

Human biotechnology also might come into play; given the shortage of eggs available to pursue certain types of stem-cell research, it's not difficult to imagine similar incentives being offered to incarcerated women to become egg donors.

These are complicated issues with remarkably high stakes. Medical research with human subjects can retain its legitimacy only if it recognizes its deep kinship with human rights. Regrettably, the impending Health and Human Services decision to loosen restrictions on prison research leads us in the wrong direction.

31. The Body Hunters

Marcia Angell

Shortly before I started work on my book *The Truth About the Drug Companies: How They Deceive Us and What to Do About It*,[1] a friend gave me John le Carré's new novel, *The Constant Gardener*, and urged me to read it right away. I did as I was told, and found the tale apposite, to put it mildly.

The villain is a global pharmaceutical company called Karel Vita Hudson (KVH). The heroine, Tessa Quayle, is the wife of a low-level British diplomat stationed in Nairobi, Kenya. She stumbles across evidence that KVH is testing a dangerous tuberculosis drug, called Dypraxa, on powerless and unsuspecting poor Africans, and not reporting the resulting deaths. When she threatens to expose the company, she is brutally murdered, and the British government colludes in the cover up. Her husband, Justin Quayle, a seemingly docile civil servant at first, becomes obsessed with finding out why his wife was murdered and by whom. He finally does, and at the end of the book, he too is murdered. In between the deaths, we follow Justin's gradual awakening to the ruthless activities of a corporation too powerful to be accountable to anyone.

Now *The Constant Gardener* has been released as a film, starring Ralph Fiennes and Rachel Weisz and directed by Fernando Meirelles. It is both better and worse than the book. Visually, it is stunning. The many aerial shots of Kenya show the stark beauty and sweep of the African countryside, and the film also conveys in its urban scenes the miserable overcrowding and hopeless poverty in Nairobi, something the book only suggests.

Where the film most improves on the book is in its treatment of the main characters. Fiennes and Weisz portray the relationship between Tessa and Justin as touching and believable, something the book fails to do. Le Carré presents Justin as self-contained to the point of inertness and seemingly with no serious interests beyond his garden. It hardly seems plausible that such a man would throw over his career, and risk his life, to investigate the death of his wife. In this film, Justin is revealed as not so much passive and narrow as controlled and quietly determined. And Weisz portrays Tessa, a passionately uninhibited champion of the poor and downtrodden, as a shrewder and more perceptive woman than the one we find in le Carré's book; she does not share her discoveries about the drug companies with her husband for fear of compromising him. As in the book, Tessa's murder takes place at the beginning, and we come to know her through flashbacks. But Fiennes's face at hearing of her death, controlled and virtually immobile, somehow manages to convey the enormity of his loss as well as his determination to find out the truth about her death.

The film falls far short of the book, however, in telling us what KVH (for some reason renamed KDH in the film) was up to; it never explains why every institution that might have interfered with the company, including the British government, was colluding with it. We only get hints. We are told in passing that KDH and the people it controlled coerced poor Africans into acting as guinea pigs by denying them medical care unless they took part in company experiments; but we learn little about the rules which prevent that sort of coercion in prosperous countries but not in poor ones. We're told that deaths were covered up—literally; bodies were thrown into a lime pit and their existence denied. But we learn little about why that was done, or why companies conduct clinical trials (that is, tests on human beings) in the first place, and why they find it advantageous to do so in Africa.

Since the film tells us very little about the motives of the drug company, we are left with a story that has plenty of passion and intrigue but is played out in something of a historical vacuum. In fact, most viewers would probably conclude that insofar as we do learn anything about KDH, its deadly practices are wildly implausible, in no way representative of real drug company behavior. After all, in the real world, we don't hear of pharmaceutical whistle-blowers being murdered, and there have been several whistle-blowers recently.

But le Carré himself cautions us against drawing any such conclusions. In an author's note at the end of the book he makes a grudging disclaimer to the effect that no person or organization in the book is based on an actual person or organization. He also makes it clear, however, that he is obliged to say this "in these dog days when lawyers rule the universe." He adds, "But

I can tell you this. As my journey through the pharmaceutical jungle progressed, I came to realize that, by comparison with reality, my story was as tame as a holiday postcard."

Quite so. Le Carré obviously did careful research, and the book is rich in details about ordinary drug company practices. Without being pedantic, he has his characters explain how drug companies distort research to make their drugs look safer and more effective than they are; how they can get away with this more easily in poor regions of the world; and how they use their vast wealth to influence governments and the medical profession and any other institutions that might interfere with their single-minded pursuit of profits. On the basis of the research I did for my book I believe that most of the background facts about drug company behavior in *The Constant Gardener,* however hard to believe, are correct.

Yet the story is based on the premise that a pharmaceutical company would be so threatened by disclosures of its activities that it would have someone killed. That is what is fantasy. In fact, many of the practices that so horrified le Carré's heroine are fairly standard and generally well known and accepted. They seldom provoke outrage, let alone murder. A company like KDH would not kill someone like Tessa even if it were willing to do so; it wouldn't have to. Her concerns would have seemed isolated and futile, and the companies would hardly have taken notice of them.

There is no question that the United States and other rich countries have been conducting more and more clinical research in Africa and other parts of the third world. Although exact figures are hard to come by, it is likely that tens of thousands of studies sponsored by first-world drug companies and governments are now under way in Africa, parts of Latin America and Asia, and the former Soviet Union. Most of this research is intended to find new treatments for use in well-to-do countries. After all, that is where the paying customers are. In this sense, third-world countries are being used as laboratories for first-world needs. Relatively few studies are devoted to finding treatments for the diseases that plague poor countries, such as malaria, sleeping sickness, and schistosomiasis. The big companies are more interested in the usual first-world conditions, like high cholesterol, obesity, and arthritis.

The rapid movement of drug studies to third-world countries began in 1980, when the US Food and Drug Administration (FDA), in considering applications to approve new drugs, first agreed to accept foreign trials as evidence of safety and effectiveness. Before a company can sell a drug in the United States (or market an old drug for a new use), it must get approval from the FDA, which means it must demonstrate in clinical trials that the drug is reasonably safe and effective. Nearly every large drug company, wherever it is located, wants to get into the US market, because that is the major source of profit for pharmaceuticals.

Probably close to half of all clinical trials are now conducted in the third world, although there is no way to know for sure. The reasons are clear. It is cheaper and in many respects easier and faster to do them there. A huge new industry has arisen that conducts third-world research for drug companies (like le Carré's fictional research firm, ThreeBees). These companies, called contract research organizations, or CROs, hire local doctors to find people who will take part in clinical trials, and while the payments to the doctors per patient are lower than in first-world studies, by local standards they are munificent. Doctors can multiply their income tenfold or more. Patients, too, are readily enticed by small amounts of money and promises of free care. In fact, as in le Carré's story, enrolling in a trial may be the only way they can get any care at all.

This system makes a mockery of the notion of informed consent—the requirement that subjects be given full information about the nature of the research and have the right to refuse to participate, without penalty or consequences for their usual health care. That requirement is enforced in the United States and other well-to-do countries, and partly for that reason, drug companies are having a hard time getting enough volunteers for the growing number of clinical trials. Not so in the third world, where authoritarian regimes and corrupt local government officials and health authorities are eager to be paid off by first-world organizations and to have good relations with them. They "encourage" entire villages or provinces to enroll in research programs, while local doctors enrich themselves by providing human subjects.

Perhaps the most important reason for conducting human research in Africa and other poor regions outside the United States is that it is a way of circumventing FDA regulations. In the United States, drug companies are required to file "investigational new drug applications" (INDs) with the FDA before they begin human testing of a drug they hope to get approved. The applications give detailed descriptions of the proposed research, including plans for obtaining informed consent and for monitoring the progress of the study. Companies must also provide evidence that ethics committees (called institutional review boards, or IRBs) have been set up to review each clinical trial. These committees are supposed to ensure that risks to human subjects are, in the words of the applicable federal regulations, "reasonable in relation to anticipated benefits, if any, to subjects, and the importance of the knowledge that may reasonably be expected to result," and further, that all risks are "minimized." The FDA can deny approval of the IND or request changes in the proposed research. It may also conduct on-site inspections of the trials.

The requirements for foreign research are much looser. In fact, the FDA may not even know about such trials until after they are completed, when the company applies for final approval of a new drug. Only then—when there is no longer an opportunity to verify the information—does the company have to describe the way in which the research was conducted, or say whether there was ethics committee approval and informed consent. Furthermore, the FDA rarely conducts on-site inspections abroad. While it conducts very few in the United States, there is always the possibility that it will decide to do so. For research done in the third world, the agency simply takes the word of the sponsors of the research.

When research does not require FDA approval, there may be no oversight at all. Companies can conduct preliminary studies of drugs in poor countries before formal testing even begins. Quite literally, the participants in their studies are used as guinea pigs, subjects of research that really should be done on experimental animals. That was the case in le Carré's fictional account. Although some research in the United States and other wealthy countries also escapes formal oversight, there are generally more restrictions on what researchers can get away with.

Several real third-world clinical trials were described in detail in a six-part *Washington Post* series in 2000, called "The Body Hunters." One of them has parallels to le Carré's story. In 1996, Nigeria was in the grip of a widespread epidemic of bacterial meningitis, which eventually claimed over fifteen thousand lives. Pfizer, the world's biggest drug company, was at that time conducting the largest research program it had ever undertaken to get FDA approval for a new antibiotic called Trovan. Eventually the drug was tested on thirteen thousand people in twenty-seven countries. When one of Pfizer's doctors heard about the epidemic in Nigeria, he immediately got approval from Nigerian authorities to bring a team to Kano, a city of two million people in northern Nigeria, to test Trovan in children with meningitis. The aim was to demonstrate that oral Trovan would work as well in these children as an established fast-acting intravenous antibiotic.

Within six weeks, the Pfizer team had set up its program in the squalid Kano Infectious Diseases Hospital at the center of the epidemic. A local doctor was named as principal investigator, and the team rapidly enrolled two hundred children for the study. Half were given Trovan, many of them in pill or drink form. The other half were given injections of ceftriaxone, an antibiotic known to be effective against epidemic meningitis. Two weeks later, at the end of the trial, an equal number of children had died in both groups. The new drug was apparently just as effective as the old one, and it could be given in oral form. On that basis, Pfizer applied to the FDA for approval to market oral Trovan for use in children with meningitis.

Those are the bare outlines of what happened. But critics such as Médecins Sans Frontières, the Nobel Prize–winning international medical relief organization, charge that this was exactly the sort of study that would not have been permitted in the United States. To these critics, it was unethical to test an experimental drug orally in the midst of an epidemic. The usual treatment for meningitis in such urgent conditions would be intravenous antibiotics. In fact, the Pfizer doctor who organized the study told *The Washington Post* that antibiotics "would never be used like that in the United States. The standard is IV therapy." It was also charged that there had not been adequate preliminary research into how Trovan is absorbed and metabolized by children or how effective it is against meningitis.

Furthermore, to lessen the pain of injections, the dose of intravenous ceftriaxone given for comparison was much smaller than originally planned. But if the dose of the comparison drug were inadequate, that would make Trovan look better than if it were compared with a full dose of ceftriaxone. Pfizer maintained that the smaller dose was still more than sufficient, but the medical director of Hoffmann-La Roche, the manufacturer of ceftriaxone, was quoted as saying, "A high dose is essential."

Questions were also raised about whether the subjects had given informed consent and whether an ethics committee had approved the trial. The families did not sign informed consent documents, but the company maintained they had consented orally. However, some doctors and family members disputed this. A laboratory technician in Kano said, "The patients did not know if it was research or not. They just knew they were sick."

Even more troubling was the issue of ethics committee approval. A Pfizer spokeswoman told *The Washington Post* that the research had been approved by a Nigerian ethics board. But a month later, the *Post* found that the lead Nigerian researcher admitted to creating and backdating the approval document. According to the *Post*, the document was typed on the letterhead of the Aminu Kano Teaching Hospital and dated March 28, 1996 (six days before the trial began), but the researcher said that he actually wrote it about a year later. Pfizer reportedly gave the document to the FDA in 1997 during an audit of records supporting its application for approval of Trovan. The hospital's medical director told the *Post* the document was "a lie." In fact, he said, the hospital didn't even have an ethics board at the time the trial was done.

In 1997, Trovan was approved by the FDA to treat certain infections, but not for children and not for epidemic meningitis. The FDA found dozens of discrepancies in the documents from Nigeria. Trovan quickly became a highly profitable antibiotic widely used against a variety of infections. However, after less than two years on the market, there were over a hundred reports that the drug produced liver toxicity, causing several deaths, and it is no longer sold.

In 2001, the families of thirty Nigerian children who either died or suffered serious injury in the Kano trial filed suit against Pfizer in a New York federal district court. They alleged that the company increased the risk of death and injury by failing to provide a treatment of proven efficacy for children who did not respond to Trovan and by giving the patients used for comparison a weakened version of ceftriaxone. They also complained that they had not given informed consent. According to the complaint,

> Pfizer took the opportunity presented by the chaos caused by the civil and medical crises in Kano to accomplish what the company could not do elsewhere—to quickly conduct on young children a test of a potentially dangerous antibiotic.

During the following four years, Pfizer argued that the case should not be heard in a US court at all. In August [2005], Southern District of New York judge William H. Pauley III agreed, ruling that Nigeria, not the United States, was the proper place to try a lawsuit over Pfizer's conduct of the Trovan trial. The families plan to appeal.

Drug companies are not the only sponsors of research in the third world that wouldn't be allowed at home. In the 1990s, two government agencies, the National Institutes of Health (NIH) and the Centers for Disease Control (CDC), sponsored some nine clinical trials in the third world in which thousands of HIV-infected pregnant women under study were given a placebo (or sugar pill) instead of the drug AZT, even though the latter had been shown to cut by 70 percent the risk of transmission of HIV/AIDS from mother to infant. The aim of the studies was to see whether a shorter, simpler course of treatment might be as effective. The standard course required taking oral AZT for the last trimester of pregnancy, an intravenous infusion of the drug during labor and delivery, and oral treatment of the newborn for six weeks. There was preliminary evidence that oral treatment limited to the last few weeks of pregnancy and the first few days of the newborn's life might also be effective.

But instead of comparing the transmission rate in women who received an experimental short course of treatment with that in women receiving the standard course, the researchers compared it with the transmission rate in women who received only a placebo—thus consigning many babies in their care to be born with HIV/AIDS that could have been prevented. This was justified as being the fastest way to show whether a short course was reasonably effective, but designing the trials in that way certainly wasn't scientifically necessary, and, in any case, it would never have been permitted in the United States.

In 1997, in an article in *The New England Journal of Medicine*, Peter Lurie and Sidney M. Wolfe of Public Citizen's Health Research Group protested that the trials should have compared short courses of treatment with the standard long one, not a placebo.[2] As executive editor of the journal, I wrote an accompanying editorial in support of their view ("The Ethics of Clinical Research in the Third World"). The public reaction was intense. Many in the US research establishment, including the directors of the NIH and CDC, vigorously defended the trials, pointing out that the women denied AZT probably wouldn't have been able to obtain it where they lived anyway. But that argument, which was meant to mitigate criticism of the trials, simply underscored the fact that the NIH and CDC were willing to take advantage of the women's poverty and vulnerability. The researchers could easily have supplied the drug to all the women they enrolled (and for whom they thereby assumed responsibility), even if it wasn't widely available in the region.[3]

Some writers who comment on medical ethics are not so much concerned with the design and conduct of particular trials in the third world as with the legitimacy of carrying on research there in the first place. They believe it is virtually impossible to conduct medical research ethically in poor countries, because it is inherently exploitative. Although this seems to me too broad a judgment, I believe the amount of research sponsored by the first world in the third world should be sharply curtailed. It is driven too much by the search for profits. It offers quick answers precisely because it is so easy to cut corners.

Before a study is exported to the third world, two important questions should be asked. First, would it be possible to do the research in the first world? And second, why is it being diverted to poor countries? It is sometimes claimed that research should be done where health needs are greatest—and that is certainly the case in the third world. But this view confuses research with treatment. There is a great need to apply the *results* of research, wherever it is conducted, to the treatment of people in the third world. Unfortunately, that is not what happens. Research findings are applied predominantly in well-to-do countries even when the research is done in poor ones. The only clinical research that clearly needs to be conducted in the third world is research on third-world diseases. Such work is amply justified, and far more of it is needed. Unfortunately, it is not a high priority either for the pharmaceutical industry or the National Institutes of Health.

In my view, research should not be done in the third world unless it concerns diseases that are virtually confined to those regions. And regulations governing research in poor countries should be every bit as stringent—and enforced just as vigilantly—as in well-to-do countries. There is no justification for the present situation in which the standards are looser precisely where human subjects are most vulnerable.

Before research on human subjects is undertaken anywhere in the world, there should be adequate animal studies and preliminary tests on normal subjects to eliminate all unnecessary risks. Consent should be truly informed, and there should be no penalties for refusing to participate or undue inducements to do so. It is not enough to claim that informed consent was oral. It should be documented. If there is any doubt about whether the information given to subjects was understood, subjects should be asked to repeat their understanding of the research. To be on the safe side, that conversation and their consent could be videotaped. Companies should no longer be allowed to conduct research in the third world that they would not be permitted to do at home.

Le Carré seems bleak about the chances of any such reform. At the end of the novel, both Quayles are dead, no one is called to account, and KVH and all the people who serve its interests in and out of government presumably continue undeterred. The film, however, adds a Hollywood-style hint of justice to come. At Justin's funeral in London, Tessa's cousin, in whom she had confided, reveals in a eulogy for the couple just why they were killed. Reporters scribble in their notebooks and race for phones, while Sir Bernard Pellegrin, the unctuous and complicit director of affairs for Africa of the British Foreign Office, looks increasingly uneasy and finally flees in consternation. In adding this unlikely scene, the film writers did a disservice to le Carré's book, but that is a small fault. The larger one is in not making it clear, as le Carré did so well, exactly what Tessa Quayle was unhappy about.

NOTES

1. Random House 2004. See also my essay "The Truth about the Drug Companies," *New York Review,* July 15, 2004.

2. "Unethical Trials of Interventions to Reduce Perinatal Transmission of the Human Immunodeficiency Virus in Developing Countries," *New England Journal of Medicine,* September 18, 1997.

3. This controversy was discussed in detail in David Rothman's "The Shame of Medical Research," *New York Review,* November 30, 2000.

32. Guinea-Pigging

Carl Elliott

On September 11, 2001, James Rockwell was camped out in a clinical-research unit on the eleventh floor of a Philadelphia hospital, where he had enrolled as a subject in a high-paying drug study. As a rule, studies that involve invasive medical procedures are more lucrative—the more uncomfortable, the better the pay—and in this study subjects had a fiber-optic tube inserted in their mouths and down their esophagi so that researchers could examine their gastrointestinal tracts.

Rockwell had enrolled in many previous studies at corporate sites at places like Wyeth and GlaxoSmithKline. But the atmosphere there felt professional, bureaucratic, and cold. This unit was in a university hospital, not a corporate lab, and the staff had a casual attitude toward regulations and procedures. "The Animal House of research units" is what Rockwell calls it. "I'm standing in the hallway juggling," he says. "I'm up at five in the morning watching movies." Although study guidelines called for stringent dietary restrictions, the subjects got so hungry that one of them picked the lock on the food closet. "We got giant boxes of cookies and ran into the lounge and put them in the couch," Rockwell says. "This one guy was putting them in the ceiling tiles." Rockwell has little confidence in the data that the study produced. "The most integral part of the study was the diet restriction," he says, "and we were just gorging ourselves at 2 A.M. on Cheez Doodles."

On the morning of September 11th, nearly a month into the five-week study, the subjects gathered around a television and watched the news of the terrorist attacks through a drug-induced haze. "We were all high on Versed after getting endoscopies," Rockwell says. He and the other subjects

Original publication by Carl Elliott. "Guinea-Pigging." *The New Yorker.* January 7, 2008. Reprinted by permission of © Condé Nast.

began to wonder if they should go home. But a mass departure would have ruined the study. "The doctors were, like, 'No, no!'" Rockwell recalls. "'No one's going home, everything's fine!'" Rockwell stayed until the end of the study and was paid seventy-five hundred dollars. He used the money to make a down payment on a house.

Rockwell is a wiry thirty-year-old massage-therapy student with a pierced nose; he seems to bounce in his seat as he speaks, radiating enthusiasm. Over the years, he estimates, he has enrolled in more than twenty studies for money. The Philadelphia area offers plenty of opportunities for aspiring human subjects. It is home to four medical schools and is part of a drug-industry corridor that stretches into New Jersey. Bristol-Myers Squibb regularly sends a van to pick up volunteers at the Trenton train station.

Today, fees as high as the one that Rockwell received aren't unusual. The best-paying studies are longer, in-patient trials, where subjects are often required to check into a research facility for days or even weeks at a time, so that their diet can be controlled, their blood and urine checked regularly, and their medical status carefully monitored. Occasionally, they also undergo invasive procedures, like a bronchoscopy or a biopsy, or something else unpleasant, such as being deprived of sleep, wearing a rectal probe, or having allergens sprayed in their faces. Because such studies require a fair amount of time in a research unit, the subjects are usually people who need money and have a lot of time to spare: the unemployed, college students, contract workers, ex-cons, or young people living on the margins who have decided that testing drugs is better than punching a clock with the wage slaves. In some cities, like Philadelphia and Austin, the drug-testing economy has produced a community of semiprofessional research subjects, who enroll in one study after another. Some of them do nothing else. For them, "guinea-pigging," as they call it, has become a job. Many of them say that they know people who have been traveling around the country doing studies for fifteen years or longer. "It's crazy and it's sad," one drug-trial veteran told me. "For me, this is not a life. But it is a life for a lot of these people."

· · ·

Most drug studies used to take place in medical schools and teaching hospitals. Pharmaceutical companies developed the drugs, but they contracted with academic physicians to carry out the clinical testing. According to the *New England Journal of Medicine,* as recently as 1991 80 percent of industry-sponsored trials were conducted in academic health centers. Academic health centers had a lot to offer pharmaceutical companies: academic researchers who could design the trials, publications in academic journals that could help market the products, and a pool of potential subjects on whom the drugs could be tested. But, in the past decade, the pharmaceutical industry has been testing more drugs, the trials have grown more complex, and the financial pressure to bring drugs to market swiftly has intensified. Impatient with the slow pace of academic bureaucracies, pharmaceutical companies have moved trials to the private sector, where more than 70 percent of them are now conducted.

This has spurred the growth of businesses that specialize in various parts of the commercial-research enterprise. The largest of the new businesses are called "contract research organizations," and include Quintiles, Covance, Parexel, and PPD (Pharmaceutical Product Development), a company that has operations in thirty countries, including India, Israel, and South Africa. (About 50 percent of clinical trials are now conducted outside the United States and Western Europe.) These firms are hired to shepherd a product through every aspect of its development, from subject recruitment and testing through FDA approval. Speed is critical: a patent lasts twenty years, and a drug company's aim is to get the drug on the shelves as early in the life of the patent as possible. When, in 2000, the Office of the Inspector General of the Department of Health and Human Services asked one researcher what sponsors were looking for, he replied, "No. 1—rapid enrollment. No. 2—rapid enrollment. No. 3—rapid enrollment." The result has been to broaden the range of subjects who are used and to increase the rates of pay they receive.

Most professional guinea pigs are involved in phase I clinical trials, in which the safety of a potential drug is tested, typically by giving it to healthy subjects and studying any side effects that it produces. (Phase II trials aim at determining dosing requirements and demonstrating therapeutic efficacy; phase III trials are on a larger scale and usually compare a drug's results with standard treatments.) The better trial sites offer such amenities as video games, pool tables, and wireless Internet access. If all goes well, a guinea pig can get paid to spend a week watching "The Lord of the Rings" and playing Halo with his friends, in exchange for wearing a hep-lock catheter on one arm and eating institutional food. Nathaniel Miller, a Philadelphia trial veteran who started doing studies to fund his political activism, was once paid fifteen hundred dollars in exchange for three days and two GI endoscopies at Temple University, where he was given a private room with a television. "It was like a hotel," he says, "except that twice they came in and stuck a tube down my nose."

The shift to the market has created a new dynamic. The relationship between testers and test subjects has become, more nakedly than ever, a business transaction. Guinea pigs are the first to admit this. "Nobody's doing this out of the goodness of their heart," Miller says. Unlike subjects in later-stage clinical trials, who are usually sick and might enroll in a study to gain access to a new drug, people in healthy-volunteer studies cannot expect any therapeutic benefit to balance the risks they take. As guinea pigs see it, their reason for taking the drugs is no different from that of the clinical investigators who administer them, and who are compensated handsomely for their efforts. This raises an ethical question: what happens when both parties involved in a trial see the enterprise primarily as a way of making money?

· · ·

In May of 2006, Miami-Dade County ordered the demolition of a former Holiday Inn, citing various fire and safety violations. It had been the largest drug-testing site in North America, with 675 beds. The operation closed down that year, shortly after the financial magazine *Bloomberg Markets* reported that the building's owner, SFBC International, was paying undocumented immigrants to participate in drug trials under ethically dubious conditions. The medical director of the clinic got her degree from a school in the Caribbean and was not licensed to practice. Some of the studies had been approved by a commercial ethical-review board owned by the wife of an SFBC vice president. (The company, which has since changed its name to PharmaNet Development Group, says that it required subjects to provide proof of their legal status, and that the practice of medicine wasn't part of the medical director's duties. Last August, the company paid $28.5 million to settle a class-action lawsuit.)

"It was a human-subjects bazaar," says Kenneth Goodman, a bioethicist at the University of Miami who visited the site. The motel was in a down-trodden neighborhood; according to later reports, paint was peeling from the walls, and there were seven or eight subjects in a room. Goodman says that the waiting area was filled with potential subjects, mainly African American and Hispanic; administrative staff members worked behind a window, like gas-station attendants, passing documents through a hole in the glass.

The SFBC scandal was not the first of its kind. In 1996, the *Wall Street Journal* reported that the Eli Lilly company was using homeless alcoholics from a local shelter to test experimental drugs at budget rates at its testing site in Indianapolis. (Lilly's executive director of clinical pharmacology told the journal that the homeless people were driven by "altruism," and that they enrolled in trials because they "want to help society." The company says that it now requires subjects to provide proof of residence.) The Lilly clinic, the journal reported, had developed such a reputation for admitting the down-and-out that subjects traveled to Indianapolis from all over the country to participate in studies.

How did the largest clinical-trial unit on the continent recruit undocumented immigrants to a dilapidated motel for ten years without anyone noticing? Part of the answer has to do with our system of oversight. Before the 1970s, medical research was poorly regulated; many phase I subjects were prisoners. Reforms were instituted after congressional investigations into abuses like the four-decade Tuskegee syphilis studies, in which researchers studied, instead of treating, syphilis infections in African American men. For the past three decades, institutional review boards, or IRBs, have been the primary mechanism for protecting subjects in drug trials. FDA regulations require that any study in support of a new drug be approved by an IRB Until recently, IRBs were based in universities and teaching hospitals, and were made up primarily of faculty members who volunteered to review the research studies being conducted in their own institutions. Now that most drug studies take place outside academic settings, research sponsors can submit their proposed studies to for-profit IRBs, which will review the ethics of a study in exchange for a fee. These boards are subject to the same financial pressures faced by virtually everyone in the business. They compete for clients by promising a fast review. And if one for-profit IRB concludes that a study is unethical, the sponsor can simply take it to another.

Moreover, because IRBs scrutinize studies only on paper, they are seldom in a position to comment on conditions at a study site. Most of the standards that SFBC violated in Miami, for example, would not be covered in an ordinary off-site ethics review. IRBs ask questions like "Have the subjects been adequately informed of what the study involves?" They do not generally ask if the sponsors are recruiting undocumented immigrants or if the study site poses a fire hazard. At some trial sites, guinea pigs are housed in circumstances that would drive away anyone with better options. Guinea pigs told me about sites that skimp on meals and hot water, or that require subjects to bring their own towels and blankets. A few sites have a reputation for recruiting subjects who are threatening or dangerous but work cheaply.

Few people realize how little oversight the federal government provides for the protection of subjects in privately sponsored studies. The Office for Human Research Protections, in the Department of Health and Human Services, has jurisdiction only over research funded by the department. The FDA oversees drug safety, but, according to a 2007 HHS report, it conducts "more inspections that verify clinical trial data than inspections that focus on human-subject protections." In 2005, FDA inspectors were finally given a code number for reporting "failure to protect the rights, safety, and welfare of subjects," and an agency spokesman says that they plan to make more human-subject safety inspections in the future, but so far they have cited only one investigator for a violation. (He had held a subject in his research unit against her will.) In any case, the FDA inspects only about one per cent of clinical trials.

Most guinea pigs rely on their wits—or on word of mouth from other subjects—to determine which studies are safe. Some avoid particular kinds of studies, such as trials for heart drugs or psychiatric drugs. Others have developed relationships with certain recruiters, whom they trust to tell them which studies to avoid. In general, guinea pigs figure that sponsors have a financial incentive to keep them healthy. "The companies don't give two shits about me or my personal well-being," Nathaniel Miller says. "But it's not in their interest for anything to go wrong." That's true, but companies also have an interest in things going well as cheaply as possible, and this can lead to hazardous tradeoffs.

The most notorious recent disaster for healthy volunteers took place in March 2006, at a testing site run by Parexel at Northwick Park Hospital, outside London; subjects were offered two thousand pounds to enroll in a phase I trial of a monoclonal antibody, a prospective treatment for rheumatoid arthritis and multiple sclerosis. Six of the volunteers had to be rushed to a nearby intensive-care unit after suffering life-threatening reactions—severe inflammation, organ failure. They were hospitalized for weeks, and one subject's fingers and toes were amputated. All the subjects have reportedly been left with long-term disabilities.

The Northwick Park episode was not an isolated incident. Traci Johnson, a previously healthy nineteen-year-old student, committed suicide in a safety study of Eli Lilly's antidepressant Cymbalta in January of 2004. (Lilly denies that its product was to blame.) I spoke to an Iraqi living in Canada who began doing trials when he immigrated. He was living in a hostel and needed money to buy a car. A friend told him, "This thing is like fast cash." When he enrolled in an immunosuppressant trial at a Montreal-based subsidiary of SFBC, he found himself in a bed next to a subject who was coughing up blood. Despite his complaints, he was not moved to a different bed for nine days. He and eight other subjects later tested positive for tuberculosis.

• • •

A decade ago, shortly after I began teaching bioethics and philosophy at the University of Minnesota, I got a phone call from a psychiatrist named

Faruk Abuzzahab. He wanted to know if he could sit in on an ethics class that I was teaching. There had been some trouble in a research study that he had conducted, it seemed, and the state licensing board had ordered him to take a class in medical ethics.

Despite some misgivings about my class being used as an instrument of punishment, I agreed. He seemed affable enough on the phone, explaining that he had been a faculty member at the university before going into private practice, and had once chaired the Minnesota Psychiatric Society's ethics committee.

I did not give much more thought to Abuzzahab until about three years ago, when a for-profit testing site called Prism Research opened in St. Paul. Prism was advertising for healthy subjects in a local alternative weekly. I discovered, on the company's website, that Abuzzahab was one of its researchers. A few more clicks revealed that he was also conducting studies at his private practice, Clinical Psychopharmacology Consultants. I began to wonder what, exactly, the incident was that had brought him to my class.

As it turned out, the disciplinary action was a response to the injuries or deaths of forty-six patients under Abuzzahab's supervision. Seventeen of them had been research subjects in studies that he was conducting. These were not healthy-volunteer studies. According to the board, Abuzzahab had "enrolled psychiatrically disturbed and vulnerable patients into investigational drug studies without ensuring that they met eligibility criteria to be in the study and then kept them in the study after their conditions deteriorated." The board had judged Abuzzahab a danger to the public and suspended his license, citing "a reckless, if not willful, disregard of the patients' welfare."

One case, which was reported in the *Boston Globe*, concerned a forty-one-year-old woman named Susan Endersbe, who had struggled for years with schizophrenia and suicidal thoughts. She had been doing well on her medication, however, until Abuzzahab enrolled her in a trial of an experimental antipsychotic drug. In the trial, she was taken off her regular medication and became suicidal. When Abuzzahab gave her a day pass to leave the hospital unsupervised, she threw herself into the Mississippi River and drowned. In another case cited by the board, Abuzzahab had prescribed a "large supply of potentially lethal medications" to a woman with a history of substance abuse, "shortly after a serious suicide attempt." She committed suicide by taking an overdose.

The public portion of Abuzzahab's disciplinary file is freely available from the Minnesota licensing board, and has been posted on the website of Circare, a watchdog group that documents research abuse. When I ran a Google search on "Faruk Abuzzahab," the first hit I got was a 1998 article in the *Globe* on his trial disasters. Yet none of this seems to have derailed Abuzzahab's research career. Even after his suspension, the *Times* has reported, he continued to supervise drug trials, and to receive payments from at least a dozen drug companies. In 2003, the American Psychiatric Association awarded him a Distinguished Life Fellowship.

The US regulatory system is built on the tacit assumption that the main threat to research subjects comes from overly ambitious academic researchers, who might be tempted to gamble with subjects' health in the pursuit of medical knowledge or academic fame. The system was intended to check this sort of intellectual ambition, mainly by insuring that studies are reviewed in advance by boards made up of the researcher's academic peers. But, like most physicians supervising clinical trials today, Abuzzahab does not work in an academic setting. The studies conducted at for-profit sites such as Prism are not the natural domain of academically ambitious researchers. They are rarely published and, even if they were, would bring little intellectual credit to the physicians carrying them out, because they are designed by the industry sponsor. A researcher like Abuzzahab would not become famous by supervising subjects in studies like these. But he might become rich.

Abuzzahab represents a new, entrepreneurial breed of physician-researcher; in fact, many of his colleagues have moved even farther from the academic realm. In 1994, according to the Tufts Center for Drug Development, 70 percent of clinical researchers were affiliated with academic medical centers. By 2006, that figure had dropped to 36 percent. The work can be lucrative, and some sponsors offer researchers additional financial incentives to recruit subjects. One doctor told the Department of Health and Human Services that he was offered twelve thousand dollars for each subject that he could enroll in a trial, plus a thirty-thousand-dollar bonus and an additional six thousand dollars per subject after the first six.

Some of the people conducting clinical trials have little training in how to conduct research. And, as the Abuzzahab case suggests, not all drug companies are especially selective about the researchers they hire. In 2001, the FDA asked the pharmaceutical company Sanofi-Aventis to perform new studies of the antibiotic Ketek, which was suspected of causing liver failure. Reports later revealed that the top-recruiting investigator hired by PPD, the firm contracted to conduct the studies, was a graduate of an offshore medical school who tested the antibiotic on clients in an obesity clinic she ran in Alabama. She was sentenced to five years in federal prison for fraud. Another top recruiting investigator was arrested when the police found him carrying a loaded semiautomatic handgun, and hiding cocaine in his underwear.

. . .

In early December of 2002, a man named Bob Helms took part in an industry-sponsored "drug delivery" study. Helms and his fellow guinea pigs were required to take a new antianxiety drug and, later, to defecate into a small basket. The unfortunate clinic staff members then searched for the remains of the tablet to determine how much had been absorbed by the body.

The guinea pigs were paid thirty-three hundred dollars and were required to live in the unit for five periods of four days each. But before the end of the first period, Helms says, the guinea pigs decided that they were getting a raw deal. The process of fecal collection was smelly and unpleasant; the amount of time allowed outside the unit had been shortened from three days to thirty-six hours; and the subjects were required to abstain from alcohol, even though the study—because of unexpected delays—was taking place over the Christmas and New Year's holidays. The guinea pigs wanted a raise.

Since the staff was collecting their feces, Helms suggested that the guinea pigs all swallow notes that said "More money." This idea was rejected. Instead, they presented a one-page memo to the staff, detailing their concerns and requesting a pay increase of eleven hundred dollars. When the memo was ignored, they began hinting that they might decamp for a better-paying study at another site. Eventually, the clinic agreed to pay each subject an additional eight hundred dollars.

Helms is a pioneer in the world of guinea-pig activism. A fifty-year-old housepainter and former union organizer, he has a calm, measured demeanor that masks a deep dissident streak. Before he started guinea-pigging, in the nineteen-nineties, he worked as a caregiver for mentally retarded adults living in group homes. There Helms began to understand the difficulties in organizing health care workers who were employed by the same company but in far-flung locations—in this case, group homes that were spread over two hundred miles of suburbs. "The other organizers told me right off the bat that I could not organize workers who might meet each other once a year at best," Helms says. "How could we ask them to take risks together? They were strangers."

Helms saw that guinea pigs faced a similar problem, and, in 1996, he started a jobzine for research subjects called *Guinea Pig Zero*. With a mixture of reporting, advocacy, and dark humor (a cartoon in an early issue shows a young man surrounded by IV bags and syringes, exclaiming, "No more fast food work for me—I've got a career in science!"), *Guinea Pig Zero* published the sort of information that guinea pigs really wanted to know—how well a study paid, the competence of the venipuncturist, the quality of the food. It even published report cards, grading research units from A to F. "Overcrowding, no hot showers, sleeping in an easy chair, incredibly cheap shit for dinner, creepy guys from New York jails—all these are a poor man's worries," Helms says. "Where are these things in the regulators' paperwork?" *Guinea Pig Zero* was not aimed at sick people who sign up for studies in order to get new treatment. It was aimed at poor people who sign up for studies in order to get money.

And here is where its perspective diverged most radically from the traditional ethical perspective. *Guinea Pig Zero* assumed that subjects should get more money, while many ethicists and regulators argued that they should get none at all. The standard worry expressed by ethicists is that money tempts subjects to take part in dangerous, painful, or degrading studies against their better judgment. FDA guidelines instruct review boards to make sure that payment is not "coercive" and does not exert an "undue influence" on subjects. It's a reasonable worry. "If there were a study where they cut off your leg and sewed it back on and you got twenty thousand dollars, people would be fighting to get into that study," a Philadelphia activist and clinical-trial veteran who writes under the name Dave Onion says.

Of course, ethicists generally prefer that subjects take part in studies for altruistic reasons. Yet, if sponsors relied solely on altruism, studies on healthy subjects would probably come to a halt. The result is an uneasy compromise: guinea pigs are paid to test drugs, but everyone pretends that guinea-pigging is not really a job. IRBs allow sponsors to pay guinea pigs, but, consistent with FDA guidelines, insist on their keeping the amount low. Sponsors refer to the money as "compensation" rather than as "wages," but guinea pigs must pay taxes, and they are given no retirement benefits, disability insurance, workmen's compensation, or overtime pay. And, because so many guinea pigs are uninsured, they are testing the safety of drugs that they will probably not be able to afford once the drugs have been approved. "I'm not going to get the benefit of the health care that is developed by this research," Helms says, "because I am not in the economic class to get health insurance."

Guinea pigs can't even count on having their medical care paid for if they are injured in a study. According to a recent survey in the *New England Journal of Medicine,* only 16 percent of academic medical centers in the United States provided free care to subjects injured in trials. None of them compensated injured subjects for pain or lost wages. No systematic data are available for private testing sites, but the provisions typically found in consent forms are not encouraging. A consent form for a recent study of Genentech's immunosuppressant drug Raptiva told participants that they would be treated for any injuries the drug caused, but stipulated that "the cost of such treatment will not be reimbursed."

Some sponsors withhold most of the payment until the studies are over. Guinea pigs who drop out after deciding that a surgical procedure is too disagreeable, or that a drug seems unpleasant or dangerous, must forfeit the bulk of their paycheck. Two years ago, when SFBC conducted a two-month study of the pain medication Palladone, it offered subjects twenty-four hundred dollars. But most of that was paid only after the last of the study's four confinement periods. A guinea pig could spend nearly two months in the study, including twelve days and nights in the SFBC unit, and get only six hundred dollars. SFBC even reserved the right to withhold payments from subjects whom it dropped from the study because of a drug's side effects.

Guinea-pig activists recognize that they are indispensable to the pharmaceutical industry; a guinea-pig walkout in the middle of a trial could wreak financial havoc on the sponsor. Yet the conditions of guinea-pigging make any exercise of power difficult. Not only are those in a particular trial likely to be strangers, if they complain to the sponsor about conditions, they risk being excluded from future studies. And, according to *Bloomberg*, when illegal-immigrant guinea pigs at SFBC talked to the press, managers threatened to have them deported.

Lawsuits on behalf of injured subjects are growing, though, and they have begun to target not just research sponsors but also institutional review boards and bioethicists. Alan Milstein, an attorney in Philadelphia, has pioneered this area of law, most notably with successful litigation against the University of Pennsylvania on behalf of the family of Jesse Gelsinger, who died in a gene-therapy trial in 1999. Milstein has represented volunteers injured at commercial sites, but most guinea pigs are in no position to hire a lawyer. "This is not something you or I do," Milstein says. "This is something the poor do so that the rich can get better drugs."

· · ·

During our early years of medical school, my classmates and I were given a course in physical diagnosis. Usually, we practiced on one another. Each of us would percuss a classmate's chest, or listen to his heart with a stethoscope. But some procedures were considered too personal to practice on a classmate. For some of these, we were assigned a "model patient"—someone from the community who was "compensated" in exchange for undergoing an examination.

This was how I performed my first rectal exam. A large group of us were led into a room, where our model patient was bent over an examining table with his pants around his ankles. One by one, we approached him nervously from behind, inserted a gloved, lubricated finger into his rectum, and felt around for the prostate. "Thank you," we all said politely to the model patient as we removed our index fingers from his anus. The model patient stared straight ahead, saying nothing.

What made the experience oddly disturbing was not just the forced, pseudo normality of the instruction, or the fact that the exam could have been done more privately, but the instrumentality of the encounter: a pretend "patient" bending over naked for anonymous strangers in exchange for money. The fact that the model patient had been paid did not make his work seem any less degrading. (Tipping him would have made it even worse.)

Perhaps there is something inherently disconcerting about the idea of turning drug testing into a job. Guinea pigs do not do things in exchange for money so much as they allow things to be done to them. There are not many other jobs where that is the case. Meanwhile, our patchwork regulatory system insures that no one institution is keeping track of how many deaths and injuries befall healthy subjects in clinical trials. Nobody appears to be tracking how many clinical investigators are incompetent, or have lost their licenses, or have questionable disciplinary records. Nobody is monitoring the effect that so many trials have on the health of professional guinea pigs. In fact, nobody is even entirely certain whether the trials generate reliable data. A professional guinea pig who does a dozen drug-safety trials a year is not exactly representative of the population that will be taking the drugs once they have been approved.

The safety of new drugs has always depended on the willingness of someone to test them, and it seems inevitable that the job will fall to people who have no better options. Guinea-pigging requires no training or skill, and in a thoroughly commercial environment, where there can be no pretense of humanitarian motivation, it is hard to think of it as meaningful work. As Dave Onion puts it, "You don't go home and say to yourself, 'Now, that was a good day.'"

33. Human Enhancement and Experimental Research in the Military

Efthimios Parasidis

INTRODUCTION

This article examines experimental research in the US military through the eyes of the human subject. It explores the egregious legal and ethical violations committed by military researchers in the mid-to-late twentieth century and evaluates investigational studies that have shadowed military medicine for the past two decades. At a time when the US military is actively pursuing transformative biomedical and technological innovations, analyzing the history of misfeasance in military research informs contemporary discussion as to the extent to which legal and regulatory reforms are desirable.

Modern military medicine has evolved from its traditional role of "preserving the fighting force,"[1] to enhancing it through application of novel biotechnologies. Current research sponsored by the US Department of Defense (DoD) and the Defense Advanced Research Projects Agency (DARPA) includes drugs that can keep soldiers awake for seventy-two hours or more, a nutraceutical that fulfills a soldier's dietary needs for up to five days, and sophisticated brain-to-computer interfaces that endeavor to permit human-to-human and human-to-computer communication via thought alone.[2]

In addition to increased health risks associated with enhancement techniques, a number of challenging legal and bioethical issues have been insuf-

This article was first published in volume 44, issue 4, of the *Connecticut Law Review* (April 2012): 1117–1132. Reprinted by permission of the author. Due to space limitations, portions of this chapter have been deleted or edited from the original.

ficiently explored. Should enhancements be a mandatory aspect of military service? Who determines the parameters for an acceptable risk-benefit profile? What remedies should be available for service members who experience adverse health effects? Adequately addressing these concerns, particularly in the context of military hierarchy and demography, provides a sociomedical framework that facilitates sensible harmonization of national security interests with fundamental notions of human dignity and patient autonomy.

A HISTORY OF UNCONSCIONABLE RESEARCH

The atrocities committed by German military researchers during World War II challenged the international community to directly address safeguards governing experimental research on human subjects. While the US military played an integral role in the prosecution of the German researchers and the drafting of the Nuremburg Code, the US government failed to publicly disclose its involvement in unethical, if not illegal, experimental research on American civilians and service members.[3] Three examples include studies related to mustard gas, nuclear weapons, and psychotropic drugs.

The mustard gas experiments involved approximately sixty thousand American soldiers in "race-based human experimentation" that sought to determine whether race or skin complexion influences one's susceptibility to injuries from mustard gas.[4] Researchers created "man-break" tests whereby service members were locked in gas chambers that were inundated with mustard gas until the point that the men became incapacitated.[5] During the experiments, some soldiers were exposed to gas levels that were equivalent to those reported on World War I battlefields.[6]

The "man-break" tests caused severe injuries to the service members. Soldiers experienced "immediate and severe eye injuries" and "enormous, grotesque blisters and oozing sores" on their "face, hands, underarms, buttocks, and genitals."[7] Exposure to mustard gas also caused blindness, intense vomiting, internal and external bleeding, and damage to the lungs and respiratory system.[8] Many soldiers suffered long-term health effects that included cancer, asthma, and psychological disorders.[9]

For decades, the US government refused to acknowledge the existence of the studies or provide injured service members with compensation or long-term health care. It was not until 1991—nearly five decades after the first studies began—that the government officially admitted to the use of soldiers in experimental research. The government also admitted that it did not fully disclose safety risks or obtain informed consent from the research participants, and that the service members may have suffered adverse health effects as a result of their participation in the studies.[10]

Contemporaneous with the mustard gas experiments, the US military conducted radiation experiments on American soldiers and civilians.[11] In addition to testing the destructive capabilities of nuclear weapons, military researchers examined the effects of nuclear warfare on humans, animals, and the environment.[12] As early as 1942, the military understood that exposure to radiation was likely to be quite dangerous, since "the deleterious effects of radiation could not be seen or felt and the results of overexposure might not become apparent for long periods after such exposure."[13]

After years of detonating atomic weapons in the South Pacific, the military began open-air testing of nuclear weapons on American soil in the 1950s.[14] Thousands of soldiers were placed, without protective clothing, in the immediate vicinity of atomic detonations. The military did not inform the soldiers of potential health risks or seek to obtain informed consent prior to participation in the trials.

While the military publicly denied any potential harm to humans, plants, or animals, internal documents indicate that government officials had determined that there existed a causal relationship between radiation exposure and serious adverse health effects.[15] Despite the health and environmental hazards, the commissioner of the US Atomic Energy Commission privately asserted that "we must not let anything interfere with this series of tests—nothing."[16] It was later revealed that radiation exposure at the test sites was comparable to that of Hiroshima and Nagasaki.[17]

Coupled with the open-air nuclear tests, the military funded studies at a number of well-respected American universities, including the University of Chicago and the University of California, whereby researchers injected unsuspecting civilians with radioactive elements that included plutonium, uranium, and polonium.[18] This work continued through the 1970s, with researchers targeting the elderly, patients in mental institutions, prisoners, and others "who did not have full faculties for informed consent."[19] A congressional investigation later found that "no evidence was elicited that informed consent was granted in any of the cases," and that "the government covered up the nature of the experiments and deceived the families of deceased victims. . . ."[20]

In the 1990s, the government acknowledged that hundreds of thousands of American service members had been involved in at least 1,400 radiation projects over a thirty-year period during and after World War II.[21] These figures do not include exposure suffered by American civilians in connection with hundreds of "intentional radiation releases," where researchers deliberately emitted radioactive substances into densely populated cities and other locations to test human response and environmental contamination.[22] Although the government was aware that the radiation releases were likely to contaminate food and water supplies, many of the releases "took place with no public awareness or understanding."[23] Within ten years after the commencement of the detonations in America, childhood leukemia deaths and diagnoses, as well as adult cancer deaths and diagnoses, were exponentially higher in several detonation regions.[24]

Along with the mustard gas and radiation experiments, the US military engaged in decades of classified research, beginning in the 1940s and continuing through the 1970s, to ascertain whether psychotropic drugs could be used as chemical weapons or interrogation-facilitating agents.[25] The products under investigation included lysergic acid diethylamide (LSD), synthetic mescaline, synthetic marijuana, and over a dozen other drugs. During the early stages of the research, the US military recruited Nazi scientists who had studied and participated in torture and brainwashing.[26] Several of the Germans had been recently identified as war criminals, and the United States falsified documents to conceal their true identities.[27] The military later justified its actions by arguing that national security interests far outweighed any ethical concerns.[28]

The psychotropic drugs were given to service members and civilians without their knowledge or consent.[29] Studies were conducted in military facilities and university medical centers, and many human subjects experienced serious adverse side effects.[30] Internally, the military justified the secret testing on "'unwitting, nonvolunteer' Americans" by arguing that national security interests permit "a more tolerant interpretation of moral-ethical values, but not legal limits."[31] The military went on to argue that legal liability could be avoided by covering up the experiments.[32]

PRESERVING AND ENHANCING THE FIGHTING FORCE

There is nothing to suggest that the US military is currently supporting research that utilizes methods similar to those employed during the mustard gas, radiation, or psychotropic drug experiments. However, recent controversies have highlighted the military's efforts to mandate use of medical products for off-label or investigational purposes, and its emphasis on developing biotechnologies that seek to facilitate the cognitive and physical enhancement of service members.[33] This part will focus on these two areas of research.

Investigational and Off-Label Use of Medical Products

Since at least the 1990s, the US military has required service members to subject themselves to both investigational and off-label use of medical products.[34] Both off-label and investigational use involve utilization of a medical product for an indication that has not earned FDA approval. While each is properly characterized as experimental research because the FDA has not found that the underlying product is safe and effective for the stated use, there is an important distinction between the two categories. For products that are used off-label, the FDA has determined that the product is safe and effective for at least one indication.[35] Investigational medical products, on the other hand, have not been approved for any indication.[36]

Nonconsensual use of off-label or investigational medical products raises a number of serious concerns. While physicians may prescribe drugs for off-label indications or investigational purposes, the decision to do so must be based on an evaluation of a patient's particular health condition and risk factors, and should only occur where medical data reflect meaningful evidence that the potential benefits are likely to outweigh the known or expected risks and the patient provides informed consent to the treatment. In a number of instances, the military has made off-label and investigational use of medical products compulsory for service members as a whole, and has not sought to obtain informed consent or provide adequate risk disclosures to individual soldiers. The discussion below explores four recent examples—pyridostigmine bromide (PB), the botulinum toxoid (BT) vaccine, the anthrax vaccine, and selective serotonin reuptake inhibitors (SSRIs).

After petitioning the FDA to establish a new rule that waives informed consent requirements for investigational use of medical products in times of existing or anticipated combat activities, the DoD sought and obtained permission from the FDA to use PB and the BT vaccine pursuant to the new regulation.[37] Fearing use of chemical weapons during the Gulf War, the military decided to administer PB and the BT vaccine to all soldiers.[38] At the time, the FDA was evaluating the safety and efficacy of both products as pretreatments for chemical warfare.[39]

In its informed consent waiver request to the FDA, the DoD argued that it would not be feasible to obtain informed consent because a soldier's "personal preference" does not supersede the military's view that the drug and vaccine would contribute to the "safety of other personnel in a soldier's unit and the accomplishment of the combat mission."[40] The DoD also argued that "obtaining informed consent in the heat of imminent or ongoing combat would not be practicable."[41]

The FDA granted the DoD's requests, but the decision was not without controversy. The DoD claims that it trusted that the FDA had granted permission to use the investigational drug without informed consent because the FDA believed that the drug was deemed to be safe.[42] The FDA, on the other hand, claims that it granted the waiver because it believed that the DoD determined that military necessity required an informed consent waiver.[43]

Regardless of the reason why the FDA granted the waiver, as a condition of the FDA's permission to use the investigational medical products without informed consent, the DoD agreed to: (1) provide information on PB to all service members; (2) collect, review, and make reports of adverse events related to PB; (3) label PB as an investigational product that was solely for "military use and evaluation"; (4) ensure that each dose of the BT vaccine was recorded in each service member's medical record; and (5) maintain adequate records related to the receipt, shipment, and disposition of the BT vaccine.[44] The DoD failed to comply with each of these requirements.[45]

Following use of PB and the BT vaccine during the Gulf War, veterans began suffering from serious health problems that include cognitive difficulties, chronic headaches, widespread pain, skin rashes, respiratory and gastrointestinal problems, and other chronic abnormalities.[46] Gulf War veterans have been diagnosed with amyotrophic lateral sclerosis (ALS) at a much higher rate than that of the general population or veteran populations from other wars.[47] Children of Gulf War veterans are also born with birth defects at an alarming rate.[48] Commonly referred to as Gulf War illness, these health problems affect over 175,000 Gulf War veterans, which amounts to more than 25 percent of the fighting force during the war.[49] PB is included in the list of factors that are most likely to be a contributing factor to Gulf War illness.[50]

The military's off-label use of vaccines continued after the Gulf War. In 1998, the DoD implemented the Anthrax Vaccine Immunization Program (AVIP), which requires the anthrax vaccine for all service members who are deemed by DoD to be at risk for anthrax exposure.[51] Although the vaccine had earned FDA approval to protect against cutaneous anthrax, the military sought to use the vaccine as a pretreatment for inhalation anthrax.[52]

In 2003, six service members filed a lawsuit seeking to enjoin the military from continuing AVIP because the military did not obtain informed consent prior to inoculations, nor did the DoD obtain a waiver for the informed consent requirements.[53] A federal district court issued a preliminary injunction that halted AVIP.[54] Days later, the FDA approved the anthrax vaccine "independent of the route of exposure," which captured the indication of inhalation anthrax.[55] The court then vacated the FDA's decision on procedural grounds because the agency did not follow its requirement to certify that the vaccine was safe and effective against inhalation anthrax.[56] In essence, the court found it impossible for the FDA to have adequately evaluated the products, pursuant to the statutory requirements, in such a short time period.

Congress stepped in to aid the DoD by enacting the Project BioShield Act of 2004,[57] which granted the FDA the ability to permit off-label or investigational use of medical products during a declared emergency.[58] In turn, the FDA used its newfound power to grant the DoD the ability to continue using the anthrax vaccine.[59] During the time that the DoD was permitted to continue with AVIP pursuant to the emergency order, the FDA approved the vaccine regardless of the route of exposure.[60] Although service members once again challenged the FDA's decision, the Court of Appeals for the DC Circuit dismissed the action because it found that the FDA had not acted arbitrarily or capriciously in approving the new indication during its second review.[61] Since March 1998, over 2,700,000 service members have received the anthrax vaccine.[62] For a number of service members, however, the administration of the vaccine is not reflected in the official medical records maintained by the military.[63]

Today, some of the most pressing medical issues facing service members include traumatic brain injury (TBI), posttraumatic stress disorder (PTSD), and other mental health issues.[64] A decade of intense fighting in Afghanistan and Iraq has resulted in a "substantial mental health burden for war veterans and their families."[65] Blast-related TBI has been labeled the signature injury of the wars, and countless soldiers have reported postconcussive symptoms.[66] Veterans of these wars have required mental health treatment for serious mental disorders much more than veterans of previous wars, and suicide rates for enlisted service members and veterans are at an all-time high.[67]

Increasingly, treatments for depression, TBI, PTSD, and anxiety disorders utilize newer psychotropic medications, particularly selective serotonin-reuptake inhibitors.[68] Military psychiatrists have recommended that physicians in war zones have SSRIs "in large quantities, to be used for both depressive disorders and anxiety disorders."[69] However, a number of studies have questioned the safety and efficacy of SSRIs. Off-label use of SSRIs is particularly troubling, with some studies finding no meaningful clinical benefit and long-term adverse health effects. To the extent that SSRIs are the standard of care for both on-label and off-label indications, and service members are not provided with accurate risk-benefit profiles, such use may place service members at a heightened risk for both short-term and long-term health problems.[70]

Physical and Cognitive Enhancement of Service Members

The fundamental goal of military training is to enhance service members—to make them smarter, stronger, and more able fighters. Increasingly, enhancement techniques have sought to leverage innovative medical products and technologies. As the director of DARPA explains, the agency's goal is to exploit "the life sciences to make the individual warfighter stronger, more alert, more endurant, and better able to heal."[71]

Such endeavors have raised a number of challenging questions. Is there a valid distinction between "artificial" and "natural" enhancement? Under what circumstances should enhancements that are under development be administered to service members? Should medical enhancements ever be a required aspect of service in the military? Examining current enhancement projects helps frame these concerns.

DARPA's "Persistence in Combat" program aims to create soldiers who are "unstoppable because pain, wounds, and bleeding are kept under their control."[72] This program includes research directed at developing a vaccine that will block intense pain within seconds, use of photobiomodulation to accelerate wound healing, and the creation of a chemical cascade to stop bleeding within minutes.[73] The agency's Metabolic Dominance program seeks to create a "'nutraceutical,' a pill with nutritional value that would vastly improve soldiers' endurance."[74] DARPA's vision is "to enable superior physical and physiological performance by controlling energy metabolism on demand. An example is continuous peak physical performance and cognitive function for 3 to 5 days, 24 hours per day, without the need for calories."[75]

Coupled with these programs, "the security establishment's interest and investment in neuroscience, neuropharmacology . . . and related areas [are] extensive and growing."[76] Under the Augmented Cognition program, DARPA seeks to "develop the technologies needed to measure and track a subject's cognitive state in real-time."[77] Another goal is to create brain-to-computer interfaces, whereby soldiers can communicate by thought alone.[78] This includes systems that can relay messages, such as images and sounds, between human brains and machines, or even from human to human.[79] Service members can receive commands via electrodes implanted in their brains, or be wired directly into the equipment they control.[80]

Through implanted electrodes, DARPA is researching whether neurostimulation can improve impaired cognitive performance and reduce the effects of sleep deprivation on soldiers.[81] This research dovetails with two other DARPA endeavors, the Continuous Assisted Performance program and the Applications of Biology to Defense Applications program.[82] The former is "investigating ways to prevent fatigue and enable soldiers to stay awake, alert, and effective for up to seven days straight without suffering any deleterious mental or physical effects and without using any of the current generation of stimulants."[83] The latter incorporates neuroscientific studies such as

> biological approaches for maintaining the warfighter's performance, capabilities and medical survival in the face of harsh battlefield conditions, biological approaches for minimizing the after-effects of battle injuries, including neurotrauma from penetrating and non-penetrating injuries as well as faster recuperation from battlefield injury and wounds[,] . . . biomolecular motors and devices[,] . . . micro/nano-scale technologies for non-invasive assessment of health[,] . . . techniques for the decoding of neural signals in real time[,] . . . novel interfaces and sensor designs for interacting with the central . . . and peripheral nervous systems[,] . . . [and] new approaches for understanding and predicting the behavior of individuals and groups, especially those that elucidate the neurobiological basis of behavior and decision making.[84]

Though DARPA-funded research is often cutting-edge and visionary, about 90 percent of its projects fail.[85] Those that succeed, however, often prove transformative for both military and civilian life.[86] DARPA-funded research has resulted in the creation of the Internet (initially called the Darpanet), the computer mouse, the stealth fighter, and unmanned aerial vehicles.[87] As one DARPA official explains, "DARPA is about trying to do those things, which are thought to be impossible, and finding ways to make them happen."[88]

CONCLUSION

I have focused my discussion in this article on military medicine and research methods employed by the US military in furtherance of its mandate to protect national security interests. One need not question the validity of the government's motivations to conduct experimental research to understand that current and past research methods run contrary to fundamental constitutional liberties and well-established research protocols governing human subjects research.

Importantly, the unique relationship between a service member and his or her commanding officer, and in turn, between the commanding officer and his or her superiors, creates an environment with enormous potential for abuse from a sociomedical context. Service members are legally obligated to submit to biomedical treatments deemed necessary for the good of the armed forces, even in instances where the treatments are purely investigational or involve unapproved uses of FDA-approved medical products.[89] Refusing "treatment" may be viewed as disobeying an order, which can result in punitive measures that include a court-martial and dishonorable discharge from the military. Coupled with the threat of punitive measures, military hierarchy often compels soldiers to submit to experimental treatment in instances where they otherwise may not have provided consent.[90]

The risks to service members are compounded when one considers the broad legal immunities that shield military researchers and the US government from civil claims. Under the *Feres* doctrine, service members are precluded from raising tort claims against the government, government employees, or third-party contractors working in furtherance of governmental research, if the underlying injury is sustained "in the course of activity incident to service."[91] Service members are also precluded from raising tort claims against the United States when the underlying injury relates to a "discretionary function" of military policy.[92] The US Supreme Court has interpreted the *Feres* doctrine broadly to encompass claims that arise from experimental research, even in instances where the government covertly experimented upon soldiers and civilians, or intentionally disregarded legal requirements and informed consent protocols.[93]

Understanding the history and dynamics of experimental research in the military, along with the legal and regulatory framework that facilitates such research, informs contemporary discussion of how best to harmonize national security interests with fundamental notions of human dignity and patient autonomy. While the goal of this article has been to use human enhancement and experimental research as paradigms to highlight the legal and regulatory shortcomings of the current framework, proposals for reform measures addressing these concerns will be the subject of future scholarship.[94]

NOTES

1. Mike Mitka, "US Military Medicine Moves to Meet Current Challenge," *JAMA* 286 (2001): 2532–33. . . .

2. See Jonathan D. Moreno, *Mind Wars: Brain Research and National Defense* (New York: Dana Press, 2006), 11–13, 51 . . .; Catherine L. Annas and George J. Annas, "Enhancing the Fighting Force: Medical Research on American Soldiers," *Journal of Contemporary Health Law and Policy* 25 (2009): 283, 287, 285–86 . . .;

Hannah Hoag, "Remote Control," *Nature* 423 (2003): 796, 796 ...; Noah Shachtman, "Darpa Offers No Food for Thought," *Wired*, February 17, 2004, www .wired.com/print/medtech/health/news/2004/02/62297. . . .

3. See David P. Rall et al., preface to *Veterans at Risk: The Health Effects of Mustard Gas and Lewisite*, ed. Constance M. Pechura and David P. Ralls (Washington, DC: National Academy Press, 1993), v–vii . . . (hereinafter IOM Report). In the 1940s and 1950s, articles in the popular press "suggest[ed] some tension between the [American] words at Nuremberg and the practices in America." *Final Report of the Advisory Committee on Human Radiation Experiments* (New York: Oxford University Press, 1996), 87 (hereinafter Human Radiation Experiments Report).

4. IOM Report, supra note 3, v; Susan L. Smith, "Mustard Gas and American Race-Based Human Experimentation in World War II," *Journal of Law, Medicine, and Ethics* 36 (2008): 517, 517. Researchers suspected that nonwhites would have a different response than whites (Smith 518).

5. Smith, supra note 4, 518. . . .

6. IOM Report, supra note 3, vii.

7. Smith, supra note 4, at 518.

8. Id.

9. Id.

10. IOM Report, supra note 3, v–vi. . . .

11. Human Radiation Experiments Report, supra note 3, xxx, 14. . . .

12. Leonard W. Schroeter, "Human Experimentation, the Hanford Nuclear Site, and Judgment at Nuremberg," *Gonzaga Law Review* 31 (1996): 147, 213.

13. Human Radiation Experiments Report, supra note 3, 6 (citation omitted).

14. See Howard Ball, "Downwind from the Bomb," *New York Times*, February 9, 1986 (§ 6), 33 . . .

15. Id.

16. Id.

17. See Schroeter, supra note 12, 213.

18. See Human Radiation Experiments Report, supra note 3, 160–61.

19. See Schroeter, supra note 12, 157–58. . . .

20. Id., 157–58.

21. Schroeter, supra note 12, 151.

22. Human Radiation Experiments Report, supra note 3, 318–22.

23. Id., 318.

24. Ball, supra note 14, 33.

25. See Paul J. Amoroso and Lynn L. Wenger, *The Human Volunteer in Military Biomedical Research*, vol. 2 of *Military Medical Ethics*, ed. Thomas E. Beam and Linette R. Sparacino (Washington, DC: Office of the Surgeon General, 2003), 563, 570 . . .

26. See John Gimbel, "German Scientists, United States Denazification Policy, and the 'Paperclip Conspiracy,'" *International History Review* 12 (1990): 441, 441–42; Peter A. Masley, "The Paperclip File: America's Secret Agenda to Import Nazi Intelligence," *Washington Post*, August 16, 1991, D3.

27. See Gimbel, supra note 26, 441–42.

28. Gimbel, "German Scientists," 441–42; Andrew Walker, "Project Paperclip: Dark Side of the Moon," *BBC News*, November 21, 2005, http://news.bbc.co.uk/2 /hi/uk_news/magazine/4443934.stm.

29. David H. Price, "Buying a Piece of Anthropology," *Anthropology Today* 23 (2007): 9.

30. Id., 9–11.

31. United States v. Stanley, 483 U.S. 669, 686, 688 (1987) . . . (citations omitted).

32. Id., 689.

33. See Annas and Annas, supra note 2, 301–4. . . .

34. See Stuart L. Nightingale et al., "Emergency Use Authorization (EUA) to Enable Use of Needed Products in Civilian and Military Emergencies, United States," *Emerging Infectious Diseases* 13 (2007): 1046, 1047. . . .

35. Nightingale et al., "Emergency Use Authorization," 2007; Randall S. Stafford, "Regulating Off-Label Drug Use—Rethinking the Role of the FDA," *New England Journal of Medicine* 358 (2008): 1427.

36. See Susan Okie, "Access before Approval—A Right to Take Experimental Drugs?," *New England Journal of Medicine* 355 (2006): 437, 439. . . .

37. Doe v. Sullivan, 938 F.2d 1370, 1372 n.1, 1374 (D.C. Cir. 1991).

38. Id. 1371–72, 1372n1.

39. Id., 1372n1.

40. Id., 1373.

41. Id.

42. Annas and Annas, supra note 2, 301–2.

43. Id., 302.

44. Revocation of 1990 Interim Final Rule, 64 Fed. Reg. 54180, 54184 (Oct. 5, 1999). . . .

45. Id.

46. Research Advisory Committee on Gulf War Veterans' Illnesses, *Gulf War Illness and the Health of Gulf War Veterans* (2008) [hereinafter Gulf War Illness Report].

47. Id., 6. . . .

48. Id.

49. Id., 4.

50. Id., 7–10. . . .

51. Rempfer v. Sharfstein, 583 F.3d 860, 863 (D.C. Cir. 2009).

52. Id., 863–64.

53. Id.

54. Id., 864.

55. Id., 863–64.

56. Id., 864.

57. Project BioShield Act, 21 U.S.C. § 360bbb-3 (2006).

58. See Nightingale et al., supra note 34, 1046. . . .

59. Id., 1050.

60. See Rempfer, 583 F.3d at 864. . . .

61. Id., 867–68.

62. "Why Get Vaccinated," *Biothrax*, accessed February 16, 2002, www .biothrax.com/whatisbiothrax/whygetvaccinated.aspx. . . .

63. See Military Vaccine (MILVAX) Agency, *Anthrax Vaccine Immunization Program: Questions and Answers Anthrax and the Persian Gulf War*, accessed March 30, 2012, www.anthrax.osd.mil/resource/qna/qaAll.asp?cID = 313. . . .

64. See Charles W. Hoge, "Interventions for War-Related Posttraumatic Stress Disorder," *JAMA* 306 (2011): 549, 549. . .; Charles W. Hoge et al., "Mild Traumatic Brain Injury in U.S. Soldiers Returning from Iraq," *New England Journal of Medicine* 358 (2008): 453, 454.

65. Hoge, supra note 64, 549.

66. Hoge et al., supra note 64, 454.

67. Annas and Annas, supra note 2, 304.

68. Id. Examples of SSRIs include Prozac, Paxil, and Zoloft.

69. Id.

70. See Efthimios Parasidis, "Patients over Politics: Addressing Legislative Failure in the Regulation of Medical Products," *Wisconsin Law Review* 2011, no. 5 (2013): 929, 987, https://papers.ssrn.com/sol3/papers.cfm?abstract_id=1964025. . . .

71. Moreno, supra note 2, 11 (internal quotation marks omitted).

72. Annas and Annas, supra note 2, 286 (internal quotation marks omitted).

73. Id.

74. Moreno, supra note 2, 121.

75. Id.

76. Id., 4.

77. Id., 51.

78. Hoag, supra note 2, 798.

79. Id., 796, 798.

80. Id., 796.

81. Moreno, supra note 2, 127.

82. Id., 11–13.

83. Id., 11 (internal quotation marks omitted).

84. Id., 12–13.

85. Id., 12.

86. See id., 12–13. . . .

87. Id., 12.

88. Id. (internal quotation marks omitted) (quoting a DARPA official).

89. Doe v. Sullivan, 938 F.2d 1370, 1372–74 (D.C. Cir. 1991).

90. See IOM Report, supra note 4, v–vii . . .; Moreno, supra note 2, 134; Annas and Annas, supra note 2, 308 . . .; Smith, supra note 5, 518.

91. Feres v. United States, 340 U.S. 135, 146 (1950). . . .

92. Boyle v. United Tech. Corp., 487 U.S. 500, 511 (1988).

93. United States v. Stanley, 483 U.S. 669, 671, 683–84 (1987).

94. See Efthimios Parasidis, "Justice and Beneficence in Military Medicine and Research," *Ohio State Law Journal* 73 (2012): 723.

34. Non-Consenting Adults

Harriet A. Washington

Sixty-five years ago in Nuremberg, Germany, American prosecutors confronted the Nazi physicians who had subjected Jews and others to a murderous regime of medical research. The "doctors' trial" was the first of the war crimes trials; one of its outcomes was the famous Nuremberg Code, a set of ethical guidelines for human experimentation.

The first tenet of the code is very clear: "The voluntary consent of the human subject is absolutely essential."

Today, the Nuremberg Code is the most important influence on US law governing human medical research. Even so, marginalized groups have frequently been coerced into studies that violate their right to consent. A recent review of the bioethics of human research in the United States offers little prospect for change.

. . . In 1994, for example, the Medical University of South Carolina in Charleston was accused of enrolling poor black women into narcotic-treatment research without their knowledge. The next year in Los Angeles, an experimental measles vaccine was tested on children, mostly black and Hispanic, without their parents' consent. In 1994 and 1995, New York City law enforcement officials helped researchers coerce black parents into enrolling their boys into a study that sought to establish a genetic propensity for violence, again without their consent. And in 2001, the Kennedy Krieger Institute in Baltimore was found guilty by a Maryland court of encouraging black families to move into lead-contaminated housing as part of a study on lead levels in children (the verdict was later overturned).

This scourge has spread beyond racial minorities. . . . Since the 1980s, around twenty US research projects have won legal waivers allowing them to bypass any form of consent. From 1990 until 2005, for example, the Department of Defense obtained a waiver that allowed it to force 8.9 million ground troops to accept inoculation with experimental anthrax vaccines.

Civilians' rights are violated too. In 1996, US law was changed to permit nonconsensual research on trauma victims on the pretext that they are unconscious and unable to give their consent.

Private companies, for whom time is money when seeking approval to sell their products from the US Food and Drug Administration (FDA), were quick to see the advantages in enrolling subjects who could not refuse. In 2003, biotechnology company Northfield Laboratories set up a nationwide trial of its blood substitute PolyHeme. The substance was randomly administered by ambulance crews to unconscious victims of car accidents, shootings, and cardiac arrests.

The law required the researchers to offer a means for opting out. Northfield's answer was to provide plastic bracelets inscribed with the words "I decline the PolyHeme study." Crucially, to opt out, one first had to know the study existed—a challenge, because the obligatory community notification tended to be desultory. Ross McKinney Jr., vice dean for research at Duke University School of Medicine in Durham, North Carolina, estimated that consultation in his area reached about 450 out of a possible 267,000 people.

In 2009 the FDA reviewed the trial. It concluded that there were more heart attacks and deaths in subjects who had received PolyHeme than those who had not and rejected Northfield's application to license PolyHeme. The company went into liquidation later that year.

A larger number of nonconsensual studies is still ongoing. The fifty-million dollar Resuscitation Outcomes Consortium aims to recruit around twenty-one thousand subjects to test the safety and effectiveness of various emergency treatments for severe injury and cardiac arrest. ROC is being conducted at eleven trauma centers in the United States and Canada. As in the PolyHeme study, subjects are enrolled at random and no consent is sought.

One ROC experiment infused concentrated saline into trauma victims' blood vessels to test its effect on traumatic brain injuries. Doctors are well aware of the dangers of administering such a highly concentrated solution and these concerns proved well founded in August 2008 when the study was suspended over concerns about patient safety.

Once news coverage alerted some residents of targeted areas to the study, the researchers were overwhelmed by demands for opt-out bracelets. One study continued even after high demand made the bracelets unavailable.

Prospects for progress appear minimal. [In late 2011,] . . . the Presidential Bioethics Commission issued a report on protecting human research subjects. Called Moral Science, it made much of the US's "robust" protections—the very rules that permit and legitimize breaches of informed consent.

The failure to elicit consent is not confined to the United States. One in every three US corporate medical studies is now carried out abroad, usually in places where trials can be conducted more cheaply than in the United States. Subjects are often unaware that the treatments are experimental.

In 2011, drug giant Pfizer paid seventy-five million dollars to settle claims that children in Kano state, Nigeria, were injured or killed by non-consensual administration of its experimental meningitis drug Trovan. Just as US physicians demanded justice at Nuremberg, Nigerian parents stormed courts in Kano and Manhattan to demand that we live up to our stated ideals. Sixty-five years on, it is high time we did.

PART VII

Baby-Making in the Biotech Age

．　．　．　．　．　．

Since the first success of in vitro fertilization (IVF) in 1978, when a child named Louise Brown was born after the creation of an embryo outside her mother's body, assisted reproduction has brought some five million children into the world for people affected by medical and social infertility. Unfortunately, millions of others have attempted to form families through assisted reproduction but have not succeeded.

Among those who have become parents through IVF, some have relied both on medical assistance and on the biological participation of sperm providers, egg providers, and/or gestational surrogates who do not intend to participate in raising the children. And in recent years, young women who have no known fertility problems, but who may decide to postpone childbearing, are being urged to have their own eggs surgically extracted and frozen for later use, as we saw in part 4.

Despite the large numbers of people whose lives have been touched by assisted reproduction—prospective parents, resulting children, and third parties who provide gametes or gestational services—adverse consequences often remain underexamined. In the United States, there is very little public policy to provide guidelines or oversight for the fertility industry. Further, assisted reproduction and surrogacy are now global enterprises. In pursuit of what is known as "cross border reproductive care" or "reproductive tourism," people hoping to become parents travel abroad, often to avoid regulations in their home countries or to save money. In recent years, a flurry of problems and scandals in international surrogacy has shut the practice down in a number of countries, sometimes in ways that discriminate against same sex or unmarried people and sometimes across the board.

Contributors in this section use a social justice lens to examine troubling aspects of assisted reproduction, focusing on the United States and on cross border fertility arrangements.

The section begins with Miriam Zoll's "Generation I.V.F.: Making a Baby in the Lab—10 Things I Wish Someone Had Told Me," an account of her own four-year odyssey into the world of assisted reproductive technology. Zoll, who winds up adopting a son, wants to communicate the conclusions she and her partner reached to others considering IVF: "Most of the information circulating about IVF predominantly focuses on its *successes;* there is virtually no counterbalance to inform us about high failure rates, its devastating effect on couples, or its bioethical conundrums."

Laura Mamo's "Queering the Fertility Clinic" traces sexual minorities' use of assisted reproduction, beginning with the "lesbian baby boom" of the 1990s and 2000s and moving to the growing number of parents who are gay or transgender. Examining the role of information technologies, fertility biomedicine, and recent social movements around queer rights, she ends with questions about the persistent implications and dilemmas of technology-mediated queer family formation.

Lisa Ikemoto's "Reproductive Tourism: Equality Concerns in the Global Market for Fertility Services" describes the cross border fertility market and examines the equality concerns it raises—issues that have received relatively little attention in scholarly and public discussions of reproductive tourism.

Another contribution on transnational assisted reproduction, Douglas Pet's "Make Me a Baby as Fast as You Can," highlights the role of companies that serve as brokers mediating between commissioning parents in the United States and surrogacy clinics in other countries. Pet describes a program offered by one such firm that is meant to accelerate the process: "the option of having embryos implanted into two surrogates at the same time."

And finally, while children are the goal of all assisted reproduction arrangements, their well-being is, paradoxically, too often ignored. One aspect of this is the desire of many children conceived through third-party gametes to know about their biological heritage. In "Let's Get Rid of the Secrecy in Donor-Conceived Families," Naomi Cahn and Wendy Kramer point out that many parents, especially heterosexual ones, don't tell their children that they were born as a result of using third-party eggs or sperm and warn of the problems that often causes.

35. Generation I.V.F.

Making a Baby in the Lab—10 Things I Wish Someone Had Told Me

Miriam Zoll

At age fifty, I am an official member of Generation IVF, having grown up after the pill and Baby Boomer feminists revolutionized women's reproductive choices and lives. We watched as millions of American women infiltrated formerly closed-to-females professions, and as home and office politics, the economy, and relations between the sexes radically shifted.

My generation also came of age alongside reproductive technologies: in vitro fertilization (IVF), frozen sperm, donor eggs, and surrogacy. I vividly remember reading front-page stories about the first test-tube baby born in Britain (in 1978) and about the first donor egg baby born (in 1984). These advances were so extraordinary that my girlfriends and I began to believe that almost anything would be possible by the time we were ready to have kids . . . that is, if we chose to have kids.

I can recite from memory the names and ages of the celebrities who first seemed to beat the biological clock: photographer Annie Leibovitz (twins at 52), supermodel Cheryl Tiegs (baby at 52), actress Geena Davis (twins at 48). Now there are hordes more, and some of these have revealed they used IVF: Mariah Carey, Celine Dion, Courtney Cox, J-Lo, and, it seems, just about every other forty-plus female in Hollywood.

Bombarded by these relentless endorsements for older motherhood, many middle-class, educated, Gen IVF women like myself started thinking, "Wow, science is finally beating Mother Nature." We reassuringly told each other, "It's okay to delay motherhood while we pursue our careers. If we

Original publication by Miriam Zoll. "Generation I.V.F.: Making a Baby in the Lab: 10 Things I Wish Someone Had Told Me." *Lilith.* Fall 2013. Reprinted by permission of Miriam Zoll. Due to space limitations, portions of this chapter have been deleted or edited from the original.

run into trouble, well, there are always fertility treatments." Science and technology were our new God.

My own fertility story has four chapters.

Chapter One: Ambivalent about motherhood and thrilled to be ensconced in a meaningful career, I married in my midthirties and five years later started trying to make a baby the old-fashioned way.

Chapter Two: a four-year odyssey into assisted reproductive technology (ART) (which 99 percent of the time means IVF). With ART, the sperm and egg are handled *outside of the womb* (ART does not include straight hormone therapy or intrauterine insemination). Most ART requires very expensive injections, via needles to the abdomen, thigh, or buttock, aimed at controlling your body's hormones and stimulating your egg production. Notoriously, they can also fuel horrible mood swings and alter your personality.

During the years I underwent IVF, I met women so addicted to the hope the science offered that they'd endured as many as eighteen rounds of treatment. My husband Michael and I abandoned the IVF treadmill after four failed cycles, including one emotionally devastating miscarriage and another in which my ovaries produced no eggs at all.

For the most part, this four-billion-dollar-a-year biotech industry is not invested in providing appropriate patient education that can help women and couples determine when to cease treatments, or support them in coming to terms with the indescribably painful fate of their biological childlessness. Still reaching for the stars ourselves, Michael and I continued on to *Chapter Three:* The Donor Egg Phase.

At this point a third party was introduced into the sacredness of the conception process, deeply challenging not only our values, but also our core sense of identity. We talked endlessly about ethics, spirituality, and the excruciating realization that we were actually considering *buying* another woman's eggs, commodifying her.

Still, sad to say, despite trepidation, disabling bouts of insomnia, and self-flagellation, our obsession to procreate prevailed. Slowly but surely, we became that thing: Fertility Junkies. We began working with a donor egg agency, spending endless hours online, often surreptitiously logging on at 3 am, addictively "shopping" for the perfect egg mother: maybe one who would look a little bit like me.

After several more months of tears and insomnia, we chose an attractive twenty-one-year-old who then had to undergo rigorous testing. When the nurse finally called to report that this donor was infertile, we were stunned. . . .

At that point, maimed and almost immobilized by grief, I slunk into a depression that kept me in bed for months. Somehow, I don't know how, we initiated *Chapter Four* of our saga: the path to parenthood through adoption. . . .

Michael's and my seven-year-long journey to have a child ended the moment we laid eyes on our newborn son, Sammy. As he snuggled next to his birth mother in a hospital bed, a maternal force as great as a tsunami welled up inside of me, and I gave thanks that he had finally arrived through the spiritual cocoon of my long-recited prayers. At the same time, I couldn't quite comprehend that he was actually "my son" and I was now "his mother." In an instant, we all assumed new identities that would bind us together for life.

Like wide-eyed pioneers, Michael and I had ventured into a wildly unregulated and subterranean branch of medicine. Most of the information circulating about IVF predominantly focuses on its *successes;* there is virtually no counterbalance to inform us about high failure rates, its devastating effect on couples, or its bioethical conundrums. So herein, abridged, are ten things I wish someone had told me before I embarked on my ride through hell.

1. **You Will Probably Be Physically and Emotionally Traumatized.** I found that the side effects of the drugs, the constant prodding and probing below my hips, and the repeated failures, miscarriage, and devastatingly dashed hopes brought me to the point where I sought treatment for post-traumatic stress disorder (PTSD).

A study by Allyson Bradow, *Primary and Secondary Infertility and Post-Traumatic Stress Disorder,* confirms that women who experience failed fertility treatments often exhibit symptoms of PTSD. Close to 50 percent of 142 participants in Bradow's study met the official criteria for the disorder; that's about six times higher than its prevalence in the general population.

Those of us who bump into age-related infertility end up confronting two tragedies: the loss of our deep primal desire to birth a baby and the realization that we guzzled the Kool-Aid: we built our entire "women-can-finally-have-it-all" adult life on an illusion.

2. **Your Sexuality Will No Longer Belong to You.** In order to endure the physical and emotional strain of multiple IVF cycles, you will eventually detach from your body and your sex drive. This is almost inevitable, as the doctors will control, through drugs and technology, what used to be controlled by nature.

By the time I reached the Donor Egg Phase, sex equaled stress. It meant needles and petri dishes, stirrups and vaginal probes. It was associated with disappointment and guilt and pain. Most nights I cried myself to sleep.

3. **You Will Blame Yourself.** In 2012, the European Society for Human Reproduction and Embryology reported that the global ART failure rate was as high as 77 percent. In the United States, treatments fail close to 60 percent of the time among women younger than thirty-five, and 88 to 95 percent of the time among women older than 40. This glaring omission of information from most mainstream media results in women blaming themselves for failed cycles rather than understanding that this fragile science has consistently missed its mark two-thirds of the time or more since 1978.

In my case, as cycle after cycle failed, I buried myself in a tomb of self-blame so disabling that I was unable to work for one full year. *It was my fault* my ovaries weren't producing enough quality eggs. *It was my fault* we waited too long. It was my fault we had a miscarriage. I was an expert when it came to contraception, but I was embarrassed about my ignorance regarding reproduction, and angry with myself for how blindly I entrusted doctors to work their magic in a laboratory. I was a failure in every way. *If I had only tried harder . . . .*

Fortunately, former infertility patients and advocates are beginning to talk about all of this publicly. . . .

4. **The Absence of the Sacred Will Deplete You.** Fertility clinics and their staff are focused on manufacturing embryos, *not* on counseling patients compassionately after miscarriages, stillbirths, and negative pregnancy tests. I often wondered what the doctors and nurses thought about me, the human being, as I lay on the gurney, and when I eagerly signed up for another cycle only days after my miscarriage. Did they feel sorry for my desperation, which kept them employed? Hooked into stirrups, did I have a face, a husband, and a life, or was I just another older woman trying to have a kid?

A few months after our second donor was diagnosed as being infertile, we finally, for the first time, sat in a room with other couples in the same situation. A minister's wife told the tale of how she'd adopted four children whose mother could no longer care for them.

Only minutes into her story, the dam inside of me broke loose and a river of tears began streaming down my face. This was the first occasion since we'd begun the arduous baby-making process that we were communing with people who actually talked about the sacredness of the path toward parenthood. Never once during treatments had clinic staff even mentioned the beauty or spirituality of creating and stewarding new life.

5. **Treatments Involve Health Risks.** In a branch of medicine that is still very much experimental, I injected into my body whatever drugs the

doctors thought might help me become pregnant. I am an educated woman, a researcher and writer by trade, a feminist, and yet I became an obedient guinea pig.

When I finally stopped treatments and was invited to join the board of *Our Bodies Ourselves,* I learned that there is scant evidence-based research about the *long-term* effects of treatments on women's and infants' health. Existing data does show an increased risk between certain fertility interventions and breast, ovarian, and endometrial cancers, among other side effects, and a 26 percent increased risk of birth defects in IVF babies. The common practice of implanting multiple embryos is known to pose serious health risks to mothers and infants, including preterm delivery, low birth weights, and costly hospitalizations.

The effect of treatments on egg donors has been even less studied, yet we do know that side effects can include blood clotting, infertility, and ovarian hyperstimulation syndrome, and in some rare instances, death. Potent drug regimens can create as many as thirty to sixty eggs in one cycle, as opposed to the solo egg a woman naturally produces during her period. (You can learn more in the film *Eggsploitation* and from the group *We Are Egg Donors.*)

On the positive side, there is now the Infertility Family Research Registry (ifrr-registry.org) that invites women going through ART to submit information about their health and that of any offspring. Of the roughly five hundred clinics in the United States, however, fewer than a hundred have signed up to promote it.

6. **Treatment Costs a Fortune. Be Prepared to Confront Your Privilege.** One average IVF cycle in the United States costs between $12K and $15K; a donor egg cycle, $30K; and surrogacy anywhere from $75K to $150K. Around the globe, the greatest cause of infertility is untreated sexually transmitted diseases; these hit poor women the hardest. Needless to say, fertility treatments are largely unavailable to them.

Only fifteen US states offer insurance policies that cover fertility procedures, compared to Britain, Israel, and many countries in Europe that subsidize some citizens' fertility treatments. In Sweden, France, and Italy, single women, and lesbians and gay men, are often barred from accessing them at all.

7. **Fertility Clinics Are Big Business.** Most clinic staff wear two incompatible hats: a medical one and a business one, so this means their advice might include steering you toward trying new technologies and drugs. Our reproductive endocrinologist told us honestly that our chances of IVF success were low, but he also said, "It only takes one good egg to

make a baby." Michael and I were awash in yearning and denial; the doctor knew that. "New techniques and protocols are constantly being developed," he said. *"You just never know what can happen."*

The world's first fertility company, Virtus Health, went public [in 2013] to the tune of almost half a billion dollars. Its CEO is quelling investors' fears that improvements in ART might mean fewer cycles for clients. Uh-oh—dwindling revenues.

8. **Fertility Clinics Are a "Wild West."** There is only one piece of US federal legislation, loosely enforced, that requires clinics to self-report their annual success rates: the 1992 Fertility Clinic Success Rate and Certification Act. Apart from this, the industry operates below the public radar.

Activities that are stunningly unregulated include: implanting multiple embryos that may increase rates of success but also endanger women's and infants' health; engineering and selling anonymous embryos in the marketplace; prescribing off-label drugs that have not been approved by the FDA for fertility use; marketing donor embryos or donor egg treatments to postmenopausal women; and offering expensive procedures—such as egg freezing—that have no proven track record in efficacy or safety.

The hype around egg freezing is a good example of the clinics going rogue. This newest technology is being marketed as though it's as revolutionary and reliable as the pill. I know older women who view it as a kind of magical insurance policy that will ensure their chances to birth babies safely when they are older. But there is virtually no long-term, evidence-based research to back up these claims.

9. **Your Treatment Options May Exploit Poor Women.** Patients wrestling with the pain of infertility and considering options like surrogacy and egg donation need to understand and connect the dots between their treatment choices and these women's lives. Take a look at the recent documentaries *Made in India* and *Show Me the Baby Bump, Please.* Many surrogates, in India and elsewhere, are illiterate, extremely poor, and often not informed about what they've consented to. They can be separated from their children for up to a year, relegated to "surrogacy dormitories," and, if their pregnancy fails, compensation and follow-up health care may be withheld. As health consumers, patients can play an important role promoting greater health and human rights protections for all parties involved in reproductive technology treatments.

Commercial surrogacy and egg vending are booming businesses. As someone who has studied the link between poverty and gender, I would much rather see women and girls acquire economic security through better access to educational, constitutional human rights protections, and sustainable employment opportunities, not by a singular focus on their gonads and wombs.

10. **You Will Dislike Yourself.** Entering the world of ART will challenge you to reassess much of what you thought you knew about yourself. Long-held beliefs about right and wrong begin to flake off your psyche like old paint on a windblown house. Moral dilemmas about eugenics and cloning invade your dreams.

For me, deciding to use donor eggs was much more difficult than choosing IVF. I was averse to how unnatural it was, and I felt deep shame for my conspicuous conception, paying another woman to risk her health and possibly deplete her own egg reserves on my behalf. How and why do these young women decide to sell their eggs to someone like me? How does a donor agency determine that one woman's eggs are worth eight thousand dollars, but another's only five thousand dollars? Blonde, svelte donors seem to get paid more than brunette, overweight ones. And Caucasian, Asian, and African American eggs carry very different price tags. Ivy League egg donors with high SAT scores and 36–24–26 body measurements have been paid as much as a hundred thousand dollars for their eggs.

While searching for a donor, Michael and I were aghast at how judgmental we became. *This one's eyes are too close together. I don't like her teeth. She looks bipolar. She looks uneducated.*

Like any protective parent, you want to be discriminating when choosing the genetic code and physical traits of someone whose egg will form half the DNA structure of your potential offspring; that's understandable. Still, it did not sit well. And even though we are now the proud and grateful parents of the most delicious little four-year-old ever, our ART ordeal may have scarred us for life.

What's redemptive for me now is my mission to reveal the hidden side of treatments, and to caution women intent on birthing babies to avoid making the same irrevocable decisions so many educated, middle-class women in Gen IVF made when we delayed childbearing. . . . If you're reading this, you doubtless know some of us. You may be one of us.

There's a global epidemic of misinformation about the age when women's fertility naturally declines and about the power of modern medicine to reverse this. If you have experienced treatments, or know someone who has, I invite you to cast off your silence and contribute to expanding an open and honest consumer-driven discussion about these life-altering technologies.

36. Queering the Fertility Clinic

Laura Mamo

> Not all bodies have sperm, some do. . . . Not all bodies have eggs,
> some do. . . . And not all bodies have a uterus, some do. . . . Who
> helped bring together the sperm and egg that made you?
>
> CORY SILVERBERG, 2012

It was in the summer of 2008 that the first pregnant man was introduced to the world (Beatie 2008). Thomas Beale's gender identification as a man coupled with his pregnant female body became an immediate cultural frenzy. Was it his masculinity, his gender nonconformity, his reproductive transgression, or something else that ignited such fascination? . . .

This article is not about Thomas Beale, trans parents, or trans family forms specifically; instead, it is about how these lived experiences of pregnancy and family formation inform and constitute a queering of the fertility clinic. More specifically, this article is about the intersections of gender, sexuality, and reproductive biomedicine for those who "choose" clinical biomedicine to achieve their pregnancy goals and the fertility clinics they encounter along the way. . . .

Through this conceptual frame, I analyze the queering of reproduction in three acts that together constitute both a queering of the fertility clinic and the bioethical tensions produced as a result. Act I briefly looks back to my earlier work and my analysis of the "lesbian baby boom"; Act II and III, in contrast, take stock of subsequent developments and scholarship that have forced me to rethink some of my earlier assertions and to attend to both new queerings and new bioethical dilemmas.

Journal of Medical Humanities. "Queering the Fertility Clinic." 34.2 (March 2013): 227–239. Laura Mamo. doi: 10.1007/s10912–013–9210–3. Springer Science+Business Media New York 2013. Reprinted with permission of Springer and authorization from Laura Mamo. Due to space limitations, portions of this chapter have been deleted or edited from the original.

ACT I: THEORIZING THE LESBIAN BABY BOOM

The emergence of lesbian reproduction was enabled through the meeting of assisted reproductive technologies with vibrant women's and lesbian health movements organized around issues of reproductive rights. Advanced, high-tech biomedical options were becoming routine, standard practices. . . . These options were constructed as not only the "best" option but as the only valid approach with a new grounding assumption, "If you can achieve pregnancy, you must procreate."

[In my 2007 book *Queering Reproduction*,] I argued that the social designations and identities of LGB or T had been transformed into a fertility "risk" factor, a biomedical classification, and a source for biomedical intervention. [These i]dentities were not relevant medically, but they were highly relevant legally, social-culturally, and in everyday practices that structured possibilities and pathways. "Risk" factors turned people to biomedicine and transformed them from parents-in-waiting to enterprising health consumers of the fertility clinic, writ large as including sperm banks, egg brokers, endocrinologists, hormone therapies, and fertility specialists.

Another central finding posed in *Queering Reproduction* was the emergence of compulsory reproduction for LGBT lives as they sought inclusions in normativity, demanding sexual citizenship. . . . [F]amily formation, including children, has become a cultural expectation for many LGBT people. Yet, full legal, social, and biomedical inclusion remains constrained. Attaining inclusion via parenting and children was constrained by economic and cultural capital. As lay health pathways declined, accessing biomedical services was available only for those with the "right" health insurance, contacts to health care providers, and ability to pay. . . .

As I completed the book much caught my attention: the fight for gay marriage was snowballing into a core political strategy for LGBT political organizations as well as communities. Gay men were launching their own baby boom, and the new world order of assisted reproduction had fast burgeoned into a global, largely unregulated set of techno-scientific practices with many groups, especially feminist social justice ones, calling for regulations. As a result, I concluded the book . . . suggest[ing] that there was much to celebrate but some to caution.

First, I . . . argued that lesbian reproduction as well as the expansion of Fertility Inc. contributes to racialized, stratified possibilities of childbearing and motherhood. Access is restricted by who can pay; by classificatory categories such as th[at] . . . of "infertility"; by legal structures; and by continued forms of pronatalist, heterosexist, racist, and neoliberal exclusions.

Second, I proposed that two-father family forms share ideological space with what is now the more domesticated lesbian mother. I advised against the domestication and displacement of lesbian mothers in favor of only seeing the radical in gay men parenting or a fetishization of masculine pregnancy and parenting. Why not see both the radical in all queer family forms as these traverse the fertility clinic (or choose not to and seek adoptions, foster parenting, and other family formations) and theorize how these at once queer and also perpetuate normativity. . . .

Further, I was struck by the rapid changes in online media and the ways these aligned with neoliberal rhetoric, and the multiple and profound structural inequalities that neoliberal policies produce and maintain. I argued that seeking fertility services constituted new subjectivities: lesbian mothers, gay fathers, and queer families. I asserted that through the construction of "affinity-ties," lesbians were using Web 2.0 to form relatedness. Reproduction had become another "do-it-yourself" self-project—a way to transform oneself and one's identity. . . . Institutions, including Fertility, Inc., served as mediators structuring possibilities, intimate and otherwise. "Choice" was constrained by medical labels, insurance codes, costs, institutional heteronorms, everyday homophobia, and transphobia. These constraints continue to be obfuscated by a message that individuals with agency can overcome them with the right attitude, knowledge, and now Web 2.0 savvy. . . .

ACT II: PARENTS-IN-WAITING, QUEER INTIMACIES IN WEB 2.0

Parents-in-waiting constitute a large slice of queer intimacies in the twenty-first century. And, similar to lesbian motherhood, the emergence of gay and trans fatherhood and motherhood are produced through expansions in information technologies, fertility biomedicine, and recent social movements around queer rights and gay and lesbian inclusions in social benefits, including access to individual liberty to achieve families. Although these intimacies are likewise constituted through the meeting of sex without reproduction and reproduction without sex, their distinctiveness requires research and theorization that engages techno-science: assisted reproductive technologies, markets in eggs and sperm, online information communities, and the fast-paced cultures of biomedicine.

The stories of new queer family forms have continued to capture the media's attention. . . . Each media story begins with a nod to the World Wide

Web: "Primarily using the Web, they found an egg donor service and, based on donor profiles, selected and purchased donor eggs (at a cost of several thousand dollars). . . . " These are stories of multiple collaborators: sellers, buyers, parents-in-waiting, donators, brokers, reproductive clinics, etc. . . .

But more than buying and selling, the Internet is today producing and expanding the possibilities for the queer intimacies that consolidate into new family forms. That is, it is producing social relationships that may not have existed materially (although, relations do exist in imaginaries). . . . From virtual spaces, parents-in-waiting let their fingers take them to real and imagined social possibilities. Connections are made and imagined, bricks and mortar fertility services are identified, and reproductive practices engaged, allowing previously obfuscated intimate social possibilities to join visible ones. . . . While gender and sexual norms have largely populated these online spaces, identifications previously marginalized occupy spaces that were once less visible to the mainstream. Parents-in-waiting sites such as "I want to be a (gay) Dad," and "It's conceivable," a site dedicated to providing "clear, no frills, pregnancy and parenting information for the LGBT community" are easily found with a simple Google search. . . .

LGBT reproduction has almost fully moved from a do-it-yourself alternative practice to complex engagements with, and consumption of, a panoply of biomedical services that rely on third and fourth parties. "Choosing" clinical biomedical for reproductive purposes continues to be shaped by and through structural intimacies (Mackenzie 2013). These intimate social forms continue to be structured by the legal gaps, discriminations, and resulting vulnerabilities whether or not queers turn to biomedicine to seek pregnancies. For example, the most recent guidelines of the World Professional Association for Transgender Health (WPATH, version 7) include in their first section on "Reproductive Health" acknowledgment that many transgender, transsexual, and gender nonconforming people will want to have children and recommending appropriate health care practices and consumer information (WPATH 2011, 50).

Yet, the turn to biomedicine continues to be a false "choice," as negotiating structural inequalities to ensure legal protection under the US family law punctuates many of these practices. It is also a means of legitimacy driven by cultural assumptions of "recognizable" familial ties and how much kids "look like" and act like their parent(s). For those who seek pregnancy outside the clinical framework, they do so with legal vulnerability. . . .

ACT III: FERTILITY TRAVELS: BIOETHICAL TENSIONS AND THE "WILD WEST" OF FERTILITY INC.

. . . Today, the US fertility industry is often referred to as the "Wild West" of assisted reproduction (Dresser 2000). Unlike most other industrialized countries, the United States relies on professional, voluntary guidelines to circumscribe its boundaries.

Fertility Inc., an economic market, consists of a growing constellation of medical practices shaped within a context of a corporate, mostly for-profit, health care system expanding its technological offerings and geographical bounds and promoting individual choice (Clarke et al. 2003). Fertility travels in what can seem to be an unregulated imperial expansion (Krolokke, Foss, and Pant 2012) following the global capitalist imperative to constantly expand one's market and diversify services. As Francine Coeytaux, Marcy Darnovsky, and Susan Berke Fogel (2011, 1) state, "While assisted reproductive technologies have increased parental options for those who can afford them, they pose numerous ethical challenges that the reproductive rights, health, and justice communities are only beginning to address."

Border crossing is frequent in this marketplace. Patients seek reproductive services in the United States that are not available elsewhere, and patients from the United States do the same. Lesbians, for example, from outside the United States, buy sperm online to be overnight expressed to their own home, a local fertility clinic, or a fertility clinic in a country where services can be rendered legally. At the same time, "outsourcing" surrogate service has fast become big business with the world's most vulnerable filling this need. Ideologies of normative gender, relatedness, and family organize the consumption intersections where buyers and sellers meet (Rudrappa 2010).

As we buy eggs, locate surrogates, and surf our way to locate and produce "donor families," the global economy is ever present. It is heterosexual couples who constitute the vast majority of users of this expanded offering of techno-sciences that push the bounds of reproductive possibilities. When considering the queering of reproduction today, while there is much to celebrate as users on the margins join the center, there is also much from which to raise concern: participation in normativity includes participation in the global trafficking in human sperm, eggs, and wombs.

Who will provide the eggs and the wombs necessary to enable these family forms? From what towns, communities, and countries will the biomaterials be drawn? From whose gendered, raced, and classed bodies will they be drawn? Will these services follow capitalism from the West to the rest to secure the bodies and labor necessary to fulfill our American dreams? How can we be accountable to these collaborative reproducers? In all, questions of how reproductive technologies should be developed and used, and by whom, include questions of LGBT actors and queer reproductive practices. As calls for further regulation sound, where are queer practices, queer bodies in these debates?

As fertility biomedicine captures queer users from the margins, implications reverberate in multiple directions. It is neither a simple celebration of expanded rights nor a further exploitation of global inequality. Yet both coexist. . . . Biomedicalization is stratified, bringing forward certain users and uses along familiar lines of economic, social, and cultural lines. As the most fortunate have their rights expanded, the least fortunate are often called forward to supply those rights. Gender, class, and race stratifications shape two-dad, transmen, and transwomen families just as they do two-mom families. . . .

In cases of reprogenetics, such as cloning techniques, these may soon permit same-sex biological parents. Embryonic stem cell research and assisted reproduction research is a potential means of filling the static market in "donated" gametes, especially eggs, for research and procreation. Most agree that any shift to the genetic line is dangerous and must be regulated. Some advocates remain and do so under the banner of reproductive liberty. LGBT appeals are often used to advocate that queer communities support this line of research. Artificial gametes and cloning would not, however, do anything to help queer communities but would have negative effects on all future generations, physically and socially. Cloning, eugenic technologies, and other genetic reproductive technologies are part of the Wild West of reproductive medicine and many are demanding regulations (see Center for Genetics and Society). While these possibilities might seem to be the realm of science fiction, the knowledge needed to realize them is already in development. More so, they are part of a longer cultural imagination. That is, one can see ideological continuities found in shifts from eugenics to genetic medicine, from infertility to fertility problems, and from nonprocreation as a product of unnatural gender to procreation for every woman and man. . . .

The commercialization of the human exists as human beings are rendered perfectible through the market. Examples are numerous, including the rise of sex selection, the sale of organs from the poor to the rich, the boom in enhancement technologies such as cosmetic surgery, and gene doping for athletes. Furthermore, social issues are increasingly being defined as strictly genetic or biomedical problems, not social or environmental phenomena. These include disability, obesity, sexual orientation, gender variance, poverty, violence, breast cancer, osteoporosis, and rickets. In all, existing social divisions are exacerbated.

In the United States, neoliberal ideals of ownership and individualism punctuate reproductive practices and services, as reproduction becomes another do-it-yourself self-project enabling us to transform our selves, identities, and social lives through consumption. The subjectivities produced and intimacies enabled are products, in part, of consumption: lesbian mothers, gay fathers, and new family arrangements brought into being through consumption.

CONCLUSIONS: BIOETHICAL DILEMMAS AND QUEER INTIMACIES IN THE TWENTY-FIRST CENTURY

Pregnant men, butch pregnancies, and other gender queer embodiments are today active participants in the queering of reproduction. While often controversial on grounds of moral panics, ethical dilemmas, and other debates now understood as part of the "culture wars," Fertility Inc. shapes these engagements be it through the "choice" to participate, the stratifications that disallow and disavow doing so, and the many possibilities that lie between. . . .

In many ways these shifts in *queer intimacies* [of the past forty years] have consolidated around regenerative possibilities offered through technoscience. And these are part of LGBTQ challenges to and claims upon the entitlements and benefits of state-sanctioned marriage, adoption, and reproductive rights. While these provoke a rethinking of kinship markers, they also raise questions about belonging and recognition. . . . Today's children are likely to live in various family forms with multiple parents, aunts, uncles, and grandparents. Children may live in a situation where they have more than one woman who operates as "mother," and/or more than one man who operates as "father," and/or no mother or father. Straight and gay families are often blended, and queer families emerge in nuclear and non-nuclear forms.

Increasingly, they also raise bioethical dilemmas: how can we account for these scenarios and be accountable to the people involved? How can we be accountable to the third and fourth party "donators" of gametes, wombs, and services? How can we be accountable to our own critical engagements with the local and global fields of ARTs and genetic technologies? As LGBTQ voices push for inclusion and are brought into biomedical contexts, what are the many implications? What boundaries are drawn around inclusions and exclusions?

To conclude, I return to a central question posed at the beginning of the article: in what ways is Fertility Inc. queered? And with what social implications? . . .

REFERENCES

Almeling, Renee. 2011. *Sex Cells: The Medical Market for Eggs and Sperm.* Berkeley: University of California Press.

Beatie, Thomas. 2008. "Labor of Love: Is Society Ready for This Pregnant Husband?" *Advocate*, April 8, 24.

Becker, Gay. 2000. *The Elusive Embryo: How Men and Women Approach New Reproductive Technologies.* Berkeley: University of California Press.

Boellstorf, Tom. 2005. *The Gay Archipelago: Sexuality and Nation in Indonesia.* Princeton, NJ: Princeton University Press.

Boodman, Sandra. 2005. "Fatherhood by a New Formula Using an Egg Donor and a Gestational Surrogate, Some Gay Men Are Becoming Dads—and Charting New Legal and Ethical Territory." *Washington Post*, January 18, HE01.

Califia, Patrick. 2000. "Family Values: Two Dads with a Difference—Neither of Us Was Born Male." *Village Voice*, June 20. Accessed April 6 2012. www .villagevoice.com/2000–06–20/news/familyvalues/.

Clarke, Adele E, Janet K. Shim, Laura Mamo, Jennifer R. Fosket, and Jennifer R. Fishman. 2010. "Biomedicalization: A Theoretical and Substantive Introduction." In *Biomedicalization: Technoscience, Health, and Illness in the U.S,* edited by A. E. Clarke et al., 1–46. Durham, NC: Duke University Press.

———. 2010. "Biomedicalization: Theorizing Technoscientific Transformations of Health, Illness, and U.S. Biomedicine." *American Sociological Review* 68:161–94.

Coeytaux, Francine, Marcy Darnovsky, and Susan Berke Fogel. 2011. "Assisted Reproduction and Choice in the Biotech Age: Recommendations for a Way Forward." *Contraception* 83:1–4.

Daniels, Cynthia R. 2006. *Exposing Men: The Science and Politics of Male Reproduction.* Oxford: Oxford University Press.

Daniels, Cynthia R., and J. Golden. 2004. "Procreative Compounds: Popular Eugenics, Artificial Insemination and the Rise of the American Sperm Banking Industry." *Journal of Social History* 38:5–27.

Darney, P.D. 2008. "Hormonal Contraception." In *Williams Textbook of Endocrinology,* edited by H.M. Kronenberg, S. Melmer, K.S. Polonsky, and P.R. Larsen, 615–44. 11th ed. Philadelphia: Saunders.

De Sutter, P., K. Kira, A. Verschoor, and A. Hotimsky. 2002. "The Desire to Have Children and the Preservation of Fertility in Transsexual Women: A Survey." *International Journal of Transgenderism* 6. Research in the Sociology of Work 20. Donor Sibling Registry Website. Accessed November 9, 2012. https://www.donorsiblingregistry.com/.

Dresser, R. 2000. "Regulating Assisted Reproduction." *Hastings Report* 30:26–27.

Duggan, Lisa. 2002. "The New Homonormativity: The Sexual Politics of Neoliberalism." In *Materializing Democracy*, edited by R. Castronovo and D. Nelson, 173–94. Durham, NC: Duke University Press.

———. 2003. *Twilight of Equality? Neoliberalism: Cultural Politics and the Attack on Democracy.* Boston, MA: Beacon.

Eng, David, Jose Munoz, and Judith Halberstam, eds. 2005. "What's Queer about Queer Studies Now?" In *Social Text* 23, no. 3–4 (Fall–Winter): 1–17. Durham, NC: Duke University Press.

Finegold, Wilfred J. (1964) 1976. *Artificial Insemination*, 2nd ed. Springfield: Thomas.

Giddens, Anthony. 1991. *Modernity and Self Identity.* Oxford: Polity Press.

Hertz, Rosanna, and Jane Mattes. 2011. "Donor-Shared Siblings or Genetic Strangers: New Families, Clans, and the Internet." *Journal of Family Issues* 32 (9): 1–27.

Hunter, M.H., and J.J. Sterrett. 2000. "Polycystic Ovary Syndrome: It's Not Just Infertility." *American Family Physician* 62 (5): 1079–95.

Inhorn, Marcia C. 2006. "Making Muslim Babies: IVF and Gamete Donation in Sunni versus Shi'a Islam." *Culture, Medicine, and Psychiatry* 30 (4): 427–50.

Kelly, Jen. 2001. "Gay Men May Make Babies." *Herald Sun*, January 4, News 15.

Kolata, Gina. 2002 "Fertility Inc.: Clinics Race to Lure Clients." *New York Times*, January 1, D1, D7.

Krolokke, Charlotte. 2009. "Click a Donor: Viking Masculinity on the Line." *Journal of Consumer Culture* 9 (1): 7–30.

Krolokke, Charlotte, Karen A. Foss, and Saumya Pant. 2012. "Fertility Travel: The Commodification of Human Reproduction." *Cultural Politics* 8 (2): 273–82.

Lambda Legal Defense Fund. 2012. Accessed November 9, 2012. www.lambdalegal.org/in-court/cases/benitez-v-north-coast-womens-care-medical-group.

Larsen, Brad. 2011. "The Phenomenological Experience of Gay Male Couples Pursuing Parenthood." PhD diss., New York University.

Lewin, Ellen. 2009. *Gay Fatherhood: Narratives of Family and Citizenship in America.* Chicago: University of Chicago Press.

Luce, Jacqueline. 2010. *Beyond Expectation: Lesbian/Bi/Queer Women and Assisted Conception.* Toronto: University of Toronto Press.

Mackenzie, Sonja. 2013. *Structural Intimacies: Sexual Stories in the Black AIDS Epidemic.* Livingston, NJ: Rutgers University Press.

Mamo, Laura. 2007. *Queering Reproduction: Achieving Pregnancy in the Age of Technoscience.* Durham, NC: Duke University Press.

———. 2010. "Fertility Inc.: Consumption and Subjectification in Lesbian Reproductive Practices." In *Biomedicalization: Technoscience, Health, and*

Illness in the U.S., edited by A.E. Clarke et al., 82–98. Durham, NC: Duke University Press.

Moore, Lisa J. 1997. "'It's Like You Use Pots and Pans to Cook, It's the Tool': The Technologies of Safer Sex." *Science, Technology and Human Values* 22 (4): 343–471.

———. 2008. *Sperm Counts: Overcome by Man's Most Precious Fluid.* New York: New York University Press.

Moore, Lisa, and Mariane Grady. Forthcoming. "Putting 'Daddy' in the Cart: Ordering Sperm Online." In *Reframing Reproduction,* edited by Meredith Nash. London: Palgrave.

Moore, Mignon. 2011. *Invisible Families: Gay Identities, Relationships, and Motherhood among Black Women.* Berkeley: University of California Press.

Nationwide News Party Limited. 2011. "No-Dad Babies Seen as Boon for Lesbians." *Mercury, Hobart,* July 12.

Pfeffer, Carla. 2012. "Normative Resistance and Inventive Pragmatism: Negotiating Structure and Agency in Transgender Families." *Gender and Society* 26 (4): 574–602.

———. Forthcoming. *Postmodern Partnerships: Women, Transgender Men, and Twenty-First Century Queer Families.* Oxford: Oxford University Press.

Plummer, Ken. 1995. *Telling Sexual Stories: Power, Change, and Social Worlds.* New York: Routledge.

———. 2001. "The Square of Intimate Citizenship: Some Preliminary Proposals." *Citizenship Studies CISDFE* 5 (3): 237–53.

Rudrappa, Sharmila. 2010. "Making India the 'Mother Destination': Outsourcing Labor to Indian Surrogates." In *Gender and Sexuality in the Workplace,* edited by Christine L. Williams and Kirsten Dellinger, 253–85. Research in the Sociology of Work 20. Bingley, UK: Emerald Group.

Schmidt, Matthew, and Lisa J. Moore. 1998. "Constructing a 'Good Catch,' Picking a Winner: The Development of Technosemen and the Deconstruction of the Monolithic Male." In *Cyborg Babies: From Techno-Sex to Techno-Tots,* edited by R. Davis-Floyd and J. Dumit, 21–39. New York: Routledge.

Silverberg, Cory. 2012. *What Makes a Baby.* Self-published.

Spar, Debora. 2006. *The Baby Business: How Money, Science, and Politics Drive the Commerce of Conception.* Cambridge, MA: Harvard Business Review Press.

Waldby, Catherine, and Robert Mitchell. 2006. *Tissue Economies: Blood, Organs and Cell Lines in Late Capitalism.* Durham, NC: Duke University Press.

Weiss, Rick. 2003. "In Laboratory, Ordinary Cells Are Turned into Eggs." *Washington Post,* May 2, A1.

World Professional Association for Transgender Health (WPATH). 2011. *Standards of Care for the Health of Transsexual, Transgender, and Gender Nonconforming People.* Version 7. Accessed November 11, 2012. www .wpath.org/.

Zhang, G., Y. Gu, X. Wang, Y. Cui, and W. K. Bremner. 1999. "A Clinical Trial of Injectable Testosterone Undecanoate as a Potential Male Contraceptive in Normal Chinese Men." *Journal of Clinical Endocrinology and Metabolism* 84 (10): 3642–47.

Zonneveldt, Mandi. 2011. "No-Dad Babies Uproar." *Herald Sun*, July 11, News 2.

37. Reproductive Tourism

Equality Concerns in the Global Market for Fertility Services

Lisa Chiyemi Ikemoto

INTRODUCTION

Assisted reproductive technology (ART) lures people across borders. The willingness to travel for ART and the practices that facilitate fertility travel are known as "reproductive tourism." In the past few years, reproductive tourism has expanded rapidly, and has acquired a public profile in the process. The news media has reported on various aspects of reproductive tourism through feature articles, health bulletins, business reports, and human interest stories. These stories illustrate, even if they do not directly address, that the underlying global inequalities between geographic regions and their residents—and local inequalities among residents based on gender, class, race, and ethnic hierarchies—enable reproductive tourism.

At the same time, news stories reported from the United States tend to focus on destination spots outside of the United States. India, in particular, has become somewhat fetishized by the media's depiction of reproductive tourism.[1] This has the effect of situating the inequalities and other problematic aspects of reproductive tourism outside of the West, in non-white jurisdictions. Yet the United States and other Western countries are active participants in reproductive tourism as both destination spots and points of departure.[2]

Excerpt of original publication by Lisa Chiyemi Ikemoto. "Reproductive Tourism: Equality Concerns in the Global Market for Fertility Services." *Law and Equality: A Journal of Theory and Practice* 27, no. 2 (August 2009): 277–309. Reprinted by permission of Lisa Chiyemi Ikemoto. Due to space limitations, portions of this chapter have been deleted or edited from the original.

Prospective fertility patients leave their home jurisdictions to use ART for a complex mix of reasons.[3] For example, laws or social rules might restrict access at home, or the cost of ART use in other countries may be lower. Some travel to bypass a local dearth of technology, and others seek to use the fruits of third parties' bodies—eggs, sperm, or wombs. Some may simply want secrecy. The supply side of reproductive tourism has formed to satisfy these needs in a sprawling commercial enterprise that is sophisticated in some respects and crude in others.

Yet, the dominant narrative of reproductive tourism, and of ART use generally, speaks of the private sphere desire for family formation. The profile of reproductive tourism, then, is that of the couple or individual who literally travels to the ends of the earth to have a child. This narrative does not mask the commerce, but characterizes it as the fortuitous means of achieving the dream of parentage. In effect, the narrative shifts the gaze from the market to the intimate aspects of family formation, [separating] . . . family and market [rather than examining] . . . how the human need of family formation interacts with commerce, and how geopolitical differences shape that interaction.

ART commerce across jurisdictional lines is fluid. The participants on both the demand and the supply side of the global fertility market change based on legal, medical, and normative innovations (or regressions). Participants include prospective patients, states, countries, providers, health care facilities, sperm banks, egg donors, surrogates, agencies, and brokers. When legal rules, technology, or social norms change, the destination spots and departure points of reproductive tourism change as well. The geographic shifts echo in the identities of the human participants. . . .

A DESCRIPTION OF REPRODUCTIVE TOURISM

Reproductive tourism is a phenomenon of moving parts. While the description is linear, the reality is interactive. That is, ART, the destination spots and points of departure, the prospective patients, and the commercial entities on the supply side are mutually responsive. As the destination spots and points of departure shift, so do the patients and those providing gametes and surrogacy as third-party participants. When new technologies or commercial practices emerge, the market realigns, expands, or contracts.[4] . . .

Because the fertility business is primarily profit-based, the supply side entities have a stake in increasing demand.[5] The growth of reproductive tourism indicates that the fertility business has successfully promoted family formation through ART use. Fertility doctors have not played the role of passive professionals surrounded by a whirl of commercial activity;[6] many have become influential stakeholders who use a combination of medical and commercial practices to enhance their market positions. The role of physician-entrepreneur is not new or unique to fertility practice, but the use of therapies that require third-party body parts for nonlifesaving procedures may be. . . .

Surrogacy, in both the domestic and international contexts, has probably triggered the greatest media attention. Surrogacy is not a technology per se. Assisted conception is used to impregnate a woman who gestates and gives birth to a child on behalf of an individual or a couple who intend to raise the child as their own.[7] IVF, with gametes provided by the intended parent(s) and/or third-party donors, results in the birth of a child who has a biological relationship with the surrogate by virtue of pregnancy and birth, but a genetic relationship with others.[8] The intentional separation of biological and social motherhood, as well as the commercial practices that accompany most surrogacy arrangements, make this practice one of the most controversial services on the market.[9] The introduction of this form of surrogacy—gestational surrogacy—has increased the demand for women to be surrogates, for third-party gametes, and for IVF.[10]

DESTINATION SPOTS AND POINTS OF DEPARTURE

Reproductive tourism spans the globe. The destination spots and points of departure in the global fertility market include both developed and developing countries. Australia, Canada, Germany, India, Israel, and South Africa are among the many destination spots. Fertility travelers hail from a range of countries such as Costa Rica, Japan, Mexico, and the United Kingdom.[11]

Many jurisdictions, such as Australia, Italy, and Germany, are both destination spots and points of departure. The United States is also a major presence on both the supply and the demand side of the market. United States residents travel within the United States to access ART that is not readily available in their home states or regions.[12] United States suppliers also serve many who enter the United States in order to access ART.[13] At the same time, many United States residents cross borders to seek fertility services outside the United States.[14]

Often, fertility travelers have racial or gender preferences that inform their choice of fertility destination. Many seeking third-party gametes want gametes from individuals whose race matches their own.[15] As a result, Romania and the Ukraine are among the popular destination spots for egg donation among whites.[16] Because of its racial diversity, the United States is popular among travelers from Asia and Latin America, as well as countries with predominantly white populations.[17] Jurisdictions that permit preimplantation genetic diagnosis (PGD) use for sex selection attract those who wish to choose the sex of their child.[18]

Some destinations are attractive because of their religious or social norms. . . . In some jurisdictions, either legal regulation or clinic-imposed rules restrict ART access based on marital status and/or sexual orientation. On the other hand, some clinics have formed for the express purpose of providing access to single people and lesbian and gay couples. Some regions or clinics offer services on an inclusive basis because it is good for business.[19] . . . As a result, there are destination spots for single women and men, and lesbian and gay couples who want access to ART, such as Belgium, Finland, Greece, India, and the United States.[20] . . .

COMPARING MEDICAL TOURISM

. . . [R]eproductive tourism is distinct from other types of medical tourism in at least three ways. First, when successful, reproductive tourism results in the birth of children. Risks arising from legal uncertainty over the parent-child relationships formed by ART use are qualitatively different from quality of care issues. The patient's interests are not the only ones at stake. The child's status and future are also at stake. If gamete donors or surrogates were involved, then their interests are also implicated.

Second, reproductive tourism relies on third parties to contribute their gametes and wombs, and to take health risks in order to enable ART use by others. Traveling to access gamete donors and/or surrogates is similar in some ways to traveling in order to access organs for transplantation. Both rely on the availability of persons who provide the raw materials of therapy from their own bodies. Both raise issues about the source and means of procuring use of others' bodies. Yet, while payment to procure gametes and surrogacy has provoked controversy, it has so far proven less controversial than payment for organ procurement.[21]

Finally, to the extent that reproductive tourism relies on the use of others' bodies, it relies primarily on women's bodies—those of egg donors and surrogates. ART is gendered technology.[22] Regardless of the cause of infertility, it is the woman who is seen as infertile,[23] [and] . . . most of the treatments are administered to the woman even in cases of male infertility.[24] ICSI is used to overcome a common cause of infertility—low sperm motility.[25] Yet it is the woman who undergoes IVF in order to retrieve the eggs that are injected with sperm, and who undergoes egg transfer to achieve pregnancy.[26] Thus, women bear most of the health risks, as egg donors, IVF patients, and surrogates. For many patients, egg donors, and surrogates, there are social risks as well.[27] . . .

EQUALITY CONCERNS ARISING FROM REPRODUCTIVE TOURISM

. . . Fertility travelers often go to less-developed countries, . . . [raising] questions about reproductive tourism's impact on health care access in those countries. One question is whether the effort and resources that are put into fertility clinics and hospitals, in order to attract foreign patients, divert resources to private facilities that provide care for the elite.[28] In many countries, this could reinforce a preexisting two-tiered health care system.[29] One could argue that the extra revenue generated by reproductive tourism can be used to expand health services for people who are dependent on public health care. There is, however, no evidence that this is occurring.[30]

A closely related question is whether a focus on ART use resets health care priorities without regard to domestic health care needs. In both developed and less developed . . . jurisdictions where ART use is not substantially covered by insurance, . . . persons with low incomes and much of the middle class cannot afford fertility treatment.[31] . . .

A third question is whether reproductive tourism lures physicians to private clinics in a well-paying field of practice and away from public hospitals, creating a shortage where health care is most needed.[32] . . .

UNEQUAL ALLOCATION OF HEALTH RISKS TO WOMEN

[As a] gendered technology, [ART] allocates most of the health risks to women. Some of the highest risks arise from egg retrieval and surrogacy.[33] Both require the administration of pharmaceutical hormones, which creates short-term risks, and for which there is little data on long-term risks.[34] Both require invasive procedures that produce real, but low, levels of risk.[35]

Surrogacy also places women at risk. The majority of women who give birth experience no adverse health effects.[36] Pregnancy, however, does present significant risk of a wide range of adverse health consequences.[37] In addition, most women who carry children for others do so as gestational surrogates.[38] Typically, a woman undergoing IVF as a gestational carrier takes hormones in order to coordinate her cycle with the embryo transfer, thus increasing the chances that the embryo transfer results in a pregnancy.

While these drugs do not prompt multiple egg production, they may produce side effects and pose long-term health risks.[39] . . .

A more subtle dynamic may also increase risk to egg donors and surrogates. Egg donors and surrogates provide the valuable raw materials of a for-profit industry. Women can provide eggs and gestational services precisely because of their sex. In addition, while egg donors and surrogates are undergoing medical procedures, they are not the patients undergoing fertility treatment. They are the means to fertility treatment. The resulting interplay between biological essentialism and commodification of the women who are the means to the end may permit a laxness in minimizing risk to those women. It may also foster a willingness to violate good medical practice in order to get results for foreign patients. For egg brokers, clinics, and fertility tourists, the more eggs retrieved per cycle, the better. Higher dosages of ovarian stimulation drugs increase the chances of multiple egg production in women. But higher dosages also increase the risk of ovarian hyperstimulation stress syndrome, which includes nausea, vomiting, accumulation of fluid in the abdomen, kidney and liver dysfunction, and even kidney failure among its symptoms.[40] This normative dynamic creates an inverse relation between the egg donor's intrinsic worth and her extrinsic value in the fertility industry.

REPRODUCTIVE ENTITLEMENT IN REPRODUCTIVE TOURISM

Reproductive tourism for two high-demand fertility services—IVF with third-party eggs and surrogacy—depends heavily on a lack of comparable economic alternatives for the women who provide eggs and surrogacy, and significant wealth disparity between those women and fertility travelers. But for those inequalities, fewer women would become egg donors and surrogates, [and] . . . the fees for third-party eggs and surrogacy might be too high to justify travel for all but a few individuals.

The causal links between these inequalities and fertility services would seem to indicate that egg procurement and surrogacy are exploitative practices. While some academics assert that exploitation is serious enough to justify regulation of these practices,[41] two other responses have so far prevailed. One argument locates egg procurement and surrogacy in the intimate sphere of family, [where the] . . . main characters are the infertile couples who desperately want children.

Within the narrative, their ability and willingness to pay for ART use evidences both their suitability and deservedness as parents. Closely examined, the narrative reveals a market-based test for parental suitability, but the narrative's trick is to shift the gaze to the yearning and need—the story's emotional content. That content is compelling because it is real. Infertility is not simply a medical condition for many. It causes emotional pain and loss of self-esteem and breaks up relationships.[42] Yet, centering the yearning for family also elides the commercial nature of the practices that enable family formation through ART use. Hence, women who provide eggs for others' use are called "donors," even though they receive thousands of dollars for doing so. They provide a "gift" that is priceless, and yet has been carefully priced.

The second response simply accepts the inequalities on which reproductive tourism depends as natural features of a market economy. After all, these inequalities preexist reproductive tourism and would persist without [it]. The free market narrative directly counters the claim of exploitation in a way that the family formation narrative does not. In the free market narrative, women who provide eggs and bear children for others are free agents. The women who participate opt into the market. . . .

Both narratives express a sense of entitlement. . . . The use of women for their biological capacity to reproduce, in a context in which geopolitical differences between departure points and destination spots often account for the wealth disparities, supports a sense of entitlement. In addition, well-established narratives that describe low-income women, women of color, and women in less developed nations as "too fertile" nourish the claim of the infertile to ART use as their due.[43] In other words, the identity of those who provide eggs and gestation for others overlaps with those deemed in need of population control.[44] The population control narrative has already cast the women sought for eggs and surrogacy as subalterns. In doing so, the population control narrative has prepared the ground for claims of entitlement in the family formation and market narratives.

RACIAL ALIGNMENT AND RACIAL DISTANCING IN REPRODUCTIVE TOURISM

Racial identity figures significantly in ART use. Those who seek third-party gametes often have specific racial preferences.[45] . . . In the gamete market, racial preferences seem natural and unobjectionable because they enable the resulting family to look like a biologically related family. The goal of a racially matched family in a commercial context may have two problematic effects. First, it makes race a commodity in the gamete market. Second, the naturalizing effect of the racial preference is elastic. It makes other genetic preferences seem both natural and acceptable, thus clouding what might otherwise seem to be obvious eugenic preferences.

When gestational surrogacy is used, the intended parents or third parties provide the gametes. As a result, the race of the surrogate does not inform the race of any child born from gestational surrogacy. Race matching plays a smaller role in the preferences of fertility travelers who seek surrogates. Hence, racial differences between fertility travelers and surrogates are more common than they are in the egg market.[46] This enables sites such as India to flourish as destination spots for white fertility travelers from Western Europe and the United States.

The racial difference might even make a destination spot more attractive. The character of the mainstream media's attention to India's surrogacy industry illustrates the reasons for this point. The media stories represent the surrogacy business in India as exotic because of racial, cultural, and economic differences between the fertility tourists from the United Kingdom and United States and the Indian surrogates.[47] At the same time, the stories make India's role in reproductive tourism seem like a predictable aspect of India's success in positioning itself as the destination spot for outsourcing the service economy.[48] What these stories express is the persistence of a form of racial distancing that may make hiring a woman to gestate, give birth to, and give up a child psychologically comfortable. It is a postindustrial form of master-servant privilege. In effect, it makes the nonwhite woman in the nonwhite country a marketable source of surrogacy.

CONCLUSION

Equality concerns arise from the very way that reproductive tourism works. It might be accurate to say that inequality is a driver of reproductive tourism. That is probably true for other markets as well. But reproductive tourism has two elements that make acknowledging the equality concerns uncomfortable. One is the desired product: a family. Reproductive tourism is a commercial means of acquiring one of the most intimate aspects of human life—the parent-child relationship. The second element of reproductive tourism that groups it with disreputable markets (sex tourism, the organ trade) and distinguishes it from reputable ones is that it relies heavily on women's bodies, and in particular, woman's reproductive capacities. Here, the discomfort arises in part from the fact that a woman's reproductive work has become part of a formal economy.

Neither the explanatory narratives for reproductive tourism nor the existing scholarship addressing reproductive tourism seems to want to look at the ways that reproductive tourism mixes the commercial and the intimate . . . in a wide variety of settings that literally span the globe. . . .

NOTES

1. Media coverage of India's role in reproductive tourism focuses almost exclusively on surrogacy, and many stories feature the same clinic—the Akanksha Fertility Clinic in Anand, India. See, e.g., . . . Henry Chu, "Wombs for Rent, Cheap," *Los Angeles Times*, April 19, 2006, A1 . . .; Abigail Haworth, "Womb for Rent: Surrogate Mothers in India," *Marie Claire*, August 1, 2007, 124 . . .; Sarmishta Subramanian, "Wombs for Rent," *Maclean's*, July 2, 2007, 40. . . .

2. See Lisa Chiyemi Ikemoto, "Reproductive Tourism: Equality Concerns in the Global Market for Fertility Services," *Law and Inequality: A Journal of Theory and Practice* 27, no. 2 (Summer 2009): 277–309, part I., C–D. . . .

3. See infra discussion of reproductive tourism. . . .

4. Third-world countries often serve as the testing ground for new reproduction technologies, which primarily supply expanding markets in the United States and Western Europe. Janice C. Raymond, *Women as Wombs: Reproductive Technologies and the Battle over Women's Freedom*, 140–44 (North Melbourne: Spinifex Press, 1993), 1–21.

5. Id., 109. . . .

6. Id.

7. Deborah L. Spar, *The Baby Business: How Money, Science, and Politics Drive the Commerce of Conception* (Cambridge, MA: Harvard Business Review Press, 2006), 70–72.

8. Id., 79.

9. Raymond, supra note 4, xxii. . . .

10. Spar, supra note 7, 81.

11. See Brian Alexander, "Health Bulletin: How Far Would You Go to Have a Baby?," *Glamour*, May 2005, 116, 116; Felicia R. Lee, "Driven by Costs, Fertility Clients Head Overseas," *New York Times*, January 25, 2005, A1, www.nytimes.com/2005/01/25/national/25fertility.html.

12. Spar, supra note 7, 210–14. . . .

13. Debora Spar, "Reproductive Tourism and the Regulatory Map," *New England Journal of Medicine* 352 (2005): 531.

14. Lee, supra note 11.

15. Anthony Barnett and Helena Smith, "Cruel Cost of the Human Egg Trade," *Observer*, April 30, 2006, news, 6, www.guardian.co.uk/uk/2006/apr/30/health.healthandwellbeing/print. . . .

16. Id.

17. See the Reproductive Specialty Medical Center, accessed March 31, 2009, www.drary.com/DonorProfile.html. . . .

18. Spar, supra note 7, 122.

19. Pacific Reproductive Services, accessed March 31, 2009, www.parcepro.com/index.php.

20. "IFFS Surveillance 07," supplement 1, *Fertility and Sterility* 87 (2007): S17; see Amrit Dhillon, "Surrogacy Debate Rages after Gays 'Father' Twins," *South China Morning Post*, November 30, 2007, 14; Farah Farouque, "Gay Couples Shop Offshore for Surrogate Mothers," *Age*, August 16, 2003, www.theage.com.au/articles/2003/08/15/1060936056512.html.

21. See Sarmishta Subramanian, "Wombs for Rent," *Maclean's*, July 2, 2007, supra note 1, 40. . . .

22. Marcia C. Inhorn and Daphna Birenbaum-Carmeli, "Assisted Reproductive Technology and Culture Change," *Annual Review of Anthropology* 37 (2008): 177.

23. See id., 180. . . .

24. See id. . . .

25. See Melinda Beck, "Ova Time: Women Line up to Donate Eggs—for Money," *Wall Street Journal*, December 9, 2008. . . .

26. See id. . . .

27. See Inhorn and Birenbaum-Carmeli, supra note 22, 182. . . .

28. See Nathan Cortez, "Patients without Borders: The Emerging Global Market for Patients and the Evolution of Modern Health Care," *Indiana Law Journal* 83 (2008): 110.

29. See Atul D. Garud, "Medical Tourism and Its Impact on Our Healthcare," *National Medical Journal of India* 18 (2005): 319.

30. See Cortez, supra note 28, 110–11. . . .

31. Inhorn and Birenbaum-Carmeli, supra note 22, 179–80. . . .

32. See Cortez, supra note 28, 109–10; Garud, supra note 29, 319.

33. See Barnett and Smith, supra note 15. . . .

34. Commission on Assessing the Medical Risks of Human Oocyte Donation for Stem Cell Research, "Workshop Report Summary," *National Academy of Science* (2007): 2–3 (hereinafter Workshop Report Summary); Human Cloning and Embryonic Stem Cell Research after Seoul; Examination Exploitation, Fraud and Ethical Problems in the Research: Hearing before the Subcomm. on Criminal Justice, Drug Policy, and Human Resources of the H. Comm. on Gov't Reform, 109th Cong. 76 (2006) (statement of Judy Norsigian, executive director, Our Bodies Ourselves).

35. Workshop Report Summary, supra note 34, 3. . . .

36. See Lars Noah, "Assisted Reproductive Technologies and the Pitfalls of Unregulated Biomedical Innovation," *Florida Law Review* 55 (2003): 619.

37. Spar, supra note 7, 81–82.

38. Noah, supra note 36, 610–12.

39. Id., 620–22.

40. Workshop Report Summary, supra note 34, 2.

41. See Bartha M. Knoppers and Sonia LeBris, "Recent Advances in Medically Assisted Conception: Legal, Ethical and Social Issues," *American Journal of Law and Medicine* 17 (1991): 329, 356; Angie Godwin McEwen, "So You're Having Another Woman's Baby: Economics and Exploitation in Gestational Surrogacy," *Vanderbilt Journal of International Law* 32 (1999): 292–96; Richard F. Storrow, "Quests for Conception: Fertility Tourists, Globalization and Feminist Legal Theory," *Hastings Law Journal* 57, 295, 327 (2005), 327–30, https://papers.ssrn.com/sol3/papers.cfm?abstract_id=879072.

42. See Inhorn and Birenbaum-Carmeli, supra note 22, 179.

43. See Lisa Ikemoto, "The In/Fertile, the Too Fertile, and the Dysfertile," 47 *Hastings Law Journal* 1007, 1028 (1995): 1008.

44. See Inhorn and Birenbaum-Carmeli, supra note 22, 179; Florencia Luna, "Assisted Reproductive Technology in Latin America: Some Ethical and Sociocultural Issues," in *Current Practices and Controversies in Assisted Reproduction: Report of a Meeting on "Medical, Ethical and Social Aspects of Assisted Reproduction,"* edited by Effy Vanyena et al. (Geneva: World Health Organization, 2001), 32, http://apps.who.int/iris/bitstream/10665/42576/1/9241590300.pdf. . . .

45. Lizette Alvarez, "Spreading Scandinavian Genes without Viking Boats," *New York Times*, September 30, 2004, A4; Kenneth R. Weiss, "Eggs Buy a College Education," *Los Angeles Times*, May 27, 2001, A31.

46. See Spar, supra note 7, 83.

47. See sources cited supra note 1.

48. See Chu, supra note 1; Haworth, supra note 1, 124; Subramanian, supra note 1, 42.

38. Make Me a Baby as Fast as You Can

Douglas Pet

The booming business in international surrogacy, whereby Westerners have begun hiring poor women in developing countries to carry their babies, has been the subject of plenty of media buzzing over the past few years. Much of the coverage regards the practice as a win-win for surrogates and those who hire them; couples receive the baby they have always wanted while surrogates from impoverished areas overseas earn more in one gestation than they would in many years of ordinary work. Heartening stories recount how infertile people, as well as lesbian and gay couples who want to have children (and who often suffer the brunt of discriminatory adoption policies), have formed families by finding affordable surrogates abroad. The *Oprah Winfrey Show* has even portrayed the practice as a glowing example of "women helping women" across borders, celebrating the arrangements as a "confirmation of how close our countries can really be."

But make no mistake: This is first and foremost a business. And the product this business sells—third-party pregnancy—is now being offered with all sorts of customizable options, guarantees, and legal protections for clients (aka would-be parents). See for example the December 2010 *Wall Street Journal* article "Assembling the Global Baby," which focused on high-profile PlanetHospital, a Los Angeles–based medical tourism company that has become one of many one-stop-shops for overseas surrogacy and that is going to great lengths to woo customers. "We take care of all aspects of the process, like a concierge service," company founder Rudy Rupak told the *Journal*.

The *Journal* article didn't go into much detail about how surrogates' rights might figure into this "concierge service." But interviews with those running the operation, information that was available on PlanetHospital's website until it was redesigned last year, and an information packet called "Results Driven Surrogacy" that the company distributes to prospective clients, begin to fill in the picture. The version of the packet that PlanetHospital sent me in July assures clients that each surrogate is "well looked after." Surrogates spend "the entire duration of the pregnancy at the clinic or a guest house controlled by the clinic" where their habits, medications, and diets are carefully regimented and monitored. PlanetHospital promises clients that when surrogates have a history of smoking, "We make sure they do not suddenly get a craving for it during pregnancy." Like most other surrogacy clinics and brokers, PlanetHospital accepts only surrogates who already have children of their own. While the usual reasoning for this sort of requirement is that having children proves a woman can safely carry a pregnancy to term, PlanetHospital's literature notes that the policy also ensures that she does "not bond with your baby."

In addition, PlanetHospital offers customers a novel means of accelerating their bid for a family: the option of having embryos implanted into two surrogates at the same time. The selling points of this package (which was previously marketed under the name "India Bundle"): Implantation in two surrogates at a time increases the chance of immediate impregnation and decreases the waiting time for a baby. As the company's website used to explain:

> PlanetHospital innovated the idea of routinely performing IVF on two surrogates simultaneously thus increasing the odds of pregnancy by more than 60%. The notion of hiring two surrogates in the US and doing IVF on both surrogates would be financially prohibitive[.] PlanetHospital has negotiated rates with a highly reputable clinic in India that not only provide couples with two surrogates, but also four attempts.

Of course, this approach could also leave a couple with multiple babies, possibly gestating in multiple women. Until recently, if both surrogates became pregnant—or if either surrogate became pregnant with twins—clients could opt to have the extra pregnancy aborted or twins reduced to a singleton, depending on how many babies the clients wanted or decided they could afford. As PlanetHospital's website used to explain, "The simple answer to that is it is up to you to decide what you wish to do, you can choose to have all the children (which will cost slightly more of course . . .) or you can request an embryo reduction." Founder Rudy Rupak told me via email that the company no longer allows clients to elect either reductions or abortions under the advice of its lawyers, who worry that it could open up some "nasty debates" as Indian authorities discuss the possibility of surrogacy regulation. "If a client wants both surrogates then they have to accept it if both are pregnant," he wrote.

According to the pricing information PlanetHospital provided me, the company's cheapest package for a single surrogate pregnancy is $28,000. To use two surrogates simultaneously, clients pay an initial $15,500, plus $19,600 for each surrogate that gets pregnant and delivers a baby. (And if one or more surrogate ends up carrying twins, clients pay a surcharge of at least $6,000 per twin.)

Of those amounts, PlanetHospital's Indian surrogates are paid between $7,500 and $9,000. By comparison, the cost of a single surrogate pregnancy in the United States can run up to $100,000 including medical expenses, of which about $20,000 goes to the surrogate. Rupak emphasized the extent to which Indian surrogates stood to benefit from the arrangement. "While some people might scream exploitation," he wrote to me in an email, "bear in mind that the per capita average income of a typical [Indian] surrogate would be $600/annum. She is thus making close to 12x her annual salary by being a surrogate."

It is worth looking beyond economic comparisons, however, to see how such transactions may compromise surrogates' choices. For example, if one of PlanetHospital's Indian surrogates wants to abort her own pregnancy, she is out of luck. The company's vice president of corporate affairs and business development, Geoff Moss, recently told me: "If they feel like terminating the pregnancy, they can't do that; there is a legal contract." He also suggested that surrogates would not want to take that step even if it were available to them: "They have children at home," he said, "so they understand how important it is for these people to be parents."

It seems unlikely that PlanetHospital's prohibition on surrogate-elected abortions would fly in the United States. George Annas, chair of the department of Health Law, Bioethics & Human Rights at Boston University's School of Public Health, told me via email that he believes "there is no way a competent adult woman could prospectively waive her constitutional right to terminate (or not) a pregnancy (or selectively reduce one) that would be upheld by a U.S. court." It would appear, then, that Western surrogacy brokers benefit by looking across borders not just because it allows them to locate cheap "labor" but also because some arrangements may face less legal scrutiny than they would in the United States. Moss confirmed that legal differences between the two countries make India an attractive location for surrogacy. "In the United States, in many cases, there will be surrogates all of a sudden saying that they want to keep the baby," he said, "In India it's all contractual."

The surrogates' lack of control over the course of their pregnancies continues through delivery day. According to PlanetHospital's information packet, "All the surrogates will deliver the child through cesarean birth." Moss explained one reason for this policy: "Because we can time it that way for the intended parents to be there for the birth. That way if the baby is going to be born on Dec. 10, the parents can make their travel arrangements, fly to India, and be there to receive the baby when it's born." The information packet adds another reason: C-sections are "much safer for the child and the surrogate." Rupak told me that while surrogates can refuse the operation and deliver naturally, PlanetHospital had been advised by independent obstetricians that routine cesarean delivery was the safest choice. Women's health experts and advocates would likely disagree, as many believe C-sections to be riskier to both mother and baby in the absence of other complications. The procedure also makes vaginal birth more hazardous in subsequent pregnancies and could therefore endanger the lives of low-income surrogates who may not have access to hospital care for future deliveries.

PlanetHospital's information packet ends with a note cautioning prospective clients not to make too much of any negative reviews of the company that they might find on the Internet. "Surrogacy is a very emotional matter," the packet explains. "This is not a matter of buying a car, this is a life you are asking us to help you create." But while we can probably all agree that ordering up a child is nothing like buying a car, PlanetHospital goes on to draw an equally unlikely parallel between its business and that of a well-known purveyor of mail-order shoes. "Like Zappos," the note concludes, "we too want to 'deliver happiness' and maintaining our integrity is the most important part of that mission."

Babies aren't like shoes any more than they're like cars, of course, but the comparison is telling.

Wombs are being rented in what amounts to an international marketplace. And with new cross border surrogacy operations springing up recently in countries such as Panama, Guatemala, Georgia, and Greece, the number of pregnancies involving multinational players and profit interests is likely to increase. If for-profit companies are going to continue to approach baby-making like an import-export business, maybe it's time for governments to start treating it that way, adapting oversight and protections for all parties involved. In the meantime, in the absence of meaningful regulation, the rights of surrogate mothers are being bought, sold, and signed away.

39. Let's Get Rid of the Secrecy in Donor-Conceived Families

Naomi Cahn and Wendy Kramer

From movies like *The Kids Are All Right* and *Delivery Man* to the MTV reality show *Generation Cryo*, donor-conceived children in fiction and real life are growing up and attempting to find out where they came from. But what about all of the kids who don't know how they were conceived? As we write in our new book, *Finding Our Families*, a majority of married straight couples still don't tell their children if they used donor eggs or sperm to get pregnant.

Secrecy has long been intimately intertwined with donor conception. Once upon a time, nondisclosure was standard. Almost no one talked about whether they had used a donor, and the donors themselves didn't worry that their biological offspring might come knocking. This culture of secrecy meant that many parents with donor-conceived children didn't think about disclosure, because no one ever told them that it might be the right thing to do for their children. Most children didn't know they were donor-conceived, so they never asked questions. Sperm banks and, more recently, egg donation programs drew on traditional adoption practices and beliefs—keeping health or genetic information private, never telling the adopted that they were not the biological children of their parents. Keeping the secret was seen as protecting the entire family from stress and pain.

The adoption world moved long ago toward telling children the truth about their origins, but change has been much slower for the donor-conceived. While professional organizations now advocate disclosure, some parents still struggle with whether, when, and how to tell, and many still will not do so.

Parents have a number of reasons to worry about telling, and have told us:

> I don't want to confuse my child and disrupt her normal childhood.

> I don't want to revisit those uncomfortable feelings of shame about infertility. If I tell my child, then I'll have to tell family members and friends as well.

> I know I probably should tell, but I fear my child will become angry and reject me when she finally finds out. She might feel more connected to her genetically related parent and not to me.

> I don't think my child needs to know unless the fact of her donor conception becomes important, such as for medical reasons.

> My partner and I disagree over whether to disclose.

> I worry that telling is not the end of the dialogue or the story, and that I may be unable to answer my child's questions about her genetics or ancestry, so better not to begin the conversation at all.

> I don't want my child to think that genetics are all that important.

So, yes, the decision to talk about donor conception, even if you value familial honesty, can be hard. But according to a new study of egg-donor families by researchers at Weill-Cornell Medical College, early disclosure is best for the whole family. Researchers found that parents who told their children before they turned ten reported no anxiety relating to disclosure and expressed full confidence that they had done the right thing. By contrast, among the nondisclosing families, there were high levels of anxiety as they waited for the "right time" to tell, and found themselves confronting the challenge of disclosing to teenagers or young adults.

And in a systematic review of forty-three studies on the disclosure decision-making process for heterosexual couples who had used donor eggs, sperm, and embryos, researchers found that the parents who disclosed emphasized children's best interests, their rights to know that they are donor-conceived, honesty as an essential component of the parent-child relationship, and the stress inherent in keeping a secret. By contrast, while those parents who had not disclosed also emphasized the best interests of their children, they saw no benefit from disclosure and wanted to protect the child from stigma or other damage. In some families, of course, disclosure decisions are fairly easy. Donor conception is difficult to hide in single-parent and LGBT families, and research shows that children in those families learn at an earlier age than do children in heterosexual families. It's the increasing number of these non-"traditional" families that has helped bring much more openness to the donor conception process.

As we write in our book, not telling your child does not render the fact of the conception irrelevant. Reminders about donor conception will frequently come up, perhaps as you or others try to figure out why your child is so tall, or so good at math, or so outgoing. When your pediatrician asks about your child's medical history, you will have to lie. You may feel uncomfortable when friends talk easily about how much their children look like them, or when they share with you their struggles on how to explain to their children how Daddy planted that seed in Mommy's belly.

This secrecy around donor conception is a heavy load to carry, and the layers of deception build up. The best-kept secret can warp family life, filling children with anxiety they don't understand, and parents with guilt. In an effort to protect kids they love from what parents perceive as the difficult truth of their origins, parents are hurting them—and the parent-child bond—instead.

PART VIII

Selecting Traits, Selecting Children

The increasing sophistication and spread of prenatal and prepregnancy testing means that parents are receiving more information about the genetic characteristics of their future children at earlier stages than ever before. As a result, they are facing decisions about what kinds of traits—and what kinds of children—they will bring into the world. These developments affect not just families but also cultural and societal norms and assumptions.

Prepregnancy genetic testing takes place in conjunction with in vitro fertilization. People using this type of fertility procedure can opt to screen and select embryos on the basis of the gene variants they carry and transfer only embryos with the desired characteristics into a woman's uterus to establish a pregnancy. Known as preimplantation genetic diagnosis, this embryo-testing technique was introduced in the early 1990s for people who risk passing on genes associated with a handful of severe genetic conditions such as sickle cell disease, Tay Sachs disease, and Duchenne muscular dystrophy.

Prenatal testing has been in use for much longer. Amniocentesis, for example, which tests amniotic fluid extracted from a pregnant woman's uterus, came into use in the late 1960s. The early 2010s saw the introduction of a new kind of prenatal genetic test, which analyzes fetal DNA that can be found in pregnant women's blood very early in their pregnancies. These "noninvasive prenatal tests" were quickly commercialized and have already been widely adopted, with several companies now marketing them aggressively not just to obstetricians and genetic counselors but to ordinary consumers as well.

Both prepregnancy tests of embryos and prenatal tests of fetuses are typically presented as ways to prevent the births of children affected by serious genetic conditions. But what counts as falling into that category? Put differently, the traits that are considered unacceptable in a future child are highly subjective and controversial.

In addition, prenatal and prepregnancy tests are increasingly being used to identify embryos or fetuses with genes correlated with late-onset, less serious, and treatable conditions. Down syndrome, for example, is widely targeted, yet many consider it not a tragedy to be avoided but simply a way of being human. Prenatal tests and embryo screening also enable social sex selection. And as one of the articles in this section recounts, a Los Angeles fertility clinic has offered preimplantation genetic diagnosis to perform tests that it claimed could screen embryos for eye, hair, and skin color.

The contributions in this section explore the selection technologies currently in use, how they are changing the experience of pregnancy, and the ethical and social challenges they pose. They engage the biopolitical concerns identified in the introduction to this volume as "avoiding technical developments and genetic narratives that embed social and political preferences at the molecular level" and "steering clear of a new market-driven eugenics."

In the first contribution, the late bioethicist Adrienne Asch explains and extends what has become known as the disability rights critique of prenatal testing and embryo selection. Life with a disability can be valuable to individuals, their families, and society, she argues, and those who endorse testing uncritically "act from misinformation about disability, and express views that worsen the situation for all people who live with disabilities now and in the future." Like most other disability rights advocates, Asch supports women's right to decide whether or not to have a child at a given time. Yet she is critical of making this decision on the basis of traits of the particular embryo or fetus.

Marcy Darnovsky and Alexandra Minna Stern's "The Bleak New World of Prenatal Genetics" comments on the noninvasive tests that have become widely available since 2012 and involve a simple blood draw at an initial prenatal visit. Currently marketed for chromosomal conditions such as Down syndrome, they may portend "the development of whole-genome fetal tests that can deliver far more extensive information about a child's genetic makeup and predispositions." Even now, however, this new generation of prenatal testing may not be nearly as accurate as the claims for it. This can lead to serious real world consequences, as Beth Daley explains in "Have New Prenatal Tests Been Dangerously Oversold?"

The next contribution focuses on social sex selection—that is, parents using selection technologies to have a boy or a girl for nonmedical reasons.

In "Sex Selection and the Abortion Trap," Mara Hvistendahl discusses how opponents of reproductive rights have disingenuously used the skewed sex ratios in South Asia to further their agenda of banning abortion in the United States and elsewhere. She argues that sex selection is a problem not just in other parts of the world but in the United States as well.

Gautam Naik's article reports on a controversial—and technically dubious—proposed application of embryo screening. A Los Angeles–based fertility clinic launched a program offering preimplantation genetic diagnosis, not just to choose the sex of a future child, but also to select hair, eye, and skin color. "This is cosmetic medicine," the clinic director told Naik. "Others are frightened by the criticism but we have no problems with it." Though the program was discontinued due to public backlash, the director made it clear that the setback is temporary. Trait selection "is a service," he said. "We intend to offer it soon."

40. Disability Equality and Prenatal Testing

Contradictory or Compatible?

Adrienne Asch

Is it possible for the same society to espouse the goals of including people with disabilities as fully equal and participating members and simultaneously promoting the use of embryo selection and selective abortion to prevent the births of those who would live with disabilities? As currently practiced and justified, prenatal testing and embryo selection cannot comfortably coexist with society's professed goals of promoting inclusion and equality for people with disabilities. Nonetheless, revamped clinical practice and social policy could permit informed reproductive choice and respect for current and future people with disabilities. In this . . . article, I argue that the typical justifications offered by practitioners and researchers for prenatal testing are mistaken about the implications of disability. . . . I [then] explain why I discount the claim that people with disabilities have made great progress—notwithstanding the advent of prenatal testing. I conclude by proposing reforms to our current prenatal testing practices that would meet the challenges posed by many critics.

What has become known as the disability rights critique of prenatal testing has been formulated as follows:

(1) Continuing, persistent, and pervasive discrimination constitutes the major problem of having a disability for people themselves and for their families and communities. Rather than improving the medical or social situation of today's or tomorrow's disabled citizens, prenatal diagnosis reinforces the medical model that

Original publication by Adrienne Asch, *Disability Equality and Prenatal Testing: Contradictory or Compatible?*, 30 FLA. ST. U. L. REV. 315 (2003). Reprinted by permission of *Florida State University Law Review*. Due to space limitations, portions of this chapter have been deleted or edited from the original.

disability itself, not societal discrimination against people with disabilities, is the problem to be solved.

(2) In rejecting an otherwise desired child because they believe that the child's disability will diminish their parental experience, parents suggest that they are unwilling to accept any significant departure from the parental dreams that a child's characteristics might occasion.

(3) When prospective parents select against a fetus because of predicted disability, they are making an unfortunate, often misinformed decision that a disabled child will not fulfill what most people seek in child rearing, namely, "to give ourselves to a new being who starts out with the best we can give, and who will enrich us, gladden others, contribute to the world, and make us proud." In these several contentions can be discerned two broad claims: that prenatal genetic testing followed by selective abortion is morally problematic, and that it is driven by misinformation.[1]

In what follows, I discuss these claims as applied to social institutions beyond the family, arguing that researchers, professionals, and policy makers who uncritically endorse testing followed by abortion act from misinformation about disability, and express views that worsen the situation for all people who live with disabilities now and in the future.[2] . . . My concern is to facilitate true reproductive choice for women by urging changes in the way prenatal testing occurs and the rhetoric that surrounds it.[3]

WHAT IS DISABILITY "REALLY" LIKE, OR HOW MISINFORMED ARE PEOPLE ANYWAY?

Prenatal testing, and the more recent and less common embryo screening and selection, are justified by mistaken assumptions about the quality of life of people with disabilities. [This is] demeaning to existing people with disabilities. . . .

For the past quarter century of disability scholarship and theory in the United Kingdom and North America, a significant tension has existed between what is seen as a traditional "medical model" of disability and two newer approaches, termed the "minority group model" and the "social model" of disability. Theorists with a minority group or a social model argue forcefully that clinicians, policy makers, genetic researchers, and bioethicists err in ascribing the major difficulties of people with disabilities to their physical, cognitive, or emotional make-up. Instead, the theorists assert that the difficulties should be ascribed to the mismatch between the range of people actually in the world and the institutional practices, physical structures, modes of communication, and social attitudes that assume a much narrower range of human beings than exist. . . .

Many in the field of bioethics . . . who reject the disability rights critique of prenatal testing acknowledge that a share of the problems of people with disabilities stem from life in a society that has still not made all the changes that would permit them to travel, communicate, learn, work, and play easily alongside their nondisabled peers. Yet they argue that it is better not to have a disability than to have one, and that it is preferable to select against the embryo or fetus with a disabling trait. . . .

The question keeps emerging: Just how much of the difficulty posed by disability is "socially constructed"? Could there be a social and natural world in which it would be as easy and enjoyable to live with disability as it is to live without a disability? Does the answer depend upon the particular condition under discussion—for example, should cystic fibrosis be distinguished from deafness (the former affecting needs for medical care and life expectancy, the latter affecting neither)? How much of what is negative about impairment or disability is "intrinsic" to the condition and would remain even in a society more inclusive of disability than the United States in the twenty-first century? If people prize health, and assume species-typical seeing, hearing, walking, speaking, and learning as foundational, is it not undesirable to have a condition that reduces one's general health or that limits or denies such functions as speech, hearing, cognition, or sight?

People without any disabilities naturally assume that the typical complement of human capabilities is desirable, and perhaps critical, for most plans of life. Health care is given high priority when people rank important social goods because health and species-typical functioning are taken to be essential for having a good life. Although there are variations on what "good lives" contain, many people in the United States would probably say that they would like their children's lives to include several of the following opportunities: to appreciate beauty; learn about the world; master some skills; make contributions to others; participate in satisfying relationships; live without physical or psychological pain; be safe from physical harm; develop their own interests; find satisfying work; take care of themselves; be interested in other people's welfare; and make decisions about their lives for themselves without pressure from others. This list is not meant to be exhaustive, and it is not intended to suggest that each life must contain all of these characteristics to be satisfactory to the person living it. . . .

It is possible to acknowledge that disabilities may preclude some activities that many people find worthwhile—appreciating sunsets, relishing bird songs, experiencing the interaction of body and nature in a hike through the woods. . . . Having capacities is good, but I am not sure that any capacity is an "intrinsic" good. If typical capacities and health achieve value because they enable people to participate in facets of life, it is crucial to note how much of life is open, in today's society, to people with disabilities. Brief acquaintance with people who have disabilities and who work, play, study, love, and enjoy the world should demonstrate that very few conditions preclude participating in the basic activities of life, even if some conditions limit some classes of them, or methods of engaging in them.

As a person who is blind, I cannot see a baby's smile, the antics of a friend's dog, or the paintings of Picasso. I am quite confident that I would get pleasure and satisfaction from such experiences. Nevertheless, if people who are blind cannot enjoy one class of aesthetic experiences, many others are available (weaving, sculpture, music, ocean breezes, etc.). When it is noted that people who are deaf create poetry and theater in American Sign Language, that people with mobility impairments become involved in adapted or typical athletics, that persons with autism or Down syndrome increasingly articulate their own views of their needs and experiences, it is evident that realms of activity often thought unimaginable for people with disabilities are components of many of their lives. . . .

If having a capacity is good, is not having a particular ability bad, negative, or "dis-valuable?"[4] My answer is that having a capacity can be good, but the absence of capacity is simply an absence; it need not be seen as negative, "dis-valuable" to be blind any more than it is negative or "dis-valuable" to be shorter than some people, or to be mystified by higher mathematics. . . .

If disability is a simple human variation, why do we try to promote good prenatal care in women, or to promote health in the population? There is nothing to lament about capacities to hear, speak, move, or think. The difference between selecting out fetuses and protecting them (by promoting prenatal care for women) is just that. We protect the possibility for capacity when we promote fetal health, but we refuse to acknowledge or permit the growth of people who will not have such capacity when we select against fetuses as potential people with disabling traits. Similarly, there is nothing wrong with possessing skills or aptitudes for athletics, physics, or carpentry; but the society has not yet said that only people who possess such aptitudes are welcomed. The absence of a capacity is not necessarily "bad"; the opposite of having a capacity is not having it; having it and not having it can be equally legitimate ways of living a life. . . .

THE LIMITS OF PROGRESS

Defenders of the practices of prenatal testing and embryo selection deny that these practices are incompatible with greater inclusion and participation of those with disabilities. . . . They claim that while prenatal testing for disability is becoming more widespread and routine, existing people with disabilities are making dramatic strides toward social and economic equality. I argue that the appearance of progress is illusory, or at least grossly exaggerated.

Gaps remain between people with and without disabilities in terms of education, employment, income, social life, and civic participation.[5] According to the traditional medical model of disability, those gaps are inextricably tied to the conditions themselves. With the advent of the minority group and social models, it has become possible to disentangle how factors in the built environment, modes of information dissemination, and laws and practices governing political participation, work, and education excluded, segregated, or limited the lives of people with disabilities. The richness of these latter models of disability is the legislation they helped to create, embodying a national commitment to equal opportunity in education,[6] public services,[7] employment, transportation, and places of public accommodation.[8]

Those who support vigorous efforts to reduce disabling conditions by preventing the births of people who will have them observe these legal gains and the increased presence of people with disabilities in schools and public places to demonstrate that there is no tension between prenatal selection and including those disabled people already in the population.[9] Under their view, it is possible to disvalue the disabling trait, and nonetheless to respect as social and moral equals people who exhibit these disliked traits. Prevailing social attitudes toward people with disabilities, and data about the effects of legal changes on employment, lead me to be anything but sanguine.

In passing the Americans with Disabilities Act in 1990 (ADA),[10] Congress recognized that millions of the nation's population continued to be treated differently and pejoratively by the nondisabled majority . . . and [there was a] need for the law to redress these systemic problems. . . . [However,] outlawing discrimination in public programs, employment, and places of public accommodation has not markedly altered how social science, medicine, and bioethics discuss disability when it comes to making childbearing decisions. . . .

Describing children with disabilities as children with "special needs," using the euphemism of "special needs adoption" when referring to placing children with disabilities in homes, and maintaining a system of "special education" reveal that people with disabling conditions are "others," not part of the total community. Were youth with disabling traits truly viewed as deserving of consideration when designing schools, day care centers, and afterschool programs, the programs would be created with the expectation that children differed from one another in many ways, and budget, staffing, and institutions would reflect the true diversity of the nation's youth. . . .

Unfortunately, the public consciousness of disability and the inclusion of adults who have disabilities appear more superficial than genuine. A decade after the passage of the ADA, and nearly thirty years after enactment of the Title V employment provisions of the Rehabilitation Act, people with disabilities are not succeeding in gaining access to work, and courts are frequently ruling against them when they bring cases of employment discrimination.[11] If society truly believed that people with disabilities could contribute to the nation's economy, the unemployment rate would be calculated to show that millions of the nation's disabled population of working age were not in the labor force. Considering that people with disabilities are estimated to be about 20 percent of the nation,[12] industry's failure to pursue their labor and business with accessible product design and representative advertising is astonishing, and actually detrimental to the society as a whole.

Despite the symbolic and tangible changes attributable to laws like the Americans with Disabilities Act, the nation's disabled population is still less educated, less employed, less involved in civic life, less represented in the political process, and less influential on the design of products than their numbers warrant. . . . It is in this discriminatory society in which researchers develop tests to discover disabling traits in embryos and fetuses; clinicians urge prospective parents to use these tests; government bodies endorse population screening for certain conditions, such as cystic fibrosis, and support the use of funds from public and private health insurance to pay for such tests.

THE "MESSAGE" OF SELECTING FUTURE CITIZENS

. . . Although not every difficulty of living with a disabling condition or health problem stems from society's failure to include its disabled citizens, a very large number can be traced to discriminatory attitudes and the social distance and segregated or restricted opportunities created by the nondisabled majority. . . . The societal promotion of the selection techniques is morally problematic. . . . [And] at least some of the rhetoric that endorses selecting children's characteristics conveys bias and disrespect for people with disabilities, and not merely information about the effects of a disabling trait. . . .

Elsewhere I used the following words to characterize possible reactions of the disability rights movement to the current practices of prenatal selection:

> People with just the disabilities that can now be diagnosed have struggled against an inhospitable, often unwelcoming, discriminatory, and cruel society to fashion lives of richness, of social relationships, [and] of economic productivity. For people with disabilities to work each day against the societally imposed hardships can be exhausting; learning that the world one lives in considers it better to "solve" problems of disability by prenatal detection and abortion, rather than by expending those resources in improving society so that everyone—including those people who have disabilities—could participate more easily, is demoralizing. It invalidates the effort to lead a life in an inhospitable world.[13]

. . . Clinicians providing medical services and prenatal counseling to pregnant women (whether obstetricians, nurse practitioners, midwives, or genetic counselors) obviously play crucial roles in communicating whether prenatal testing should be undertaken, what the tests reveal, and what they can mean for the health of the potential child and the life of the family. Despite the professional commitment to nondirectiveness in genetic counseling, it is clear that many professionals do not practice in a way that legitimates the choice to maintain a pregnancy of a fetus affected by a disabling trait.[14]

Counselor education contains little opportunity for contact with disabled children or adults in nonmedical settings where clinicians could observe how people with disabilities manage day-to-day life. An especially troubling example is the finding . . . that pediatric and prenatal genetic counseling gave radically different information about the same conditions to families.[15] In situations where parents were raising infants and young children with Down syndrome and cystic fibrosis, counselors stressed ways in which lives of the affected children would resemble those of nondisabled peers, focusing on capacities for education, stimulation, play, and relationships. By contrast, the stories given to prospective parents if the diagnosis was made prenatally concentrated on medical complications and differences from the lives of nondisabled children.[16] Such differences in information run afoul of nondirectiveness. . . .

If prospective parents are ever to have the opportunity to make thoughtful decisions about whether they are prepared to raise a child with a prenatally detectable disability, they need to know as much as counselors can tell them about the overall experience of children and families living with the diagnosed condition. Omitting the ways in which a child with cystic fibrosis or Down syndrome can participate in the life of family, school, and community underscores disability as a negative factor, especially if the information parents are given about what children with either condition cannot do focuses on the needs for medical follow-up or on shortened life expectancy.

Similarly, the parent learning that her or his newborn daughter or son can expect to go to school, get a job, and enjoy loving relationships with others should not be kept in ignorance regarding the need for medication, therapy, or hospitalization that may be part of her or his child's life. The premise of counseling, or of educating people about their own and their children's possible futures, is that anyone contemplating raising a child or actually involved in parenting will profit from learning about what could be in store. If professionals in one instance accentuate the negative, and in another instance accentuate the positive, they show disrespect for the intelligence and sincerity of the people who rely upon them for information and assistance.

Counselees in each situation deserve to learn as much as they can from knowledgeable professionals; and professionals betray the people they serve by slanting the information in the direction of a particular result. If counselors, midwives, and obstetricians are truly committed to patient decision making and to informed reproductive choice, they should be providing enough information about life with a disabling condition so that prospective parents can imagine the ways in which life can be worthwhile as well as those in which it can be difficult. Similarly, the parent of a toddler whose health is going to be affected by the need for medications, home-based therapy, early educational services, or hospital stays should be able to take account of those factors when deciding where to live and what job to seek. The stated neutrality and nondirectiveness of genetic counselors is very much open to question if further research determines that these differences in prenatal and pediatric counseling are the norm. . . .

If prenatal testing and embryo selection are not intended to give messages about which types of children the society will accept and welcome, proposals for "drawing lines" about the types of tests to be offered or withheld must be carefully examined and, in my view, rejected. Many clinicians and bioethicists fear that the consumerism of assisted reproduction, the anxieties of people who delay parenting and expect to have only one or two children, and the pressures parents feel to give their children "the best" start in life all contribute to a desire for "'designer' children."[17] To counter these tendencies, some are urging that researchers decline to develop, and clinicians decline to provide, tests that inform people about what professionals perceive to be traits that do not pose serious harms to the child or the family.

... I believe that ... limit[ing] diagnoses to only some, but not all, characteristics that might be determined prenatally, turns the professional assistance to reproductive autonomy into the very "message" about the badness of disability that alarms critics of the current practices. Why should parents be told by test designers: "We think that cystic fibrosis, or muscular dystrophy, or deafness, or Down syndrome should make parents think at least twice before contemplating childraising; but other conditions are too trivial for parents to object." If prenatal selection is not intended to harm existing people with the conditions that can now be diagnosed and instead is designed to give value-free information to prospective parents, creating an official list of conditions that parents should worry about will have an undesirable effect on the societal acceptance and self-esteem of those with the listed conditions. Why should it be acceptable to avoid some characteristics and not others? How can the society make lists of acceptable and unacceptable tests and still maintain that only disabling traits, and not people who live with those traits, are to be avoided? If it is legitimate to be a person with a disability, or to parent a child with such a disabling condition, should the society make a list of "serious" and "trivial" characteristics?

Endorsing testing and selecting against some traits, and refusing to let people select against other traits, will surely exacerbate the discrimination and stigmatization of future children with the listed conditions. I, and many others with a disability critique of the existing practices, find this suggestion of line-drawing clear evidence that the current arrangement and any future line-drawing reforms are much too close for comfort to running the Confederate flag up the flagpole. The flying Confederate flag tells people historically victimized by racist discrimination that racism and the history of racism is and was acceptable; enumerating a set of testable genetic diseases tells people who currently have those conditions that it would be better if prospective parents went to considerable lengths to prevent the births of children with those conditions.

Consequently, I can only urge people who support reproductive choice and also support disability inclusion and equality to oppose line-drawing efforts. It must become as acceptable to test for tone deafness or color blindness (if tests are ever developed) as it now is to test for certain forms of deafness and blindness. Undoubtedly, more prospective parents will terminate for the latter conditions than the former, but at least the decisions will be those of the people ultimately raising children, and not society, in the form of its insurance carriers and clinicians as gatekeepers. . . .

. . . I believe that it will be very difficult for most families to consider bringing children with diagnosable disabilities into the world if they know that the society believes that their births should have been prevented. . . . Raising children is work, whether or not the child has a characteristic termed a disabling trait. Virtually every parent worries about whether his son's moodiness or her daughter's adventurousness will cause problems down the line. Will children find friends, love, work, community? Will others appreciate them, warts and all? Will children grow to find a place for themselves that they will take pride in, will comfortably rest in?

If these are the anxieties of all parents raising all children, those anxieties can only be heightened if parents know and love a child whose disabling characteristics meet with aversion, social embarrassment, discrimination, and exclusion. Only when policies, laws, medical professionals, schools, and media communicate that it is respectable and legitimate to live with a disability, and only when day-to-day reality approximates the aspirations that gave rise to the Americans with Disabilities Act, will it be possible to imagine that the social problems of disability will not compound any biological limitations. Ever-increasing prenatal testing and vigorous enforcement of existing antidiscrimination laws might continue to develop along their separate tracks, because geneticists and doctors work in arenas quite different from the advocates for greater social services, increased access to education, and employment for the nation's disabled population. Yet I persist in believing that as part of the goal of creating such a welcoming society, we must persuade professionals to change what they tell prospective parents about life with disability; convince those parents to learn about how children and adults in today's world survive and thrive; and then endorse the choices people make about their reproductive and family lives. . . .

NOTES

1. Erik Parens and Adrienne Asch, "The Disability Rights Critique of Prenatal Genetic Testing: Reflections and Recommendations," in *Prenatal Testing and Disability Rights,* edited by Erik Parens and Adrienne Asch (Washington, DC: Georgetown University Press, 2000), 3, 12–13 (hereinafter Parens and Asch, "Disability Rights Critique" will refer to this specific article and *Prenatal Testing* will refer to the entire work).

2. I do not speak in this article for other members of the Hastings Center Project on Prenatal Testing for Genetic Disability, or for any advocacy group associated with the disability rights movement.

3. I, and nearly all others sharing a disability rights critique of prenatal testing, maintain an ardent pro-choice stance and assert that women should be free to make whatever decision they wish about maintaining a pregnancy or having an abortion. . . .

4. Parens and Asch, "Disability Rights Critique," supra note 1, 23–26. . . .

5. Gary L. Albrecht and Patrick J. Devlieger, "The Disability Paradox: High Quality of Life against All Odds," *Social Science and Medicine* 48 (1999): 977, 977–88.

6. See Individuals with Disabilities Education Act, 20 U.S.C. §§ 1400–1487 (2000).

7. See Rehabilitation Act of 1973, 29 U.S.C. § 794(a) (2000) (codifying Rehabilitation Act of 1973, Pub. L. No. 93–112, § 504, 87 Stat. 355).

8. See Americans with Disabilities Act of 1990, 42 U.S.C. §§ 12181–12189 (2000).

9. See Allen Buchanan et al., *From Chance to Choice: Genetics and Justice* (Cambridge: Cambridge University Press, 2000), 156; Bonnie Steinbock, "Disability, Prenatal Testing, and Selective Abortion," in *Prenatal Testing*, supra note 1, 115.

10. Americans with Disabilities Act of 1990, Pub. L. No. 101–336, 104 Stat. 327 (codified as amended at 42 U.S.C. §§ 12101–12213 (2000).

11. See 2000 N.O.D./Harris Survey of Community Participation (Harris Interactive, Inc., ed., 2000) . . .; Richard V. Burkhauser, "An Economic Perspective on ADA Backlash: Comments from the BJELL Symposium on the Americans with Disabilities Act," *Berkeley Journal of Employment and Labor Law* 21 (2000): 367–69; Ruth Colker, "Winning and Losing under the Americans with Disabilities Act," *Ohio State Law Journal* 62 (2001), 239. . . .

12. Jack McNeil, U.S. Department of Commerce, Pub. No. P70–73, Americans with Disabilities: Household Economic Studies 1 (2001), accessed December 6, 2002, www.sipp.census.gov/sipp/p70s/p70–73.pdf (on file with author).

13. Adrienne Asch, "Why I Haven't Changed My Mind about Prenatal Diagnosis: Reflections and Refinements," in *Prenatal Testing*, supra note 1, 240.

14. Parens and Asch, "Disability Rights Critique," supra note 1, 5–8.

15. Abby Lippman and Benjamin S. Wilfond, "Twice-Told Tales: Stories about Genetic Disorders," *American Journal of Human Genetics* 51 (1992): 936, 936–37.

16. Id.

17. Thomas H. Murray, "The Worth of a Child (Berkeley: University of California Press, 1996); Jeffrey R. Botkin, "Fetal Privacy and Confidentiality," *Hastings Center Report*, September–October (1995): 32.

41. The Bleak New World of Prenatal Genetics

Marcy Darnovsky and Alexandra Minna Stern

Four million American women are expecting a child this year, and many of them will encounter something entirely new in human pregnancy. Based on a simple blood draw at an initial prenatal visit, they'll be able to learn key genetic information about the fetus they're carrying—and face potentially wrenching decisions about what to do.

These noninvasive prenatal tests, called NIPTs, work by using a sample of cell-free fetal DNA circulating in the mother's blood to detect chromosomal conditions. The tests' most frequent target is trisomy 21, the genetic variation that causes Down syndrome in approximately one in every seven hundred births in the United States.

Bioethicists, genetic counselors, and advocates for disability rights have nervously anticipated the commercial rollout of these tests. Even—or perhaps especially—those who firmly support reproductive rights know that NIPTs have profound implications.

The tests have the potential to transform women's experience of early pregnancy, reduce the number of people with Down syndrome, and reinforce the assumption that Down syndrome is a dread disease to be prevented. Ultimately, these tests could dramatically reshape our understanding of what it means to be healthy and normal.

Since the end of 2011, four US companies have launched slightly different versions of NIPTs (as well as a bevy of patent-infringement lawsuits against each other). As a commercial product, NIPTs are in a rare and sought-after category: the kind whose introduction creates a new demand

Original publication by Marcy Darnovsky and Alexandra Minna Stern. "The Bleak World of Prenatal Genetics." *The Wall Street Journal.* June 12, 2013. Reprinted by permission of the authors.

and a large new consumer base. Because the predominant diagnostic prenatal test, amniocentesis, is invasive and carries some risk, only about 2 percent of pregnant women in the United States undergo it, roughly a hundred thousand per year. NIPTs could push that number into the millions in the next few years.

One of the most visible NIPTs on the market is Sequenom's MaterniT21 PLUS, a name that conflates motherhood and the trisomy 21 link to Down syndrome. In 2012, Sequenom alone processed 120,000 tests. The company has just announced that it is expanding its testing capacity to more than 300,000 samples per year.

Ideally, genetic counselors or medical professionals would work with expectant parents before they decide to take a test like MaterniT21 PLUS, giving them objective information and helping them to fully understand what could lie ahead. Because the tests are performed so early in pregnancy, NIPTs theoretically give parents more time to meet with parents who have a child with Down syndrome, and to gain balanced information about the joys and challenges of raising a child with the condition before deciding whether or not to abort.

The problem is that the surge in the number of women taking these tests is swamping the available counseling resources. The National Society of Genetic Counselors estimates that there are only about three thousand certified genetic counselors practicing in the United States. Many parents will wind up taking their cues from the glossy brochures of the NIPT companies, with their language about "risk" and "abnormalities" and their images of perfect babies with no visible disabilities. Meaningful and responsible genetic counseling is supposed to be nondirective, marketing is intended to persuade.

Several flagship medical organizations that represent obstetricians, genetic counselors, and medical geneticists—including the American Congress of Obstetricians and Gynecologists and the National Society of Genetic Counselors—have issued statements urging caution. All recommend that NIPTs be reserved for high-risk pregnancies and not used as part of routine prenatal care. And the groups concur that counseling by trained genetic-health professionals should accompany testing.

A recent statement by the American College of Medical Genetics and Genomics does a particularly good job of providing accurate and useful information for potential parents of children with Down syndrome. But the advice of these medical professionals and genetic counselors is up against a growing commercial force.

Beyond the immediate questions raised by the existing NIPTs looms an even more unsettling prospect: the development of whole-genome fetal tests that can deliver far more extensive information about a child's genetic makeup and predispositions. As with the gene tests currently available, most of the results will be presented as risk probabilities about conditions for which few, if any, therapies are available. What will parents do, for example, if they discover that their ten-week-old fetus has a five times higher-than-average chance of being diagnosed with breast cancer later in life, or a 34 percent higher-than-average risk of developing Alzheimer's seventy years from now?

Like so many other powerful technologies, fetal gene tests must be used with caution and care. Exactly what that entails is a concern not simply for medical professionals, or for people with Down syndrome and their advocates, or for companies marketing NIPTs. This portentous development in prenatal testing also raises thorny ethical problems for parents-to-be—and for everyone who cares about how we collectively understand what it means to be "healthy" and who counts as "normal."

42. Have New Prenatal Tests Been Dangerously Oversold?

Beth Daley

Stacie Chapman's heart skipped when she answered the phone at home and her doctor—rather than a nurse—was on the line. More worrisome was the doctor's gentle tone as she asked, "Where are you?"

On that spring day in 2013, Dr. Jayme Sloan had bad news for Chapman, who was nearly three months pregnant. Her unborn child had tested positive for Edwards syndrome, a genetic condition associated with severe birth defects. If her baby—a boy, the screening test had shown—was born alive, he probably would not live long.

Sloan explained that the test—MaterniT21 PLUS—has a 99 percent detection rate. Though Sloan offered additional testing to confirm the result, a distraught Chapman said she wanted to terminate the pregnancy immediately.

What she—and the doctor—did not understand, Chapman's medical records indicate, was that there was a good chance her screening result was wrong. There is, it turns out, a huge and crucial difference between a test that can detect a potential problem and one reliable enough to diagnose a life-threatening condition for certain. The screening test only does the first.

Sparked by the sequencing of the human genome a decade ago, a new generation of prenatal screening tests, including MaterniT21, has exploded onto the market in the past three years. The unregulated screens claim to detect with near-perfect accuracy the risk that a fetus may have Down or Edwards syndromes, and a growing list of other chromosomal abnormalities.

Original publication by Beth Daley. "Oversold and Misunderstood: Prenatal Screening Tests Prompt Abortion." *New England Center for Investigative Reporting.* December 12, 2014. Reprinted by permission of New England Center for Investigative Reporting.

Hundreds of thousands of women in early pregnancy have taken these tests—through a simple blood draw in the doctor's office—and studies show them to perform far better than traditional blood tests and ultrasound screening.

But a three-month examination by the New England Center for Investigative Reporting has found that companies are overselling the accuracy of their tests and doing little to educate expecting parents or their doctors about the significant risks of false alarms.

Two recent industry-funded studies show that test results indicating a fetus is at high risk for a chromosomal condition can be a false alarm half of the time. And the rate of false alarms goes up the more rare the condition, such as Trisomy 13, which almost always causes death.

Companies selling the most popular of these screens do not make it clear enough to patients and doctors that the results of their tests are not reliable enough to make a diagnosis.

California-based Sequenom Inc., for instance, promises on its Web page that its MaterniT21 blood test provides "simple, clear results." Only far down below does Sequenom disclose that "no test is perfect" and that theirs can produce erroneous results "in rare cases."

Now, evidence is building that some women are terminating pregnancies based on the screening tests alone. A recent study by another California-based testing company, Natera Inc., which offers a screen called Panorama, found that 6.2 percent of women who received test results showing their fetus at high risk for a chromosomal condition terminated pregnancies without getting a diagnostic test such as an amniocentesis.

And at Stanford University, there have been at least three cases of women aborting healthy fetuses that had received a high-risk screen result.

"The worry is women are terminating without really knowing if [the initial test result] is true or not," said Athena Cherry, professor of pathology at the Stanford University School of Medicine, whose lab examined the cells of the healthy aborted fetuses.

In one of the three Stanford cases, the woman actually obtained a confirmatory test and was told the fetus was fine, but aborted anyway because of her faith in the screening company's accuracy claims. "She felt it couldn't be wrong," Cherry said.

Companies that sell the screens stand behind their tests, saying they provide much more reliable assurance for expecting mothers than earlier screens. Some say their research focused first on how to accurately identify fetuses with potential genetic defects and only recently have they been able to get enough data to understand how often positive tests are wrong.

"The clinical performance of [noninvasive prenatal tests] has been extremely robust," Dr. Vance Vanier, vice president of marketing for reproductive and genetic health for San Diego–based Illumina Inc., which offers the Verifi prenatal screen, wrote in an e-mail statement.

The screens are not subject to approval by the Food and Drug Administration. Because of a regulatory loophole, the companies operate free of agency oversight and the kind of independent analysis that would validate their accuracy claims. Doctors often get that information from salespeople, according to doctors themselves.

And there are other emerging concerns about the new generation of prenatal tests. Two Boston-area obstetricians, with funding from a testing company, recently sent samples from two nonpregnant women to five testing companies for analysis. Three companies returned samples indicating they came from a woman who was carrying a healthy female fetus.

Meanwhile, there are a growing number of cases emerging of women told their screen shows virtually no chance of a fetus having a problem but who then deliver a child with a genetic condition.

"My son lived for four days," said Belinda Boydston, a Web/graphic designer in Chandler, Ariz., whose son was born with Edwards syndrome despite a screening test showing that he was extremely unlikely to carry the condition.

But Stacie Chapman knew nothing about these uncertainties when Dr. Sloan told her that her unborn son had screened positive for a genetic condition that was largely incompatible with life.

Hysterical with grief when she hung up, Chapman phoned her husband at a Las Vegas airport on his way home from a business trip. Together, sobbing, they concluded that their son would only suffer if he survived birth. So, that afternoon, Sloan put her in touch with a nurse who found a doctor who could do the termination the next morning.

Chapman spent the afternoon Googling the horrors of Edwards syndrome, with its heart defects, development delays, and extraordinarily high mortality. She was steeling herself for the termination when Sloan called back, urging her to wait, according to Chapman's medical record.

Chapman had a diagnostic test and learned her son did not have Edwards syndrome. A healthy Lincoln Samuel just turned one and has a wide smile that reminds Chapman of her recently deceased father.

However briefly considered, their decision to abort—informed by the MaterniT21's advertised 99 percent detection statistic—haunts them to this day.

"He is so perfect," Chapman, forty-three, said, choking up as she watched her son play with a toy lamb. "I almost terminated him."

OVERSELLING A SCREEN

Advertisements for these new prenatal screens are filled with bright skies, serene, full-bellied women, and, most of all, assurances that the tests can be trusted.

"Never maybe" promises MaterniT21 in pamphlets. Panorama states its test is "99% Accurate, Simple & Trusted" on a Web page.

The screens, conducted as early as nine weeks into a pregnancy, detect placental DNA in a mother's blood and test it for chromosomal abnormalities as well as gender.

Originally designed for older women and others at higher risk for having a problematic fetus, some of the screens are now marketed to all pregnant women. Company and analyst data indicate there have probably been between 450,000 to 800,000 tests performed in the United States since 2011 and several companies are racing to corner what one market research firm predicts will be a $3.6 billion global industry by 2019.

Independent medical experts say the new screens are far better able to rule out the possibility of certain fetal conditions, especially Down syndrome, than traditional ultrasounds and blood screenings.

That dramatically cuts down on the number of women who are incorrectly told their fetus may have a problem, and helps them avoid a more invasive follow-up test, such as amniocentesis, which carries a small risk of miscarriage.

But the new screens are not perfect, and can indicate fetuses may have serious genetic abnormalities when they do not, the industry's own research shows.

For example, among older pregnant women such as Chapman, for whom chromosomal abnormalities are more common, a positive test result for Edwards syndrome is accurate around 64 percent of the time, according to a recent study by Quest Diagnostics, a large provider of medical tests and other services. That means Baby Lincoln had about a 36 percent chance of not having the condition.

Among younger women who are at lower risk for fetal abnormalities, the error rate is even higher. Only about 40 percent of the positive tests showing high risk for Edwards syndrome turned out to be accurate, according to a recent well-regarded Illumina study of its Verifi screen.

The testing companies do recommend that women who get a positive screen result should seek additional tests to confirm it—much the way a woman whose mammogram showed potential breast cancer would undergo a biopsy to determine whether the dark spots actually are cancerous.

But some companies blur the distinction between the results of their screening tests and a true diagnosis, potentially confusing patients and doctors about the trustworthiness and meaning of their test results. Illumina, for example, claims its Verifi screen has "near-diagnostic accuracy," a term medical experts say has no meaning.

"The companies have done a very poor job of education [and] advertising this new technology, failing to make clear that it is screening testing with very good but inevitably not perfect test performance . . . and that doctors are recommending, offering, ordering a test they do not fully understand," said Dr. Michael Greene, director of obstetrics at Massachusetts General Hospital and a professor at Harvard Medical School.

Company officials say they are now focusing on the accuracy of all test results, including false positives, carrying out research on their frequency, and looking for ways to reduce the stress of these events.

Getting patients to understand limits on the accuracy of positive test results can be "extremely challenging," said Juan-Sebastian Saldivar, vice president of clinical services and medical affairs at Sequenom, which offers the screen that Chapman used. He said the company works with doctors on how to best explain it to patients.

Officials at Natera, which offers the Panorama test, insist their screens are highly accurate and produce relatively few false alarms. A company-funded study found that positive test results are correct 83 percent of the time, which is more accurate than studies of other tests have shown.

Melissa Stosic, Natera's director of medical education and clinical affairs, stresses that the company doesn't oversell the test, working to educate doctors "to be very clear it's a screening and not diagnostic."

However, the same Natera study found that some women are ignoring that advice and having abortions without getting a confirmatory diagnostic test. In its study, 22 women out of 356 who were told their fetuses were at high risk for some abnormality terminated the pregnancy without getting an invasive test to confirm the results.

"It's troubling," said Katie Stoll, a genetic counselor in Washington state who has written extensively on the marketing of noninvasive prenatal screens. "Women are getting the wrong message."

A SHORT LIFE

The field of this prenatal testing is so new, and the research so dominated by the companies selling the tests, that basic questions about their reliability and testing methods remain unresolved.

For instance, screening companies say they have had exceedingly few cases of women whose screen shows little problem, but then deliver a baby with a genetic condition. Yet such cases are popping up across the country.

It is not clear how large a problem it is, but critics say the troubling anecdotes are another example of quality control concerns that are not being independently vetted.

Belinda Boydston, forty-three, of Arizona, was stunned to give birth to a baby with Edwards syndrome after a Harmony screen result in October 2013 showed her fetus had only a 0.01 percent chance of having Edwards syndrome, also known as Trisomy18.

"I delivered prematurely, and they knew right away he had Trisomy18," Boydston said, recounting the scene in the delivery room. "I could hear them talking, 'Didn't she have that test?'"

This year alone, at the Emory Healthcare Department of Human Genetics in Georgia, officials have seen five newborns with Down syndrome whose prenatal noninvasive screens indicated that there was little risk of the condition.

And a couple in Belmont, Calif., sued Ariosa Diagnostics Inc. last year, alleging its Harmony screen falsely indicated their child would not have Down syndrome and that the company's marketing material misled patients about the screen's accuracy. The lawsuit is still pending. Ariosa declined to comment, as did Roche, which is acquiring it.

"There needs to be more transparency and accuracy from the companies about what their results mean," said Mary Norton, vice chair of genetics in the department of obstetrics, gynecology, and reproductive sciences at the University of California San Francisco. Norton, who has received funding from Ariosa and Natera, said research in the field is almost all being funded by industry. "There isn't a lot of independent research going on."

FDA REGULATORY LOOPHOLE

The loophole that allows unregulated tests in the marketplace dates back to the mid-1970s, when the FDA began overseeing diagnostic tests. The FDA exempted what was then a small group of relatively simple tests developed, manufactured, and performed all in a single lab—for example, in a hospital.

In the past decade, for-profit companies have used that regulatory running room to develop complex tests to diagnose or screen for conditions ranging from cancer to Lyme disease and now, fetal chromosomal conditions. Not all of the tests undergo robust independent review and it is challenging for the public to distinguish good and bad tests, according to medical experts.

In late September, the FDA stepped in, publishing draft regulations for the industry expected to take nine years to be fully phased in. Officials there said that the new prenatal tests would probably be among the first few groups to be regulated. Already, the industry is expected to challenge the legality of those rules and a trade group, the American Clinical Laboratory Association, has retained Harvard law professor Laurence Tribe to help represent them.

However, the FDA has made it clear that the companies don't have to wait for regulations to act.

We "have told the [prenatal] companies ... they really should bring their test into the agency," said Alberto Gutierrez, director of the FDA's office of in vitro diagnostics in the Center for Devices and Radiological Health. He indicated they could voluntarily submit to regulation, as other types of diagnostic-testing companies have done.

While the regulatory fight goes on, doctors say one immediate way to limit potential misuse of possibly inaccurate prenatal screening results is for women to receive genetic counseling before they get the screens—and after. Several women, including Chapman, told the New England Center they were offered the prenatal screens without any clear understanding of what their fetus would be tested for other than Down syndrome and to learn the sex of the baby.

Lincoln Chapman, Stacie's husband, said he will be forever grateful to Dr. Sloan, of Harvard Vanguard Medical Associates in Boston, for calling Stacie Chapman back to more strongly recommend further tests.

But Stacie Chapman remains conflicted about the doctor's first call that set in motion the near termination of her pregnancy. Sloan did not stress that the test was just a screen that could be wrong, she recalled.

"I didn't seek this test out—this test was offered to me by the doctor's office. They should know how it performs," Chapman said, adding that she would never have considered a pregnancy termination if she had better understood the odds that her result could be wrong.

A health professional involved in Chapman's case, who was not authorized by her employer to comment, said Sloan initially was unaware that the screen could be wrong a significant amount of the time.

Sloan did not return calls to discuss her understanding of the statistics, but a statement from Harvard Vanguard noted its policy recommends genetic counseling and confirmatory testing for women whose fetuses test positive for a genetic abnormality.

Chapman said enormous heartache could have been avoided in her family if companies advertised more scrupulously, and if her doctor had understood the limitations of the screen.

As millions of women in the United States and elsewhere expect babies this year, some inevitably will be in the same situation as Chapman.

"You know when I found out [the baby] was fine, my midwife said, 'You are one in a million, you are so lucky,'" Chapman recalls. "But you know? I really wasn't."

43. Sex Selection and the Abortion Trap

Mara Hvistendahl

For nearly two decades, anti-abortion activists have been at work in a disingenuous game, using the stark reduction of women in the developing world as an argument for taking away hard-earned rights. Conservative theorists have written openly about how sex-selective abortion is merely a convenient wedge issue in the drive to ban all abortions, both in the United States and abroad. And now, conservative commentators like the *New York Times'* Ross Douthat, the *Wall Street Journal's* Jonathan V. Last, and the editors of the *New York Sun* have claimed that my book, *Unnatural Selection*, strengthens their case.

This does not surprise me. One of the themes that cropped up again and again in my reporting was the extent to which American abortion politics on both sides of the question has stalled action on issues of major global importance. But it is deeply unfortunate. The American obsession with abortion does not just hinder work on maternal and child health or access to safe birth control abroad—two areas that have suffered because of domestic campaigns by anti-abortion activists. It's also distracting US policy makers from what should be the real conversation in a country that leads the world in human reproductive technology: whether to allow parents to use a growing range of methods to select for characteristics like sex (or diseases that come on late in life and, perhaps one day, IQ) in their children. Because sex selection is not just a developing-world problem—it's an American problem, too.

Original publication by Mara Hvistendahl. "The Abortion Trap: How America's Obsession with Abortion Hurts Families Everywhere." *Foreign Policy.* July 27, 2011. Reprinted by permission of The Slate Group LLC, Foreign Policy.

. . .

Western influence and technology have caused their fair share of damage abroad, but often the damage is not explicitly foreseen, so that decades later it is still possible for trouble's architects to look back and say, "We just didn't know." But with fetal sex determination and sex selection, a good number of people knew. The 1960s population control enthusiasts who supported pushing along research into sex selection methods as a way to reduce the global birth rate knew those methods would occasion a reduction in the proportion of girls born, even if they did not envision the scale of that reduction—the disappearance of 160 million females, by 2005, from Asia's population. As I delved into this history for my book, the tragedy for me became this cold foresight: the fact that some prominent activists and scientists actually anticipated the side effects of widespread sex selection—that a massively imbalanced sex ratio at birth would result in rising instability, risks for those women who are born, and a social environment bordering on what one early proponent described as a "giant boy's public school or a huge male prison"— and yet dismissed those effects as necessary ills in the quest to solve humanity's problems through technology. They knew, and still they plowed ahead.

The story behind sex selection should serve as a cautionary tale as we consider other technological quick fixes as solutions to our current global problems: dumping iron into our oceans in the quest to fight climate change, say. But the American response to my book has largely overlooked this point, glossing over the moral implications of sex determination technologies like ultrasound, amniocentesis, and add-ons to in vitro fertilization (IVF), and instead homing in on abortion.

True enough, abortion is the primary method of sex selection in the developing world today. But it is hardly the most ideal method. Given the choice between a second-trimester abortion and a cheap and easy procedure performed close to the moment of conception, most couples would choose the second. That is why activists concerned about a reduction in the number of women in the world—or in the number of disabled, red-haired, or short people—are looking ahead to emerging technologies like fetal DNA testing, sperm sorting, and embryo screening.

The high-tech sex selection front-runner in both the developing and developed worlds for the moment is a technique called preimplantation genetic diagnosis (PGD), the add-on to in vitro fertilization sometimes called embryo screening. Once a donor's eggs have been retrieved and fertilized with a man's sperm and the resulting zygotes have grown into embryos, a single cell can be removed and tested for sex.

Like ultrasound and amniocentesis before it, PGD can be used to diagnose much more than an embryo's sex. That is why most of the Western world regulates it instead of banning it outright. The United Kingdom has made perhaps the most ambitious effort at regulation in establishing the Human Fertilisation and Embryology Authority (HFEA), an agency under the Department of Health charged with regulating fertility clinics, sperm banks, and other businesses connected with assisted reproduction. Thanks to the HFEA, British couples may undergo PGD to avoid passing on debilitating diseases like cystic fibrosis—but not for what is called social sex selection, or simply because they want a boy or a girl. For that, many fly to the United States.

Because its health care system is privatized and because techniques like social sex selection are a cash cow for the clinics that provide them, the United States has become the Wild West of assisted reproduction, home to Octomom and a host of less well-known ethical blunders. America also allows nearly unfettered use of the new sex selection technologies. The American Society for Reproductive Medicine issues guidelines for fertility clinics, but they are only voluntary—and many clinics don't follow them.

But since assisted reproduction is typically carried out in private clinics that do not receive federal funding, the field gets far less oversight than many other areas of medicine. Some techniques have "moved from concept to clinic without systematic animal studies or reviews by any independent agency or committee," Marcy Darnovsky, associate executive director of the Center for Genetics and Society in Berkeley, California, wrote in a recent article for the website Science Progress. In many regards, international adoption is governed by stricter regulations than assisted reproduction.

That's why, some three decades after sex selection found its way to Asia, the United States has become a destination for couples from around the world seeking the latest techniques. Couples from China and India fly to Los Angeles to undergo PGD at Fertility Institutes, a clinic I visited last year. (A good number of healthy Americans shell out the fifteen thousand dollar–plus cost of PGD simply to get a boy or a girl as well. One out of every hundred babies in the United States is born through IVF, and fertility patients are often encouraged to tack on PGD. While clinics aren't required to report what PGD is used for, a 2006 survey of those that perform the procedure by the Genetics and Public Policy Center at Johns Hopkins University found 42 percent of responding clinics offered it for social sex selection.) The United States has also become an exporter of technologies with questionable ethical underpinnings. Jeffrey Steinberg, Fertility Institutes' founder, has opened an office in Guadalajara, Mexico, and clinics offering PGD are being set up in Asia, in some cases in countries with already severe gender imbalances.

Take South Korea, for example: Sex-selective abortion has gone out of style there, and the country's once-skewed sex ratio at birth is now balanced. But wealthy parents still turn to PGD, according to Sohyun Kim, a doctor who runs a small clinic in a tony part of Seoul. "These days, because of IVF technology, parents have a lot of different embryos to choose from," she told me. "They ask for the ones they don't want to be eliminated. The girls can be erased. And the boys remain."

. . .

In his review of my book, Last contended the work is "aimed, like a heat-seeking missile, against the entire intellectual framework of 'choice.'" It is true that I don't believe absolute choice is a great framing point for the politics of modern reproduction. The range of options available to parents today is simply too great. The rhetoric surrounding abortion in the United States sorely needs to be updated. (I am hardly alone in this thought. While the term "pro-choice" lives on, it is no coincidence that the new buzzword is "reproductive justice.")

That said, selecting for sex—or any other quality—is different from a woman's decision not to carry a pregnancy to term. When parents choose to have a child because he is male, they may do so with the expectation that their son will turn out to be an upstanding heir or that he will carry on the family line. Or, should they want a girl, they may be seeking a child who enjoys wearing pink dresses and playing with dolls. Ethically, this is worlds apart from a woman's choice not to continue a pregnancy—or not to get pregnant in the first place. One is the decision of a woman considering her own body. The other involves the creation of a new human being—and expectations for how that human being will turn out.

Sex selection, indeed, represents a form of choice that looks a lot like willful manipulation. ("Sex control," the typically bold 1960s phrase used by those who proposed sex selection as population control, has turned out to be a good term for it.) It marks the dawn of what Harvard University political philosopher Michael Sandel calls "hyperagency"—"a Promethean aspiration to remake nature, including human nature, to serve our purposes and satisfy our desires."

But the distinction between these two types of choices is not merely intellectual; it has legal grounding as well. *Roe v. Wade* came at a time when today's reproductive dilemmas were still the stuff of speculation, yet it explicitly protects a woman's right to decide whether to carry a pregnancy to term—while also acknowledging that right can have limits. In crafting a framework for prepregnancy sex selection, the United States might look to countries like France, the Netherlands, and Denmark, where abortion rights are intact and yet sex selection for social reasons is prohibited.

The anti-abortion movement should agree that sex selection, along with many other forms of selection, is wrong, and indeed many in the religious community see what happens today as tampering with God's work. (The Vatican struck out against using PGD for trait selection in 2009.) But unfortunately the same people who would ban abortion in the United States are often allied with the very ones who advocate for free-market health care and lax oversight of industry—the very ingredients that have made the United States a destination for the hypercontrolling parents of the world intent on getting a certain type of child. Conservatives who object to tampering with reproduction are rivaled in strength by those who favor letting market forces govern health care.

The result is nothing less than consumer eugenics. "Gender selection is a commodity for purchase," Steinberg, the Los Angeles clinic founder, reportedly told journalist Mimi Rohr in 2006. Sex is only the first non-medical or nondisability condition that can be tested using PGD, though the arrival of more sophisticated diagnostic techniques push the boundary of what constitutes a medical condition. (Adult-onset Alzheimer's? A future that may involve breast cancer? A propensity toward obesity?) In 2009, Steinberg announced on Fertility Institutes' website that his clinic would soon offer selection for eye color, hair color, and skin color. He retracted the announcement after a public outcry, but when I met him last year he told me it wasn't a change of heart—he's just waiting for public sentiment to come around.

Others with less of a financial stake in the matter agree that unfettered reproductive selection is the way of the future. James Watson, co-discoverer of DNA's double-helix structure, told an audience in 2003 that he supported knowledge gleaned from his discovery being used to shape the outcome of reproduction on an individual level, while Princeton University molecular biologist Lee M. Silver has written on the inevitability of a *Gattaca*-like society split between the genetic haves and have-nots.

Whatever awaits us, it's clear that the issue of abortion is increasingly a red herring. Activists on both the right and the left might now be wise to abandon the abortion fray and consider speaking out for restraint in other areas. Governments in Asia have introduced measures to address sex-selective abortion, and Western support for those measures is critical. But beyond that, the United States should now lead in addressing the new technologies emerging from within its borders. We owe as much to the world and to future generations—so that next time it can't be said that we knew and yet chose not to act.

44. A Baby, Please: Blond, Freckles—Hold the Colic

Gautam Naik

Want a daughter with blond hair, green eyes, and pale skin?

A Los Angeles clinic says it will soon help couples select both gender and physical traits in a baby when they undergo a form of fertility treatment. The clinic, Fertility Institutes, says it has received "half a dozen" requests for the service, which is based on a procedure called preimplantation genetic diagnosis, or PGD.

While PGD has long been used for the medical purpose of averting life-threatening diseases in children, the science behind it has quietly progressed to the point that it could potentially be used to create designer babies. It isn't clear that Fertility Institutes can yet deliver on its claims of trait selection. But the growth of PGD, unfettered by any state or federal regulations in the United States, has accelerated genetic knowledge swiftly enough that preselecting cosmetic traits in a baby is no longer the stuff of science fiction.

"It's technically feasible and it can be done," says Mark Hughes, a pioneer of the PGD process and director of Genesis Genetics Institute, a large fertility laboratory in Detroit. However, he adds that "no legitimate lab would get into it and, if they did, they'd be ostracized."

But Fertility Institutes disagrees. "This is cosmetic medicine," says Jeff Steinberg, director of the clinic that is advertising gender and physical trait selection on its Web site. "Others are frightened by the criticism but we have no problems with it."

PGD is a technique whereby a three-day-old embryo, consisting of about six cells, is tested in a lab to see if it carries a particular genetic disease. Embryos free of that disease are implanted in the mother's womb. Introduced in the 1990s, it has allowed thousands of parents to avoid passing on deadly disorders to their children.

But PGD is starting to be used to target less-serious disorders or certain characteristics—such as a baby's gender—that aren't medical conditions. The next controversial step is to select physical traits for cosmetic reasons.

"If we're going to produce children who are claimed to be superior because of their particular genes, we risk introducing new sources of discrimination" in society, says Marcy Darnovsky, associate executive director of the Center for Genetics and Society, a nonprofit public interest group in Oakland, Calif. If people use the method to select babies who are more likely to be tall, the thinking goes, then people could effectively be enacting their biases against short people.

In a recent US survey of 999 people who sought genetic counseling, a majority said they supported prenatal genetic tests for the elimination of certain serious diseases. The survey found that 56 percent supported using them to counter blindness and 75 percent for mental retardation.

More provocatively, about 10 percent of respondents said they would want genetic testing for athletic ability, while another 10 percent voted for improved height. Nearly 13 percent backed the approach to select for superior intelligence, according to the survey conducted by researchers at the New York University School of Medicine.

There are significant hurdles to any form of genetic enhancement. Most human traits are controlled by multiple genetic factors, and knowledge about their complex workings, though accelerating, is incomplete. And traits such as athleticism and intelligence are affected not just by DNA, but by environmental factors that cannot be controlled in a lab.

While many countries have banned the use of PGD for gender selection, it is permitted in the United States. In 2006, a survey by the Genetics and Public Policy Center at Johns Hopkins University found that 42 percent of 137 PGD clinics offered a gender-selection service.

The science of PGD has steadily expanded its scope, often in contentious ways. Embryo screening, for example, is sometimes used to create a genetically matched "savior sibling"—a younger sister or brother whose healthy cells can be harvested to treat an older sibling with a serious illness.

It also is increasingly used to weed out embryos at risk of genetic diseases—such as breast cancer—that could be treated, or that might not strike a person later in life. In 2007, the Bridge Centre fertility clinic in

London screened embryos so that a baby wouldn't suffer from a serious squint that afflicted the father.

Instead of avoiding some conditions, the technique also may have been used to select an embryo likely to have the same disease or disability, such as deafness, that affects the parents. The Johns Hopkins survey found that 3 percent of PGD clinics had provided this service, sometimes described as "negative enhancement." Groups who support this approach argue, for example, that a deaf child born to a deaf couple is better suited to participating in the parents' shared culture. So far, however, no single clinic has been publicly identified as offering this service.

Like several genetic diseases, cosmetic traits are correlated with a large number of DNA variations or markers—known as single nucleotide polymorphisms, or SNPs—that work in combination. A new device called the microarray, a small chip coated with DNA sequences, can simultaneously analyze many more spots on the chromosomes.

In October 2007, scientists from deCode Genetics of Iceland published a paper in *Nature Genetics* pinpointing various SNPs that influence skin, eye, and hair color, based on samples taken from people in Iceland and the Netherlands. Along with related genes discovered earlier, "the variants described in this report enable prediction of pigmentation traits based upon an individual's DNA," the company said. Such data, the researchers said, could be useful for teasing out the biology of skin and eye disease and for forensic DNA analysis.

Kari Stefansson, chief executive of deCode, points out that such a test will only provide a certain level of probability that a child will have blond hair or green eyes, not an absolute guarantee. He says: "I vehemently oppose the use of these discoveries for tailor-making children." In the long run, he adds, such a practice would "decrease human diversity, and that's dangerous."

In theory, these data could be used to analyze the DNA of an embryo and determine whether it was more likely to give rise to a baby of a particular hair, skin, or eye tint. (The test won't work on other ethnicities such as Asians or Africans because key pigmentation markers for those groups haven't yet been identified.)

For trait selection, a big hurdle is getting enough useful DNA material from the embryo. In a typical PGD procedure, a single cell is removed from a six-cell embryo and tested for the relevant genes or SNPs. It's relatively easy to check and eliminate diseases such as cystic fibrosis that are linked to a single malfunctioning gene. But to read the larger number of SNP markers associated with complex ailments such as diabetes, or traits like hair color, there often isn't enough high-quality genetic material.

William Kearns, a medical geneticist and director of the Shady Grove Center for Pre-implantation Genetics in Rockville, Md., says he has made headway in cracking the problem. In a presentation made at a November meeting of the American Society of Human Genetics in Philadelphia, he described how he had managed to amplify the DNA available from a single embryonic cell to identify complex diseases and also certain physical traits.

Of forty-two embryos tested, Dr. Kearns said he had enough data to identify SNPs that relate to northern European skin, hair, and eye pigmentation in 80 percent of the samples. (A patent for Dr. Kearn's technique is pending; the test data are unpublished and have yet to be reviewed by other scientists.)

Dr. Kearns's talk attracted the attention of Dr. Steinberg, the head of Fertility Institutes, which already offers PGD for gender selection. The clinic had hoped to collaborate with Dr. Kearns to offer trait selection as well. In December, the clinic's Web site announced that couples who signed up for embryo screening would soon be able to make "a pre-selected choice of gender, eye color, hair color and complexion, along with screening for potentially lethal diseases."

Dr. Kearns says he is firmly against the idea of using PGD to select non-medical traits. He plans to offer his PGD amplification technique to fertility clinics for medical purposes such as screening for complex disorders, but won't let it be used for physical trait selection. "I'm not going to do designer babies," says Dr. Kearns. "I won't sell my soul for a dollar." A spokeswoman for Dr. Steinberg said: "The relationship between them is very amicable, and this center looks forward to working with Dr. Kearns."

For trait selection, Dr. Steinberg is now betting on a new approach for screening embryos. It involves taking cells from an embryo at day five of its development, compared with typical PGD, which uses cells from day three. The method potentially allows more cells to be obtained, leading to a more reliable diagnosis of the embryo.

Trait selection in babies "is a service," says Dr. Steinberg. "We intend to offer it soon."

PART IX

Reinventing Race in the Gene Age

．　　．　　．　　．　　．

The history of science has been marred by racist ideas and practices, with low points including the eugenics movement and shockingly abusive medical experiments conducted on vulnerable populations. In the mid-twentieth century, most social scientists and many biologists rejected the idea that biological variations among humans can be used to classify people into discrete racial groups. While racism clearly has physiological consequences that create health disparities, race is a social, political, and legal category rather than a biological one.

This conclusion is substantiated by a wide variety of findings in both the life and social sciences. Studies of human genetics, for example, have demonstrated that more genetic variation occurs within population groups than between them. Anthropologists have documented that racial categories vary widely in societies around the world because of differences in how they were devised and enforced through social, political, and legal mechanisms.

Despite all this, the idea of "biological race" is now making a comeback. In recent years we have seen the use of dubiously defined racial categories in genetic and biomedical research and in health care; the introduction of race-specific drugs; and the emergence of DNA ancestry tests that purport to discover individuals' precise racial heritage. These and related developments may revive discredited ideas about race as a biological category, naturalize and normalize social and health inequities between racially defined groups, and distort our understanding of the way that racism damages health and well being. Contributors in this section explain the scientific fallacies behind the resurgence of race as a biological concept, and its dangerous social consequences.

Evelynn Hammonds discusses the current controversy about the meaning of race—which she characterizes as a "raging debate"—in "Straw Men and Their Followers: The Return of Biological Race." She begins with a British evolutionary biologist who argues that using traditional racial categories in medical genetics *is* meaningful and useful, and that those who deny this are standing in the way of scientific progress and improved medical care. In countering this perspective, which seems to be enjoying increasing "popular and political currency," Hammonds calls for scientists to "directly confront the methodological limitations, errors, and uncertainties in the way they use race constructs in their research designs and statistical analyses," and for all of us to "abandon any use of race that fails to capture the true complexity of human genetic variation."

Another prominent scholar drawing attention to the resurgence of race as a biological category is Dorothy Roberts, author of *Fatal Invention: How Science, Politics, and Big Business Re-create Race in the Twenty-First Century*. In a recent TEDMed talk titled "The Problem with Race-Based Medicine," Roberts delivers a powerful indictment of the persistence of race-as-biology in medical "diagnoses, measurements, treatments, prescriptions, even the very definition of diseases." She points out that race medicine is not only bad medicine that harms patients of color but that it also promotes "a false and toxic view of humanity." The way forward, she says, is "to affirm our common humanity by ending the social inequalities that truly divide us."

In "Race in a Bottle," Jonathan Kahn tells the story of the US Food and Drug Administration's 2005 approval of BiDil, the first "race-specific" drug claiming to treat heart failure in self-identified African American patients. As Kahn notes, BiDil was developed without regard to race or genetics, and there is no valid evidence that it works better or differently in African Americans. But the FDA's approval, and a patent award that makes the race-specific label attractive to the drug company marketing it, has "given the imprimatur of the federal government to using race as a genetic category."

The co-authors of "The Science and Business of Genetic Ancestry Testing" warn that commercial DNA tests that promise to reveal the geographic origins of one's ancestors should not be dismissed as "merely recreational." They argue that the tests, which are "less informative than many realize," can profoundly affect individuals and communities, and that they reinforce misconceptions about what is meant by *race* or *ethnicity*. Even though the websites of many genetic ancestry testing companies state that race is not genetically determined, "the tests nevertheless promote the popular understanding that race is rooted in one's DNA."

In "All That Glitters Isn't Gold," Osagie Obasogie and Troy Duster consider the emergence of DNA evidence and police DNA databases as "the evidentiary gold standard" in criminal prosecutions. What's wrong with that? While two DNA samples can be compared with high levels of accuracy and used to exonerate a suspect or convict a perpetrator, finding suspects in increasingly massive police databases based on crime-scene DNA evidence is a different story altogether. Techniques based on database searches are subject to multiple kinds of inaccuracies and raise significant civil liberties concerns. And because communities of color are disproportionately policed, arrested, and prosecuted, they will wind up being overrepresented in DNA databases and unfairly burdened by their problems.

Osagie Obasogie pursues these concerns in "High-Tech, High-Risk Forensics," a *New York Times* op-ed based on the story of a man arrested for murder when DNA found on the victim's body was submitted to a police database and his profile turned up. Although his "alibi was rock solid"—he was hospitalized at the time of the crime—he spent more than five months in jail with the possibility of death row before him. Prosecutors ultimately explained this by suggesting that the paramedics who took the man to the hospital on the night of the murder must have transferred his DNA to the victim's body when they arrived at the crime scene a few hours later. As Obasogie points out, contamination is just one of the problems that make DNA databases fallible. This is a technology, he says, "with known limitations, and susceptible to human error and prosecutorial misuse."

45. Straw Men and Their Followers

The Return of Biological Race

Evelynn M. Hammonds

Until Armand Marie Leroi's *New York Times* op-ed of March 14, 2005, it is unlikely that many Americans, even among the daily readers of the paper, knew that we are living in the midst of a raging debate over the existence of human races. This debate is occurring among and between a variety of researchers in genetics and social scientists from a range of disciplines. A number of evolutionary biologists, geneticists, biological anthropologists and medical researchers have recently challenged the view put forth by other scientists and social scientists that "Race is only a social concept, not a scientific one." They claim rather that current genetic research shows that "races are real" and that using race in genetic research has clear benefits especially for researchers who are trying to uncover the genetic basis of diseases that have a greater prevalence in certain groups. The broader claim, however, as articulated by Leroi, is that those who have argued against race as a biologically meaningful concept have based their arguments not on current genetic data but rather on political grounds. He erroneously claims that the originator of this so-called political view is the eminent Harvard geneticist, Richard Lewontin, who in 1972 first argued that since there was greater genetic variation within any given race than between races the very concept of race was not a useful way to understand genetic variation in humans. Those scientists who continued to use race, Lewontin argued, did so less for scientific reasons than for ideological ones.

For Leroi and those unnamed scientists who support his view, Lewontin's 1972 work opened the door to the politicization of race in science. They

Original publication by Evelynn M. Hammonds, "Straw Men and Their Followers: The Return of Biological Race," Is Race "Real"? Social Science Research Council, June 7, 2006. Reprinted by permission of Evelynn M. Hammonds.

have characterized those ascribing to the view that race is socially constructed as "race deniers"—people who refuse to acknowledge what any child can see—that human beings can be lumped together in groups by skin color, hair type, eye shape and color, head shape, and body type. Indeed, Leroi and others argue, these clear visible markers signal deeper differences within our bodies which are expressed in the differences in our genes. More importantly Leroi notes that, using sophisticated new technologies, "if a sample of people from around the world are sorted by computer into five groups on the basis of genetic similarity, the groups that emerge are native to Europe, East Asia, Africa, America and Australasia—more or less the major races of traditional anthropology." Therefore we can look at genes and get right back to the races we started with. Eureka! Race is real!

How could anyone argue with the story Leroi tells? To assert that race is real is not, Leroi claims, a return to the position that races are pure or that some races are superior to others. Thus there is no reason to fear that the use of race in science today could lead to the negative policies or racist attitudes of the past. Race today is a benign and beneficial concept, he says. It is merely a "shorthand that enables us to speak sensibly, though with no great precision, about genetic rather than cultural or political differences." Indeed, through the use of race, medical scientists will be able to achieve the laudable goal of improving the health and treatment of diseases that disproportionately affect African Americans and other groups in which disease has been poorly understood and treated. By acknowledging race, Leroi continues, it becomes clear why we need to value and protect some of the world's most obscure and isolated peoples. In the end, Leroi wants readers to believe that there are no dangers inherent in the use of traditional racial categories in science today and thus those who argue against its use are the ones who stand in the way of medical progress and the preservation of marginalized populations within the human species.

This is in many ways a familiar, almost biblical, competitive tale in which the righteous son speaks in the voice of "true science." In Leroi's story, his "true" science has vanquished those liberal, but gullible, scientists and other critics who have jettisoned a useful scientific concept in order to affirm a liberal political position. In his cautionary tale, he recuperates race as nothing more than a useful heuristic—a useful shorthand—for obvious phenotypic and genetic diversity. In so doing, he becomes the hero of his own tale for his obvious courage in going against the forces of political correctness in the service of addressing health disparities. In this way Leroi marks his allegiance to other conservative voices who in this politically divisive moment in American politics and life characterize those with whom they disagree as the ones who have politicized the debate.

To make this competition even clearer, with the exception of Richard Lewontin, the critics of the view that race is a useful scientific concept are not named. Indeed, the very position that Leroi argues against—that race is socially constructed rather than "real"—is never explained beyond the few quotes attributed to Lewontin. For Leroi and others of his ilk, the opposition is a straw man. There is no need to provide a careful articulation of the opposing argument because the point is not to explain to the public why race remains a complex and imprecise concept for those studying human variation within biology, genetics and medicine. Rather the point is to offer so-called irrefutable scientific confirmation of what we already know about ourselves and our differences from those who are not like us.

It is tempting for social scientists to try to argue against articles like Leroi's by carefully taking each point apart and showing how it is factually incorrect, illogical or detailing how much more complex the story is. There will be those who point to the explicit genetic determinism implicit in his argument, for example. Some geneticists will take a similar tack, reasonably pointing out the way Leroi has overstated what the data of the best genetic studies actually show about the relationship between race and genetics. Human genetic variation is essentially a continuous phenomenon, reflecting the various histories and migration patterns of groups of human beings. The fact that statistical programs can sort humans into "buckets" that very roughly correspond to "races" provides little information useful for understanding patterns of disease among human beings. They will surely note that while genetic data can be used to distinguish and allocate individuals into groups, whether or not those groups meaningfully correspond to races as defined in the United States remains an open question. These are useful and necessary points to make.

Such an exercise, however, often fails to address the larger questions that Leroi's article raises. Indeed, often such corrective exercises end up producing their own straw men, pejoratively casting scientists who believe in the usefulness of race as ignorant relics of a bygone era. Leroi's story has a certain appeal that cannot be undermined by a recitation of different facts about the meaning and uses of race in science. Rather the appeal of a story that links race to medical and scientific progress is in the way in which it naturalizes the social order in a racially stratified society such as ours. Therein lies its appeal and its popular and political currency.

Consider the controversy over the 1994 publication of Richard J. Herrnstein and Charles Murray's book, *The Bell Curve: Intelligence and*

Class Structure in American Life.[1] The book generated extensive critiques by historians, social scientists and journalists. In the collection of reviews, arguments, historical background and critiques of the work published in 1995, there are detailed criticisms against each aspect of the argument, evidence and research presented by Herrnstein and Murray. *The Bell Curve* was a serious book that generated several hundred thousand readers within months of its publication and hundreds of critiques afterward. However, I would suggest that none of the hundreds of critiques had as much appeal as the book and none of the critiques is remembered as much as the main argument of the book that "the ills of welfare, poverty, and an underclass are less matters of justice than biology."[2]

It is the power of biology as a naturalizing discourse that has to be challenged. And race is a key figure in this discourse. In the United States, race serves as a dense transfer point between nature and society. It links our social structure to our individual and group biologies and it links our biological differences back to our social structure. As one of my students quipped, "Race in America is not a biological category; it is a cosmology, an entire world view." As soon as someone shows that "intelligence" is biological and fixed and it is expressed differently in different races, you have a natural explanation for all sorts of social ills that can be linked to racial differences in intelligence. The same is true of disease. If the incidence of disease differs by race and if race is biological, then we must use race to explore the cause and treatment of disease. Race in America has always explained too much and too little. Yet, Americans are deeply attracted to and readily accept racial narratives—especially when they are produced by biology.

The biologizing of race in the United States has put biologists in a difficult position since the beginning of the twentieth century. In the period before World War II, some American biologists rushed to use the then new research in genetics to explore racial differences between African Americans and white Americans. When this project encountered methodological and political problems and its links to the horrific uses of racial science under the Nazis were exposed, such work was repudiated. In the period after World War II, some biologists thought they had successfully separated biology from race and politics. Social science buttressed this position by arguing that culture and societal structures were far more significant in human affairs than biology.

The consensus in the social sciences that race is a social concept grew so strong that by the end of the twentieth century it was rarely questioned. At the same time, a new generation of biologists armed with new technologies for understanding human genetic diversity that resulted from the Human Genome Project readily adopted common notions of race as the rubric for explaining this variation. Race is a term few of this generation of geneticists and biomedical researchers had ever questioned or even defined. It is not surprising, then, that many find it troubling that despite their best intentions, their work is used to confirm or deny lay views of race. It is even more troubling to geneticists that there is no consensus within science as to what race is, how it should be used, or its utility for predicting health outcomes in individuals. This uncertainty within science over the meaning and status of race has been politicized and exploited by a wide range of public and private entities that few Americans know about. Leroi's opinion piece is just the most recent salvo in what promises to be a growing and contentious debate.

If we are not to use our differing views about race as either a machine to produce straw men or to reproduce old narratives that naturalize the many social inequities that have produced health disparities in this country—a different approach is needed. It is time for geneticists and biomedical researchers to directly confront the methodological limitations, errors and uncertainties in the way they use race constructs in their research designs and statistical analyses. Social scientists are rightfully skeptical of scientists' use of race when the term is not clearly defined or thoughtfully employed. The system of peer review must be enlisted to ensure that scientists clearly specify why and how racial categories are used in their research. It is time too for the leading scientists in genetic research to produce a consensus document on the use of race. This cannot be done without the help of social scientists. Social scientists know that much of the power of race comes from the fact that it is open to differing and contradictory interpretations. There is a critical need for rigorous interdisciplinary work on race between geneticists and social scientists to develop new ways to analyze and explain the relations between biology and society and how they interact with each other. Unlike consensus documents on race produced in the past, resolving the issues of the meaning and uses of race today will engage a complexity not seen before. Can we use race to capture the complexity of human differences from the genetic to the social? If so, to what extent? How? And if not, why not?

Lastly, journalists and reputable news organizations have a duty to reveal the ethical, legal, financial and social implications of genetic research that invokes race. More importantly, they need to reflect in their accounts that they understand that racial narratives are always narratives about power. The need for informed and critical reporting is clear. Pharmaceutical and computer companies, politicians, and so-called independent experts are using Americans' attraction to racial narratives to obscure their own interests in the outcomes of genetic research. The *New York Times* article on the launching of a new DNA database is a case in point.[3] The project called "The Genographic Project" is a joint venture between the computer giant IBM and the National Geographic Society. The goal is to "combine population genetics and molecular biology to trace the migration of humans from the time we first left Africa, 50,000 to 60,000 years ago, to the places where we live today." The project organizers plan to invite public participation by encouraging Americans to purchase a DNA test kit for $99.95. The buyer then returns the kit with a cheek swab with their own DNA which will be analyzed and stored in a database owned by IBM and the National Geographic Society. The article in the *Times* reported little more than what was in the press release of the project. At no point were serious concerns raised about the launch of a project to create one of the largest privately owned DNA databases in the country. By marketing the project as a way for Americans to find their "true" origins and to participate in a scientific project, the report seemingly left little room for criticism. And none was found in the article with the exception of a few cautionary statements by a Yale geneticist. The public was given no information about the potential problems that this project raises, the most obvious being questions of privacy, future use of the DNA that will reside in the database, and even the waste of money that might have gone elsewhere.

We are in the middle of a debate about the power and authority of genetic information and the meaning of race. Can genetic research tell us who we really are, where we come from, who we are related to, or why we get sick without resorting to concepts of race that confound and distort these very questions? Leroi is among those who are using race as biology as a ruse for making progress on health disparities. When one scratches the surface of his argument, one sees that it is little more than a thinly guised continuation of a long tradition of using biology to explain racial differences in order to claim that such disparities are due more to genetics than to the societal forces that have historically disenfranchised people of color within the US health care system. If we want to avoid this naturalization of the inequities of our current health care system and make real progress toward understanding the underlying causes of health disparities, then we must abandon any use of race that fails to capture the true complexity of human genetic variation. In the end, there can be no simple answer to the problem of race in genetic research—until we confront the problem of race and racism in America and understand that they are not the same thing.

NOTES

1. R. Herrnstein and C. Murray, *The Bell Curve: Intelligence and Class Structure in American Life* (New York: Free Press, 1994).

2. R. Jacoby and N. Glauberman, *The Bell Curve Debate* (New York: Times Books, 1995), ix.

3. Nicholas Wade, "Geographic Society Is Seeking a Genealogy of Humankind," *New York Times,* April 13, 2005.

46. The Problem with Race-Based Medicine

Dorothy Roberts

Fifteen years ago, I volunteered to participate in a research study that involved a genetic test. When I arrived at the clinic to be tested, I was handed a questionnaire. One of the very first questions asked me to check a box for my race: white, black, Asian, or Native American.

I wasn't quite sure how to answer the question. Was it aimed at measuring the diversity of research participants' social backgrounds? In that case, I would answer with my social identity, and check the box for "black." But what if the researchers were interested in investigating some association between ancestry and the risk for certain genetic traits? In that case, wouldn't they want to know something about my ancestry, which is just as much European as African? And how could they make scientific findings about my genes if I put down my social identity as a black woman? After all, I consider myself a black woman with a white father rather than a white woman with a black mother entirely for social reasons. Which racial identity I check has nothing to do with my genes. Well, despite the obvious importance of this question to the study's scientific validity, I was told, "Don't worry about it, just put down however you identify yourself." So I check[ed] "black," but I had no confidence in the results of a study that treated a critical variable so unscientifically.

That personal experience with the use of race in genetic testing got me thinking: Where else in medicine is race used to make false biological predictions?

Well, I found out that race runs deeply throughout all of medical practice. It shapes physicians' diagnoses, measurements, treatments, prescrip-

Transcript of Dorothy Roberts, "The Problem with Race-Based Medicine." TEDMED 2015. Courtesy of TED. For more talks, visit ted.com.

FIGURE 46.1. Dorothy Roberts with her parents, Robert Roberts and Iris Roberts, at her graduation from Harvard Law School in 1980. (Photo courtesy of Dorothy Roberts.)

tions, even the very definition of diseases. And the more I found out, the more disturbed I became.

Sociologists like me have long explained that race is a social construction. When we identify people as black, white, Asian, Native American, Latina, we're referring to social groupings with made-up demarcations that have changed over time and vary around the world. As a legal scholar, I've also studied how lawmakers, not biologists, have invented the legal definitions of races.

And it's not just the view of social scientists. You remember when the map of the human genome was unveiled at a White House ceremony in June 2000? President Bill Clinton famously declared, "I believe one of the great truths to emerge from this triumphant expedition inside the human genome is that in genetic terms, human beings, regardless of race, are more than 99.9 percent the same." And he might have added that that less than 1 percent of genetic difference doesn't fall into racial boxes.

Francis Collins, who led the Human Genome Project and now heads NIH [the National Institutes of Health], echoed President Clinton. "I am happy that today, the only race we're talking about is the human race."

Doctors are supposed to practice evidence-based medicine, and they're increasingly called to join the genomic revolution. But their habit of treating patients by race lags far behind.

Component	Your Value	Standard Range
Sodium	137	135–146
Potassium	4.3	3.6–5.3
Chloride	100	96–106
Total CO_2	26	20–30
Glucose	89	65–99
Glomerular Filtration Rate	...	
GFR Est. for Non-African American	86	
GFR Est. for African American	89	

GFR > 89 Normal
GFR 60–89 Normal to mildly reduced
GFR 30–59 Moderately reduced
GFR 15–29 Severely reduced
GFR <15 Kidney failure

FIGURE 46.2.   An example of the glomerular filtration rate (GFR) estimate for black and nonblack patients. (Table courtesy of Dorothy Roberts.)

Take the estimate of glomerular filtration rate, or GFR. Doctors routinely interpret GFR, this important indicator of kidney function, by race. As you can see in this lab test, a different GFR estimate is reported depending on whether or not the patient is African American. Why?

I've been told it's based on an assumption that African Americans have more muscle mass than people of other races. But what sense does it make for a doctor to automatically assume I have more muscle mass than that female bodybuilder? Wouldn't it be far more accurate and evidence-based to determine the muscle mass of individual patients just by looking at them?

Well, doctors tell me they're using race as a shortcut. It's a crude but convenient proxy for more important factors, like muscle mass, enzyme level, [and] genetic traits they just don't have time to look for. But race is a bad proxy. In many cases, race adds no relevant information at all. It's just a distraction. But race also tends to overwhelm the clinical measures. It blinds doctors to patients' symptoms, family illnesses, their history, [and] their own illnesses—all more evidence-based than the patient's race. Race can't substitute for these important clinical measures without sacrificing patient well-being.

Doctors also tell me race is just one of many factors they take into account, but there are numerous medical tests, like the GFR, that use race categorically to treat black, white, [and] Asian patients differently just because of their race.

Race medicine also leaves patients of color especially vulnerable to harmful biases and stereotypes. Black and Latino patients are twice as likely to receive no pain medication as whites for the same painful long bone fractures because of stereotypes that black and brown people feel less pain, exaggerate their pain, and are predisposed to drug addiction.

The Food and Drug Administration has even approved a race-specific medicine. It's a pill called BiDil to treat heart failure in self-identified African-American patients. A cardiologist developed this drug without regard to race or genetics, but it became convenient for commercial reasons to market the drug to black patients. The FDA then allowed the company, the drug company, to test the efficacy in a clinical trial that only included African-American subjects. It speculated that race stood in as a proxy for some unknown genetic factor that affects heart disease or response to drugs. But think about the dangerous message it sent, that black people's bodies are so substandard, a drug tested in them is not guaranteed to work in other patients.

In the end, the drug company's marketing scheme failed. For one thing, black patients were understandably wary of using a drug just for black people. One elderly black woman stood up in a community meeting and shouted, "Give me what the white people are taking!"

And if you find race-specific medicine surprising, wait until you learn that many doctors in the United States still use an updated version of a diagnostic tool that was developed by a physician during the slavery era, a diagnostic tool that is tightly linked to justifications for slavery.

Dr. Samuel Cartwright graduated from the University of Pennsylvania Medical School. He practiced in the Deep South before the Civil War, and he was a well-known expert on what was then called "Negro medicine." He promoted the racial concept of disease, that people of different races suffer from different diseases and experience common diseases differently. Cartwright argued in the 1850s that slavery was beneficial for black people for medical reasons. He claimed that because black people have lower lung capacity than whites, forced labor was good for them. He wrote in a medical journal, "It is the red vital blood sent to the brain that liberates their minds when under the white man's control, and it is the want of sufficiency of red vital blood that chains their minds to ignorance and barbarism when in freedom." To support this theory, Cartwright helped to perfect a medical device for measuring breathing called the spirometer to show the presumed deficiency in black people's lungs.

Today, doctors still uphold Cartwright's claim that black people as a race have lower lung capacity than white people. Some even use a modern day spirometer that actually has a button labeled "race" so the machine adjusts the measurement for each patient according to his or her race. It's a well-known function called "correcting for race."

The problem with race medicine extends far beyond misdiagnosing patients. Its focus on innate racial differences in disease diverts attention and resources from the social determinants that cause appalling racial gaps in health: lack of access to high-quality medical care; food deserts in poor neighborhoods; exposure to environmental toxins; high rates of incarceration; and experiencing the stress of racial discrimination.

You see, race is not a biological category that naturally produces these health disparities because of genetic difference. Race is a social category that has staggering biological consequences, but because of the impact of social inequality on people's health. Yet race medicine pretends the answer to these gaps in health can be found in a race-specific pill. It's much easier and more lucrative to market a technological fix for these gaps in health than to deal with the structural inequities that produce them.

The reason I'm so passionate about ending race medicine isn't just because it's bad medicine. I'm also on this mission because the way doctors practice medicine continues to promote a false and toxic view of humanity. Despite the many visionary breakthroughs in medicine we've been learning about, there's a failure of imagination when it comes to race. Would you imagine with me, just a moment: What would happen if doctors stopped treating patients by race? Suppose they rejected an eighteenth-century classification system and incorporated instead the most advanced knowledge of human genetic diversity and unity, that human beings cannot be categorized into biological races? What if, instead of using race as a crude proxy for some more important factor, doctors actually investigated and addressed that more important factor? What if doctors joined the forefront of a movement to end the structural inequities caused by racism, not by genetic difference?

Race medicine is bad medicine, it's poor science, and it's a false interpretation of humanity. It is more urgent than ever to finally abandon this backward legacy and to affirm our common humanity by ending the social inequalities that truly divide us.

47. Race in a Bottle

Jonathan Kahn

[On] June 23, 2005, the US Food and Drug Administration approved the first "ethnic" drug. Called BiDil (pronounced "bye-dill"), it was intended to treat congestive heart failure—the progressive weakening of the heart muscle to the point where it can no longer pump blood efficiently—in African Americans only. The approval was widely declared to be a significant step toward a new era of personalized medicine, an era in which pharmaceuticals would be specifically designed to work with an individual's particular genetic makeup. Known as pharmacogenomics, this approach to drug development promises to reduce the cost and increase the safety and efficacy of new therapies. BiDil was also hailed as a means to improve the health of African Americans, a community woefully underserved by the US medical establishment. Organizations such as the Association of Black Cardiologists and the Congressional Black Caucus strongly supported the drug's approval.

A close inspection of BiDil's history, however, shows that the drug is ethnic in name only. First, BiDil is not a new medicine—it is merely a combination into a single pill of two generic drugs, hydralazine and isosorbidedinitrate, both of which have been used for more than a decade to treat heart failure in people of all races. Second, BiDil is not a pharmacogenomic drug. Although studies have shown that the hydralazine/isosorbidedinitrate (H/I) combination can delay hospitalization and death for patients suffering from heart failure, the underlying mechanism for the drug's efficacy is not fully understood and has not been directly connected to any specific genes. Third, and most important, no firm evidence exists that BiDil actually works better or differently in African Americans than in anyone else. The FDA's approval of BiDil was based primarily on a clinical trial that enrolled only self-identified African Americans and did not compare their health outcomes with those of other ethnic or racial groups.

So how did BiDil become tagged as an ethnic drug and the harbinger of a new age of medicine? The story of the drug's development is a tangled tale of inconclusive studies, regulatory hurdles, and commercial motives. BiDil has had a relatively small impact on the marketplace—over the past two years, only a few million dollars' worth of prescriptions have been sold—but the drug has demonstrated the perils of using racial categories to win approval for new pharmaceuticals. Although African Americans are dying from heart disease and other illnesses at younger ages than whites, most researchers believe the premature deaths result from a complex array of social and economic forces. Some medical professionals and policy experts, however, have pointed to BiDil as proof that genetic differences can explain the health disparity. Worse, some pharmaceutical companies are now using this unfounded argument to pursue other treatments targeted at various ethnic groups, a trend that may segregate medicine and fatten the profits of drugmakers without addressing the underlying causes that are killing so many African Americans before their time.

BIRTH OF BIDIL

The BiDil saga began more than twenty years ago with a pair of studies designed to gauge the effects of vasodilating drugs—which widen blood vessels—on heart failure, a debilitating and ultimately fatal disease that afflicts millions of Americans. Until then, doctors treated heart failure with diuretics (to reduce the accumulation of fluid that results from inadequate pumping) and digoxin (to increase the contraction of the heart muscle) but had little else at their disposal. In the early 1980s Jay Cohn, a cardiologist at the University of Minnesota, hypothesized that administering two vasodilators, hydralazine and isosorbidedinitrate, might ease the strain on weakened hearts by relaxing both the arteries and veins. Together with the US Veterans Administration, Cohn designed and conducted two trials to assess this theory.

The first Vasodilator Heart Failure Trial (V-HeFT I) tested the H/I combination against a placebo and a drug called prazosin, which is used to treat high blood pressure. The results seemed to show great promise for the combination. The second trial, V-HeFT II, tested H/I against enalapril, a first-generation angiotensin-converting enzyme (ACE) inhibitor. (ACE inhibitors lower blood pressure by curbing the production of vessel-constricting peptides.) As it turned out, enalapril proved more effective than H/I for treating heart failure. From that point forward, ACE inhibitors became the new first-line therapy for heart failure patients. Doctors began recommending hydralazine and isosorbidedinitrate—both available as inexpensive generic pills—for those who did not respond well to ACE inhibitors.

Cohn, however, remained committed to developing a treatment that combined hydralazine and isosorbidedinitrate because he believed in its effectiveness. In 1987 he applied for a patent on the method of using the drugs together to treat heart failure in all people, regardless of race. (He could not get a patent on the drug combination itself because both medicines were already available in generic form.) He then licensed the patent rights to Medco, a small pharmaceutical firm in North Carolina, which took steps in the early 1990s to put the H/I combination into a single pill—and BiDil was born.

Medco and Cohn brought BiDil to the FDA for approval in 1996. In early 1997 the agency refused to approve the drug. Ironically, most of the doctors on the FDA's review panel thought BiDil did in fact work and said they would consider prescribing it. The problem was not with the drug but with the statistical data from the V-HeFT trials, which were designed not to meet the regulatory standards for FDA approval but to test the hypothesis that vasodilators could treat heart failure. After the rejection, Medco's stock plummeted by more than 20 percent, and the company let the patent rights revert to Cohn. By 1997 half of the twenty-year life of the original BiDil patent had already passed, which may explain Medco's reluctance to sink more money into the drug.

BIDIL'S RACIAL REBIRTH

It was only at this point that race entered the story. After the FDA's rejection of BiDil, Cohn went back to the V-HeFT results from the 1980s and broke down the data by race, examining how well African Americans had responded to the competing treatments. Such retrospective "data dredging" can yield useful insights for further investigations, but it is also fraught with statistical peril; if the number of research subjects in each category is too small, the results for the subgroups may be meaningless. Cohn argued that H/I worked particularly well in the African Americans enrolled in the V-HeFT studies. The clearest support for this claim came from V-HeFT I, which placed only forty-nine African Americans on H/I—a tiny number considering that new drug trials typically enroll thousands of subjects. In 1999 Cohn published a paper in the *Journal of Cardiac Failure* on this hypothesized racial difference and filed a new patent application. This second patent was almost identical to the first except for specifying the use of H/I to treat heart failure in black patients. Issued in 2000, the new patent lasts until 2020, thirteen years after the original patent was set to expire. Thus was BiDil reinvented as an ethnic drug.

Race-specific patent in hand, Cohn relicensed the intellectual-property rights to NitroMed, a small Massachusetts firm. The FDA then gave NitroMed the go ahead to conduct the African American Heart Failure Trial (A-HeFT), a relatively small study involving 1,050 self-identified African Americans. In A-HeFT, half the heart failure patients took BiDil while the other half received a placebo; at the same time, the patients in both groups continued taking their already prescribed treatments for heart failure (for example, about 70 percent of the subjects in both groups were on ACE inhibitors). The results were strikingly positive: the mortality rate in the BiDil subjects was 43 percent lower than that in the placebo group. In fact, BiDil appeared so effective that A-HeFT's Data Safety Monitoring Board suspended the trial early, in July 2004, so that the drug could be offered to the subjects in the placebo group as well. NitroMed's stock surged on the news, more than tripling in value in the following days. The next June the FDA formally approved BiDil with a race-specific label, indicating that it was for use in black patients.

But researchers have good reason to believe that BiDil would also be effective in nonblack patients. Indeed, Cohn himself has said he believes the drug should work in people of all races. So why did the developers of the drug test it in only one ethnic group? The answer seems to be driven more by commerce than by science. If the FDA had approved BiDil for the general population, the patent protection for the drug's manufacturer would have expired in 2007. Restricting the clinical trial to African-Americans maximized the chances that the FDA would approve the race-specific use of BiDil, giving NitroMed an additional thirteen years to sell the H/I combination without competition.

SEGREGATED MEDICINE

Science and commerce have always proceeded together in advancing medicine, but in the case of BiDil the balance seems to have gotten out of whack. There can be no doubt that Cohn and the other medical professionals behind the drug's development sincerely want to improve the lives of the many people suffering from heart failure. In this respect, the approval of BiDil is certainly a good thing. But Cohn and NitroMed have also used race to obtain commercial advantage. The patented drug costs about six times as much as the readily available generic equivalents. The high cost has already made many insurers reluctant to cover BiDil and may place it beyond the reach of the millions of Americans without health insurance. Moreover, the unprecedented media attention to the race-specific character of the drug may lead many doctors and patients alike to think that non–African Americans should not get the drug, when, in fact, it might help prolong their lives.

Perhaps most problematically, the patent award and FDA approval of BiDil have given the imprimatur of the federal government to using race as a genetic category. Since the inception of the Human Genome Project, scientists have worked hard to ensure that the biological knowledge emerging from advances in genetic research is not used inappropriately to make socially constructed racial categories appear biologically given or natural. As a 2001 editorial in the journal *Nature Genetics* put it, "scientists have long been saying that at the genetic level there is more variation between two individuals in the same population than between populations and that there is no biological basis for 'race.'" More recently, an editorial in *Nature Biotechnology* asserted that "race is simply a poor proxy for the environmental and genetic causes of disease or drug response. . . . Pooling people in race silos is akin to zoologists grouping raccoons, tigers and okapis on the basis that they are all stripey."

The FDA's approval of BiDil was based on accepting NitroMed's argument that the drug should be indicated only for African Americans because the trial population was African American. This labeling sends the scientifically unproved message that the subject population's race was somehow a relevant biological variable in assessing the safety and efficacy of BiDil. Most drugs on the market today were tested in overwhelmingly white populations, but we do not call these medicines "white," nor should we. The FDA's unstated assumption is that a drug that proves effective for white people is good enough for everyone; the same assumption should apply when the trial population happens to be black. Otherwise, the FDA is implying that African Americans are somehow less fully representative of humanity than whites are.

In November 2004 *Nature Genetics* published an article by Sarah K. Tate and David B. Goldstein of University College London entitled "Will Tomorrow's Medicines Work for Everyone?" The paper noted that "29 medicines (or combinations of medicines) have been claimed, in peer-reviewed scientific or medical journals, to have differences in either safety or, more commonly, efficacy among racial or ethnic groups." Journalists immediately quoted the study as providing further evidence of biological differences among races; for example, an article in the *Los Angeles Times*, after discussing BiDil, referred to "a report in the journal *Nature Genetics* last month [that] listed 29 drugs that are *known* to have different efficacies in the two races." (The italics are mine.) Similarly, a story in the *Times of London* asserted that "only last week, *Nature Genetics* revealed research from University College London *showing* that 29 medicines have safety or efficacy profiles that vary between ethnic or racial groups." (Again, the italics are mine.) And a *New York Times* editorial entitled "Toward the First Racial Medicine" began with a discussion of BiDil and went on to note that "by one count, some 29 medicines show evidence of being safer or more effective in one racial group or another, suggesting that more targeted medicines may be coming."

One small problem: these newspaper stories totally misrepresented the *Nature Genetics* piece. Tate and Goldstein asserted that the racial differences in drug safety or efficacy have only been claimed, not proved, and in the next sentence they go on to say, "But these claims are universally controversial, and there is *no consensus* on how important race or ethnicity is in determining drug response." (My italics again.)

In only four of the twenty-nine medicines identified, Tate and Goldstein found evidence that genetic variations between races could possibly be related to the different responses to the drugs. (All four are beta blockers used for treating high blood pressure and other cardiovascular ills; some research indicates that these drugs work better in individuals carrying a gene variant that is more common in people of European ancestry than in African Americans.) For nine of the medicines, the authors found "a reasonable underlying physiological basis" to explain why blacks and whites may respond differently to the drugs; for example, some scientists have speculated that ACE inhibitors may be more effective in people of European descent than in African Americans because of variations in enzyme activity. (Other researchers have contested this hypothesis.) For five of the drugs, Tate and Goldstein found no physiological reasons to explain the varying responses; for the remaining eleven they concluded that the reports of differing responses may not be valid.

RACIAL INJUSTICE

Nevertheless, the appeal of race-specific drugs is growing. In 2003 VaxGen, a California biopharmaceutical company, made an abortive attempt to use a retrospective analysis of racial subgroups to salvage a proposed AIDS vaccine called AIDSVAX. Although the clinical trial for AIDSVAX showed no decrease in HIV infection rates in the study population as a whole, VaxGen claimed a significant reduction in infection among the black and Asian participants. But only a few hundred blacks and Asians were involved in the study, meaning that a handful of infections could have skewed the results. The claim of race-specific response was undercut later that year when another trial in Thailand showed that AIDSVAX was ineffective there as well. In a similar case, AstraZeneca, the British pharmaceutical firm, argued that its lung cancer drug, Iressa, worked better in the Asians enrolled in a 2004 clinical trial, which showed that the medicine did not improve survival rates overall. (Unconvinced, the FDA changed the labeling for Iressa, disallowing its use in any new patients.) More recently, AstraZeneca has conducted trials of Crestor, the company's multibillion-dollar cholesterol-lowering drug, in African Americans, South Asians, and Hispanics. Consumer groups have claimed that Crestor is less safe than other cholesterol-lowering drugs, but AstraZeneca says the race-specific studies demonstrate the safety and efficacy of the medicine.

Researchers using race to develop drugs may be motivated by good intentions, but such efforts are also driven by the dictates of an increasingly competitive medical marketplace. The example of BiDil indicates that researchers and regulators alike have not fully appreciated that race is a powerful and volatile category. When used to bolster the commercial value of a drug, it can lead to haphazard regulation, substandard medical treatment, and other unfortunate unintended consequences. The FDA should not grant race-specific approvals without clear and convincing evidence of a genetic or biological basis for any observed racial differences in safety or efficacy. Approving more drugs such as BiDil will not alleviate the very serious health disparities between races in the United States. We need social and political will, not mislabeled medicines, to redress that injustice.

48. The Science and Business of Genetic Ancestry Testing

Deborah A. Bolnick, Duana Fullwiley, Troy Duster,
Richard S. Cooper, Joan H. Fujimura, Jonathan Kahn,
Jay S. Kaufman, Jonathan Marks, Ann Morning,
Alondra Nelson, Pilar Ossorio, Jenny Reardon,
Susan M. Reverby, Kimberly TallBear

At least two dozen companies now market "genetic ancestry tests" to help consumers reconstruct their family histories and determine the geographic origins of their ancestors. More than 460,000 people have purchased these tests over the past six years[1], and public interest is still skyrocketing.[2] Some scientists support this enterprise because it makes genetics accessible and relevant; others view it with indifference, seeing the tests as merely "recreational." However, both scientists and consumers should approach genetic ancestry testing with caution because (i) the tests can have a profound impact on individuals and communities, (ii) the assumptions and limitations of these tests make them less informative than many realize, and (iii) commercialization has led to misleading practices that reinforce misconceptions.

THE IMPACT OF "RECREATIONAL GENETICS"

Although genetic ancestry testing is often described as "recreational genetics," many consumers do not take these tests lightly. Each test costs a hundred to nine hundred dollars, and consumers often have deep personal reasons for purchasing these products. Many individuals hope to identify biological relatives, to validate genealogical records, and to fill in gaps in family histories. Others are searching for a connection to specific groups or places in Eurasia and Africa. This search for a "homeland" is particularly poignant for many African Americans, who hope to recapture a history

From "The Science and Business of Genetic Ancestry Testing." *Science* 318, no. 5849 (October 19, 2007): 399–400. doi: 10.1126/science.1150098. Reprinted with permission from AAAS (see http://www.aaas.org/).

stolen by slavery. Others seek a more nuanced picture of their genetic backgrounds than the black-and-white dichotomy that dominates US racial thinking.

Genetic ancestry testing also has serious consequences. Test-takers may reshape their personal identities, and they may suffer emotional distress if test results are unexpected or undesired.[3] Test-takers may also change how they report their race or ethnicity on governmental forms, college or job applications, and medical questionnaires.[4] This could make it more difficult to track the social experiences and effects of race and racism.[5] Genetic ancestry testing also affects broader communities: tests have led African Americans to visit and financially support specific African communities. Other Americans have taken the tests in hope of obtaining Native American tribal affiliation (and benefits like financial support, housing, education, health care, and affirmation of identity) or to challenge tribal membership decisions.[6]

LIMITATIONS

It is important to understand what these tests can and cannot determine. Most tests fall into two categories. Mitochondrial DNA (mtDNA) tests sequence the hypervariable region of the maternally inherited mitochondrial genome. Y-chromosome tests analyze short tandem repeats and/or single nucleotide polymorphisms (SNPs) in the paternally inherited Y chromosome. In both cases, the test-taker's haplotype (set of linked alleles) is determined and compared with haplotypes from other sampled individuals. These comparisons can identify related individuals who share a common maternal or paternal ancestor, as well as locations where the test-taker's haplotype is found today. However, each test examines less than 1 percent of the test-taker's DNA and sheds light on only one ancestor each generation.[7] A third type of test (DNAPrint's AncestryByDNA test) attempts to provide a better measure of overall ancestry by using 175 autosomal markers (inherited from both parents) to estimate an individual's "bio-geographical ancestry."

Although companies acknowledge that mtDNA and Y-chromosome tests provide no information about most of a test-taker's ancestors, more important limitations to all three types of genetic ancestry tests are often less obvious. For example, genetic ancestry testing can identify some of the groups and locations around the world where a test-taker's haplotype or autosomal markers are found, but it is unlikely to identify all of them. Such inferences depend on the samples in a company's database, and even databases with 10,000 to 20,000 samples may fail to capture the full array of human genetic diversity in a particular population or region.

Another problem is that questionable scientific assumptions are sometimes made when companies report results of a genetic ancestry test. For instance, when an allele or haplotype is most common in one population,[8] companies often assume it to be diagnostic of that population. This can be problematic because high genetic diversity exists within populations and gene flow occurs between populations. Very few alleles are therefore diagnostic of membership in a specific population, but companies sometimes fail to mention that an allele could have been inherited from a population in which it is less common. Consequently, many consumers do not realize that the tests are probabilistic and can reach incorrect conclusions.

Consumers often purchase these tests to learn about their race or ethnicity, but there is no clear-cut connection between an individual's DNA and his or her racial or ethnic affiliation. Worldwide patterns of human genetic diversity are weakly correlated with racial and ethnic categories because both are partially correlated with geography.[9] Current understandings of race and ethnicity reflect more than genetic relatedness, though, having been defined in particular sociohistorical contexts (i.e., European and American colonialism). In addition, social relationships and life experiences have been as important as biological ancestry in shaping individual identity and group membership.

Many genetic ancestry tests also claim to tell consumers where their ancestral lineage originated and the social group to which their ancestors belonged. However, present-day patterns of residence are rarely identical to what existed in the past, and social groups have changed over time, in name and composition.[10] Databases of present-day samples may therefore provide false leads.

Finally, even though there is little evidence that four biologically discrete groups of humans ever existed,[11] the AncestryByDNA test creates the appearance of genetically distinct populations by relying on "ancestry informative markers" (AIMs). AIMs are SNPs or other markers that show relatively large (30% to 50%) frequency differences between population samples. The AncestryByDNA test examines AIMs selected to differentiate between four "parental" populations (Africans, Europeans, East Asians, and Native Americans). However, these AIMs are not found in all peoples who would be classed together as a given "parental" population. The AIMs that characterize "Africans," for example, were chosen on the basis of a sample of West Africans. Dark-skinned East Africans might be omitted from the AIMs reference panel of "Africans" because they exhibit different gene variants.[12] Furthermore, some of the most "informative" AIMs involve loci that have undergone strong selection,[13] which makes it unclear whether these markers indicate shared ancestry or parallel selective pressures (such as similar environmental exposures in different geographic regions) or both.

The problems described here are likely responsible for the most paradoxical results of this test. For instance, the AncestryByDNA test suggests that most people from the Middle East, India, and the Mediterranean region of Europe have Native American ancestry.[14] Because no archaeological, genetic, or historical evidence supports this suggestion, the test probably considers some markers to be diagnostic of Native American ancestry when, in fact, they are not.

Thus, these tests should not be seen as determining the race or ethnicity of a test-taker. They cannot pinpoint the place of origin or social affiliation of even one ancestor with exact certainty. Although wider sampling and technological advancements may help,[15] many of the tests' problems will remain.

EFFECTS OF COMMERCIALIZATION

Although it is important for consumers to understand the limitations of genetic ancestry testing and the complex relation between DNA, race, and identity, these complexities are not always made clear. Web sites of many companies state that race is not genetically determined, but the tests nevertheless promote the popular understanding that race is rooted in one's DNA[16]—rather than being an artifact of sampling strategies, contrasting geographical extremes, and the imposition of qualitative boundaries on human variation. Because race has such profound social, political, and economic consequences, we should be wary of allowing the concept to be redefined in a way that obscures its historical roots and disconnects it from its cultural and socioeconomic context.

It is unlikely that companies (and the associated scientists) deliberately choose to mislead consumers or misrepresent science. However, market pressures can lead to conflicts of interest, and data may be interpreted differently when financial incentives exist. For scientists, these incentives include paid consultancies, patent rights, licensing agreements, stock options, direct stock grants, corporate board memberships, scientific advisory board memberships, media attention, lecture fees, and/or research support. Because scientific pronouncements carry immense weight in our society, claims must be carefully evaluated when scientists have a financial stake in them. Unfortunately, peer review is difficult here, because most companies maintain proprietary databases.

As consumers realize that they have been sold a family history that may not be accurate, public attitudes toward genetic research could change. Support for molecular and anthropological genetics might decrease, and historically disadvantaged communities might increase their distrust of the scientific establishment.[17] These tests may also come up in medical settings: Many consumers are aware of the well-publicized association between ancestry and disease, and patients may ask doctors to take their ancestry tests into consideration when making medical decisions. Doctors should be cautious when considering such results.[18]

We must weigh the risks and benefits of genetic ancestry testing, and as we do so, the scientific community must break its silence and make clear the limitations and potential dangers. Just as the American Society of Human Genetics (ASHG) recently published a series of recommendations regarding direct-to-consumer genetic tests that make health-related claims,[19] we encourage ASHG and other professional genetic and anthropological associations to develop policy statements regarding genetic ancestry testing.

NOTES

1. H. Wolinsky, "Genetic Genealogy Goes Global," *EMBO Rep.* 7, 1072 (2006).

2. H. Wolinsky, "Genetic Genealogy Goes Global," *EMBO Rep.* 7, 1072 (2006); J. Simons, *Fortune* 155 (39) (2007); Thirteen/WNET New York, *African American Lives*, "Episode 2: The Promise of Freedom," press release (July 27, 2007); P. Harris, *Observer* [London], July 15, 2007, 22.

3. *Motherland: A Genetic Journey* (Takeaway Media Productions, London, 2003).

4. A. Harmon, "Seeking Ancestry in DNA Ties Uncovered by Tests," *New York Times*, April 12, 2006, A1.

5. Ibid.

6. B. Hoerner, *Wired* 13 (2005).

7. A. Yang, *Chance* 20, 32–39 (2007).

8. K. Weiss and M. Fullerton, "Racing around, Getting Nowhere," *Evolutionary Anthropology* 14 (2005), 165,

9. Ibid.

10. C. Rotimi, "Genetic Ancestry Tracing and the African Identity: A Double-Edged Sword?," *Developing World Bioethics* 3 (2003), 151–58.

11. Weiss and Fullerton, op. cit.

12. S. Tishkoff et al., "Convergent Adaptation of Human Lactase Persistence in Africa and Europe," *Nature Genetics* 39 (2006), 31–40; A. Mourant, A. Kopec, and K. Domaniewska-Sobczak, *The Distribution of the Human Blood Groups and Other Polymorphisms* (London: Oxford University Press, 1976); M. Hamblin and A. Di Rienzo, "Detection of the Signature of Natural Selection in Humans: Evidence from

the Duffy Blood Group Locus," *American Journal of Human Genetics* 66 (2000), 1669–79.

13. J. Akey et al., "Interrogating a High-Density SNP Map for Signatures of Natural Selection," *Genome Biology* 12 (2002), 1805–14.

14. AncestryByDNA, www.ancestrybydna.com/welcome/productsandservices /ancestrybydna/ethnicities.

15. M. Shriver and R. Kittles, *Nature Reviews Genetics* 5 (2004), 611.

16. DNAPrint, Frequently Asked Questions, no. 1, www.ancestrybydna.com /welcome/faq/#q1.

17. J. Reardon, *Race to the Finish: Identity and Governance in an Age of Genomics* (Princeton, NJ: Princeton University Press, 2004).

18. In contexts such as gene mapping and genome-wide associations, genetic ancestry information can protect against confounding by population stratification or provide evidence of the population origin of specific susceptibility alleles. See M. Enoch et al., "Using Ancestry-Informative Markers to Define Populations and Detect Population Stratification," *Journal of Psychopharmacology* 20 (2006), 19. These applications are much narrower than determination of individual ancestry.

19. K. Hudson et al., "ASHG Statement on Direct-to-Consumer Genetic Testing in the United States," *American Journal of Human Genetics* 81 (2007), 635.

49. All That Glitters Isn't Gold

Osagie K. Obasogie and Troy Duster

The increasing use of DNA evidence has revolutionized criminal investigations. Over the past several years, DNA forensics—once thought to be a less reliable identifier than other forensic techniques, such as latent fingerprinting—have now become the evidentiary gold standard in criminal prosecutions. At the same time, non-DNA-based forensic techniques that have incarcerated thousands are coming under fire.

The policy implications of this shifting dynamic—what Michael Lynch and colleagues call an "inversion of credibility"[1]—can be most clearly seen in the National Research Council's [NRC] 2009 report, *Strengthening Forensic Science in the United States: A Path Forward.* Conducted at Congress's request by a highly esteemed committee, this report—over three hundred pages—assesses the current state of forensic science.

The committee found remarkable shortcomings in what they call the forensic science knowledge base, noting that the scientific theories and methods used to substantiate many forensic claims frequently cannot withstand close scrutiny. They found an alarmingly "wide variability in capacity, oversight, staffing, certification, and accreditation."[2] For example, lack of transparency, susceptibility to bias, and questionable methodologies for friction ridge analyses (analyses of the prints left by fingers, palms, or soles) make it difficult for two analysts to come to the same conclusion.[3] The report's sobering message is that many forensic applications simply lack scientific rigor despite their routine use in legal proceedings.

Original publication by Osagie K. Obasogie and Troy Duster. "All That Glitters Isn't Gold." *Hastings Center Report* 41, no. 5 (September/October 2011): 15–18. doi:10.1002/j.1552146X.2011.tb00135.x. Reprinted by permission of John Wiley and Sons.

Although the committee acknowledges that DNA forensics are not always perfect, the report and its recommendations are framed by an implied yet powerful claim: non-DNA forensic techniques should live up to the gold standard created by DNA typing. But this framing has its own serious drawbacks that obscure much deeper issues concerning both technical matters related to the scientific validity of extending basic DNA identification techniques to novel applications and the ethical, legal, and social implications of DNA forensics' expanding uses.

DNA EXCEPTIONALISM?

The NRC report is replete with both general and specific declarations that frame the accuracy and reliability of DNA typing as the new standard for forensic investigations. At the broad end of the spectrum, the committee notes "DNA typing is now universally recognized as *the standard* against which many other forensic individualization techniques are judged."[4] This sentiment also shapes the analysis of specific techniques. For example, they note that "overall, the process for toolmark and firearms comparison lacks the specificity of the protocols for, say, 13 STR [short tandem repeat] DNA analysis."[5]

Few seriously doubt DNA typing's high reliability in determining whether any two isolated samples match. Yet DNA typing is only one of many ways in which DNA analyses are used in forensic investigations. For over two decades, state and federal governments have been collecting convicted felons' genetic profiles and depositing them into databases in order to be able to identify repeat offenders who leave biological samples at crime scenes.

DNA databases give rise to techniques beyond mere DNA typing that expand criminal investigations' scope and impact. This repository's growth—the federal database itself is almost at eight million profiles[6]—has given rise to three related techniques: cold hits, partial matches, and familial searches.

"Cold hits" occur when investigators are able to match unknown biological materials left at a crime scene with a known database profile. Partial matches occur when investigators identify a suspect using fewer than thirteen loci—the standard number of chromosome sites where, if identical, a "match" between two profiles can be declared. Familial searches work from the premise that relatives share many identical loci. While the number of shared markers between an unknown suspect and a database hit might not incriminate the person with the known profile, it can and has pointed to a relative, who is then the subject of a criminal investigation.

Many assume that these database-oriented techniques have the same precision as typing two individual samples. But what often gets obscured—as it does in the NRC report—is that these newer uses of DNA forensics share many of the same shortcomings that the NRC identifies with non-DNA forensic techniques.

PARTIAL MATCHES AND THE SCIENTIFIC KNOWLEDGE BASE

DNA databases gain much of their authority from the oft-repeated claim that the chance that two profiles will randomly match—even partially—is only one in several million. It has been argued that a nine-locus match can uniquely identify perpetrators,[7] and individuals have been convicted on such partial-match evidence.

However, increasing evidence suggests that the way scientists have calculated the probability for random matches may not be accurate. Data obtained from the Arizona state DNA database showed that out of 65,493 profiles, 122 pairs matched at nine loci, twenty pairs matched at ten, and two pairs (siblings) matched at eleven and twelve loci.[8] Findings from the Illinois state database yielded similar results: 903 pairs matched at nine or more loci out of a total of more than 200,000 profiles. Data from the Maryland database, with 30,000 profiles, was also surprising: thirty-two pairs matched at nine loci, and three matched at all thirteen.[9]

How could so many profiles randomly match at so many loci? No one knows for sure, which is why scientists and legal scholars are calling for more access to research government databases.[10] The FBI and several states have thus far refused to comply.

ESTABLISHING STANDARDS

The NRC report is particularly concerned with forensic science's lack of standardization; the field has few established protocols on how to accurately describe the significance of such evidence. While this is a demonstrable problem with non-DNA forensics, it is also true for DNA evidence related to database searches.

For example, with "cold-hit" evidence, where investigators run an unknown sample against thousands of database profiles in search of a "hit," any match is only as significant as the statistical probability that it might be coincidental. Yet there is no agreed-upon standard, much less a *gold standard*, for how to calculate this statistic or how to present this evidence in court.

The seemingly compelling one-in-a-million statistic that we often hear associated with cold-hit matches uses a population figure as a referent—the likelihood that an unknown profile matches a suspect (identified for other reasons) purely by coincidence. However, cold-hit matches that occur *within* databases have a substantially higher probability of being coincidental; when searching through large databases with millions of profiles (like the federal database), there are millions of chances for coincidental matches.[11] Transposing the statistical significance of the former approach to the latter is what is often referred to as the prosecutor's fallacy and has been known to impact determinations of individuals' guilt or innocence.[12]

The substantially diminished probabilities stemming from cold-hit database searches that take database size into account more accurately reflect the statistical limitations of this approach. This has led an NRC committee[13] and an FBI advisory board[14] to recommend making these database limitations part of the calculation. Yet neither recommendation has been widely implemented by authorities. While the committee briefly points to these prior recommendations, they do not fully discuss how the absence of consistently enforced standards, procedures, or oversight can lead courts to receive misleading information that can undermine DNA forensics' credibility as much as a failure to standardize non-DNA forensics affects its integrity.

QUESTIONS OF JUSTICE

While DNA typing itself is far from infallible, database-oriented DNA forensics raise a profound series of second-order questions for entire groups, in contrast to the NRC report's singular focus on individuals. Much of this stems from the unique ways in which the criminal justice system interacts with blacks and Latinos, particularly through policies regarding DNA databases. For example, the overpolicing of minority communities, along with related injustices, has led to these groups' dramatic overrepresentation in prisons; 30 percent of black males will be convicted of a felony at least once compared to 5 percent of whites, and an adult black male is eight times more likely to be incarcerated.[15] Aggressive public policies encouraging sample collections for almost any contact with law enforcement is leading to an alarming statistic: although blacks represent only 13 percent of the population, they make up an estimated 40 percent of federal DNA database profiles.[16]

Given the disparate composition of DNA databases, techniques like familial searching raise significant questions regarding systemic bias.

Leveraging the shared genetic variants and short tandem repeat lengths between relatives to find suspects will have a much larger impact on blacks and Latinos. What does it mean for government to turn people with existing stored profiles into "genetic informants" on their relatives without their knowledge or consent and with few safeguards to prevent wrongful convictions from errant cold hits? What are the ethical, political, and legal implications of placing a population under a lifetime of genetic surveillance in which each DNA fragment shared with a banked relative is screened against future crime scene evidence? And is it just for these significant civil liberties concerns to disproportionately fall upon groups already unfairly burdened by injustices linked to what we know to be selective patterns of law enforcement?[17]

WORTH ITS WEIGHT IN GOLD?

In addition to these concerns, several other technical issues and questionable practices with unresolved social and ethical implications also plague DNA forensics. On the technical side, contaminated samples can lead to false positives and false negatives, clerical errors can lead to incorrectly logged samples and poor data entry, and crime labs can misinterpret old, small, or mixed samples from multiple individuals. These and other technical issues are linked to broader social and ethical issues, such as varying practices regarding the destruction of samples after law enforcement has analyzed them, the propriety of using forensic samples for nonforensic purposes like research, the ethics of surreptitious sample acquisition by police, and a host of other privacy issues connected to the general idea of the government storing sensitive genetic information indefinitely.

Public policy regarding the use and expansion of DNA forensics is evolving as quickly as the technologies themselves. For example, a California law that went into effect in 2009 allows authorities to take and retain DNA samples from individuals merely arrested—not charged or convicted—for felonies.[18] And New York enacted a law in late 2009 allowing authorities to use DNA database partial matches to identify suspects.[19] This highlights a current trend, whereby law is being used to radically expand DNA databases to include larger, unsuspecting portions of the population.

The NRC report and its recommendations represent an important first step to putting the scientific method's rigor into forensics so that justice can prevail. If we are to take the report as seriously as it deserves, then the critiques it raises for non-DNA forensic applications must also be applied to the growing spectrum of DNA forensics. Like our prior uncritical acceptance of latent fingerprint technology, the new proverbial gold standard might, on closer inspection, have far more tarnish then we have been led to believe.

NOTES

1. M. Lynch, S.A. Cole, R. McNally, and K. Jordan, *Truth Machine: The Contentious History of DNA Fingerprinting* (Chicago, IL: University of Chicago Press, 2008).

2. National Research Council, Committee on Identifying the Needs of the Forensic Sciences Community, *Strengthening Forensic Science in the United States: A Path Forward* (Washington, DC: National Academies Press, 2009), 14.

3. Ibid., 136–45.

4. Ibid., 130.

5. Ibid., 155.

6. Federal Bureau of Investigation, www.fbi.gov/about-us/lab/codis/codis_brochure.

7. W.C. Thompson, "The Potential for Error in Forensic DNA Testing (and How That Complicates the Use of DNA Databases for Criminal Identification)" (paper presented at the Council for Responsible Genetics' national conference, Forensic DNA Databases and Race: Issues, Abuses and Actions, August 12, 2008), www.councilforresponsiblegenetics.org/pageDocuments/H4T5EOYUZI.pdf.

8. J. Jefferson, "Cold Hits Meet Cold Facts: Are DNA Matches Infallible?," *Transcript Magazine* 40, no. 1 (2008): 29–33.

9. L. Geddes, "Unreliable Evidence? Time to Open Up DNA Databases," *New Scientist* no. 2742 (January 6, 2010): 8–9, www.newscientist.com/article/mg20527424.700-unreliable-evidence-timeto-open-up-dna-databases.html?full = true&print = true.

10. D.E. Krane et al., "Time for DNA Disclosure," *Science* 326 (2009): 1631–32.

11. Thompson, "The Potential for Error in Forensic DNA Testing." The NRC report briefly cites this issue in a footnote (chapter 3, footnote 61).

12. J. Felch and M. Dolan, "DNA Matches Aren't Always a Lock," *Los Angeles Times*, May 3, 2008.

13. National Research Council, *The Evaluation of Forensic DNA Evidence. Commission on DNA Forensic Science: An Update* (Washington, DC: National Academies Press, 1996).

14. FBI DNA Advisory Board, "Statistical and Population Genetics Issues Affecting the Evaluation of the Frequency of Occurrence of DNA Profiles Calculated from Pertinent Population Databases," *Forensic Science Communications* 2, no. 3 (2000), www2.fbi.gov/hq/lab/fsc/backissu/july2000/dnastat.htm.

15. D.H. Kaye and M.E. Smith, "DNA Identification Databases: Legality, Legitimacy, and the Case for Population-Wide Coverage," *Wisconsin Law Review* 413 (2003): 413–59.

16. H.T. Greely et al., "Family Ties: The Use of DNA Offender Databases to Catch Offender's Kin," *Journal of Law, Medicine, and Ethics* 34, no. 248 (2007): 248–62.

17. New York State Assembly Committees on Codes and on Corrections, Testimony of Harry G. Levine Regarding Pending and Proposed Legislation to Collect DNA from All People Convicted of a Misdemeanor in New York State, and Also Regarding New York City's Epidemic of Marijuana Possession Arrests, Albany,

NY, May 31, 2007, www.aclu-md.org/aLegislative/Docs/Testimony-Harry_G_
Levine.pdf.

18. The DNA Fingerprint, Unsolved Crime and Innocence Protection Act
(November 2004), California Proposition 69, http://ag.ca.gov/bfs/prop69.php.

19. J.W. Peters, "New Rule Allows Use of Partial DNA Matches," *New York
Times,* January 24, 2010.

50. High-Tech, High-Risk Forensics

Osagie K. Obasogie

When the police arrived [in November 2012] at the ransacked mansion of the millionaire investor Raveesh Kumra, outside of San Jose, Calif., they found Mr. Kumra had been blindfolded, tied, and gagged. The robbers took cash, rare coins, and ultimately Mr. Kumra's life; he died at the scene, suffocated by the packaging tape used to stifle his screams. A forensics team found DNA on his fingernails that belonged to an unknown person, presumably one of the assailants. The sample was put into a DNA database and turned up a "hit"—a local man by the name of Lukis Anderson.

Bingo. Mr. Anderson was arrested and charged with murder.

There was one small problem: the twenty-six-year-old Mr. Anderson couldn't have been the culprit. During the night in question, he was at the Santa Clara Valley Medical Center, suffering from severe intoxication.

Yet he spent more than five months in jail with a possible death sentence hanging over his head. Once presented with Mr. Anderson's hospital records, prosecutors struggled to figure out how an innocent man's DNA could have ended up on a murder victim.

Late last month, prosecutors announced what they believe to be the answer: the paramedics who transported Mr. Anderson to the hospital were the very same individuals who responded to the crime scene at the mansion a few hours later. Prosecutors now conclude that at some point, Mr. Anderson's

DNA must have been accidentally transferred to Mr. Kumra's body—likely by way of the paramedics' clothing or equipment.

This theory of transference is still under investigation. Nevertheless, the certainty with which prosecutors charged Mr. Anderson with murder highlights the very real injustices that can occur when we place too much faith in DNA forensic technologies.

In the end, Mr. Anderson was lucky. His alibi was rock solid; prosecutors were forced to concede that there must have been some other explanation. It's hard to believe that, out of the growing number of convictions based largely or exclusively on DNA evidence, there haven't been any similar mistakes.

In one famous case of crime scene contamination, German police searched for around fifteen years for a serial killer they called the "Phantom of Heilbronn"—an unknown female linked by traces of DNA to six murders across Germany and Austria. In 2009, the police found their "suspect": a worker at a factory that produced the cotton swabs police used in their investigations had been accidentally contaminating them with her own DNA.

Contamination is not the only way DNA forensics can lead to injustice. Consider the frequent claim that it is highly unlikely, if not impossible, for two DNA profiles to match by coincidence. A 2005 audit of Arizona's DNA database showed that, out of some 65,000 profiles, nearly 150 pairs matched at a level typically considered high enough to identify and prosecute suspects. Yet these profiles were clearly from different people.

There are also problems with the way DNA evidence is interpreted and presented to juries. In 2008, John Puckett—a California man in his seventies with a sexual assault record—was accused of a 1972 killing, after a trawl of the state database partially linked his DNA to crime scene evidence. As in the Anderson case, Mr. Puckett was identified and implicated primarily by this evidence. Jurors—told that there was only a one-in-1.1 million chance that this DNA match was pure coincidence—convicted him. He is now serving a life sentence.

But that one-in-1.1 million figure is misleading, according to two different expert committees, one convened by the FBI, the other by the National Research Council. It reflects the chance of a coincidental match in relation to the size of the general population (assuming that the suspect is the only one examined and is not related to the real culprit). Instead of the general population, we should be looking at only the number of profiles in the DNA database. Taking the size of the database into account in Mr. Puckett's case (and, again, assuming the real culprit's profile is not in the database) would have led to a dramatic change in the estimate, to one in three.

One juror was asked whether this figure would have affected the jury's deliberations. "Of course it would have changed things," he told reporters. "It would have changed a lot of things."

DNA forensics is an invaluable tool for law enforcement. But it is most useful when it corroborates other evidence pointing to a suspect, or when used to determine whether any two individual samples match, like in the exonerations pursued by the Innocence Project.

But when the government gets into the business of warehousing millions of DNA profiles to seek "cold hits" as the primary basis for prosecutions, much more oversight by and accountability to the public is warranted. For far too long, we have allowed the myth of DNA infallibility to chip away at our skepticism of government's prosecutorial power, undoubtedly leading to untold injustices.

In the Anderson case, thankfully, prosecutors acknowledged the obvious: their suspect could not have been in two places at once. But he was dangerously close to being on his way to death row because of that speck of DNA. That one piece of evidence—obtained from a technology with known limitations, and susceptible to human error and prosecutorial misuse—might mistakenly lead to execution at the hands of the state should send chills down every one of our spines. The next Lukis Anderson could be you. Better hope your alibi is as well documented as his.

PART X

Biopolitics and the Future

This concluding section is oriented toward the future. It touches on many of the challenges explored in preceding sections as well as on the core concerns of the new biopolitics as they are articulated in the introduction to *Beyond Bioethics:*

- Reckoning with the role of commerce and markets in biomedicine and biotechnology
- Understanding the human genome as part of the common heritage of humanity
- Avoiding technical developments and genetic narratives that embed social and political preferences at the molecular level
- Ensuring democratic oversight of powerful human biotechnologies
- Steering clear of a new market-driven eugenics

The essays in this section examine contemporary social dynamics that demonstrate the need for a new biopolitics. They go on to explore paths toward opportunities for incorporating new biopolitical ways of thinking into scientific and bioethical discourses, policy debates, and public understandings of genetic and reproductive technologies. They also question outmoded paradigms of genetic science that are still too frequently encountered—especially in science journalism—and propose new metaphors and analytic approaches that acknowledge rather than erase the social and policy implications of human biotechnologies. These writings insist that these powerful tools, which have the potential both to reshape human bodies and reconfigure societies, are appropriate objects of public deliberation and democratic governance. They offer new ways to understand the social and ethical challenges of human biotechnologies and put forward broad proposals for policy and advocacy responses that can bolster rather than undermine social justice, human rights, and the public good.

"Die, Selfish Gene, Die" is David Dobbs's reconsideration of "the selfish gene," a metaphor popularized by Richard Dawkins in his 1976 book of that name, in which he argued that evolution is driven by genes rather than by organisms or populations. Dobbs asserts that the selfish gene is an out-moded metaphor; that it can't account for numerous findings over the past forty years in the fields of genomics, anthropology, and evolutionary studies; and that it "threatens to impoverish the way both scientists and the rest of us view genetics and evolution." In its place, Dobbs says, we should consider the entirety of an organism's genome "in conversation with itself, with other genomes, and with the outside environment." In other words, he argues, we should replace "the selfish gene" with "the social genome."

In "Toward Race Impact Assessments," Osagie Obasogie explores the role of public policy in relation to new biotechnologies that can encourage distorted ideas about biology and racial difference. He reviews several biotech developments—race-specific medicines, genetic ancestry tests, and DNA forensics—that fall under the purview of government agencies, including in these instances the Food and Drug Administration, the Federal Trade Commission, and the Federal Bureau of Investigation. Obasogie proposes the development and use of a new tool: race impact assessments. He argues that deliberate and systematic evaluation of human biotechnologies' potential consequences would help ensure that any claims made about the relationship between race and genetics are based on sound evidence and do not promote unfounded biological assumptions about race. In addition, race impact assessments could encourage shared responsibility among regulators, researchers, and affected communities and promote recognition of the social construction of race and the social determinants of racial disparities.

Most of those involved in the debate about new human biotechnologies acknowledge that they raise profoundly consequential social challenges. What kinds of responses are appropriate? In "Human Genetic Engineering Demands More Than a Moratorium," Sheila Jasanoff, Benjamin Hurlbut, and Krishanu Saha look at a range of reactions to the now-imminent prospect of gene editing for human reproduction and find that the moratorium approach proposed by experts in the field is entirely inadequate. "A moratorium without provisions for ongoing public deliberation narrows our understanding of risks and bypasses democracy," they write. "We need a more complex architecture for public deliberation, built on the recognition that we, as citizens, have a duty to participate in shaping our biotechnological futures, just as governments have a duty to empower us to participate in that process."

The last contributor to the volume, Marcy Darnovsky, takes up the politics of new human biotechnologies in "'Moral Questions of an Altogether Different Kind': Progressive Politics in the Biotech Age." She reviews the George W. Bush–era battles over human embryonic stem cell and cloning research and examines the shortcomings of the positions that some progressives and liberals staked out. She argues that in evaluating scientific developments that enable "unprecedented biotechnological powers," progressives must break the habit of considering only safety and efficacy, and resist the temptation to bracket moral questions. Instead, she calls for a "biopolitical imagination" that is unafraid to draw on progressive moral legacies in crafting our visions of a healthy, just, and equitable human future.

51. Die, Selfish Gene, Die

David Dobbs

A couple of years ago, at a massive conference of neuroscientists—thirty-five thousand attendees, scores of sessions going at any given time—I wandered into a talk that I thought would be about consciousness but proved (wrong room) to be about grasshoppers and locusts. At the front of the room, a bug-obsessed neuroscientist named Steve Rogers was describing these two creatures—one elegant, modest, and well-mannered, the other a soccer hooligan.

The grasshopper, he noted, sports long legs and wings, walks low and slow, and dines discreetly in solitude. The locust scurries hurriedly and hoggishly on short, crooked legs and joins hungrily with others to form swarms that darken the sky and descend to chew the farmer's fields bare.

Related, yes, just as grasshoppers and crickets are. But even someone as insect-ignorant as I could see that the hopper and the locust were radically different animals—different species, doubtless, possibly different genera. So I was quite amazed when Rogers told us that grasshopper and locust are in fact the same species, even the same animal, and that, as Jekyll is Hyde, one can morph into the other at alarmingly short notice.

Not all grasshopper species, he explained (there are some eleven thousand), possess this morphing power; some always remain grasshoppers. But every locust was, and technically still is, a grasshopper—not a different species or subspecies, but a sort of hopper gone mad. If faced with clues that food might be scarce, such as hunger or crowding, certain grasshopper species can transform within days or even hours from their solitudinous

Original publication by David Dobbs. "Die, Selfish Gene, Die." *Aeon Magazine.* December 3, 2013. Reprinted by permission of David Dobbs.

hopper states to become part of a maniacally social locust scourge. They can also return quickly to their original form.

In the most infamous species, *Schistocercagregaria,* the desert locust of Africa, the Middle East, and Asia, these phase changes (as this morphing process is called) occur when crowding spurs a temporary spike in serotonin levels, which causes changes in gene expression so widespread and powerful they alter not just the hopper's behavior but its appearance and form. Legs and wings shrink. Subtle camo coloring turns conspicuously garish. The brain grows to manage the animal's newly complicated social world, which includes the fact that, if a locust moves too slowly amid its million cousins, the cousins directly behind might eat it.

How does this happen? Does something happen to their genes? Yes, but—and here was the point of Rogers's talk—their genes don't actually change. That is, they don't mutate or in any way alter the genetic sequence or DNA. Nothing gets rewritten. Instead, this bug's DNA—the genetic book with millions of letters that form the instructions for building and operating a grasshopper—gets reread so that the very same book becomes the instructions for operating a locust. Even as one animal becomes the other, as Jekyll becomes Hyde, its genome stays unchanged. Same genome, same individual, but, I think we can all agree, quite a different beast.

Why?

. . .

Transforming the hopper is gene expression—a change in how the hopper's genes are "expressed," or read out. Gene expression is what makes a gene meaningful, and it's vital for distinguishing one species from another. We humans, for instance, share more than half our genomes with flatworms; about 60 percent with fruit flies and chickens; 80 percent with cows; and 99 percent with chimps. Those genetic distinctions aren't enough to create all our differences from those animals—what biologists call our particular phenotype, which is essentially the recognizable thing a genotype builds. This means that we are human, rather than wormlike, flylike, chickenlike, feline, bovine, or excessively simian, less because we carry different genes from those other species than because our cells read differently our remarkably similar genomes as we develop from zygote to adult. The writing varies—but hardly as much as the reading.

This raises a question: if merely reading a genome differently can change organisms so wildly, why bother rewriting the genome to evolve? How vital, really, *are* actual changes in the genetic code? Do we always need

DNA changes to adapt to new environments? Are there other ways to get the job done? Is the importance of the gene as the driver of evolution being overplayed?

You've probably noticed that these questions are not gracing the cover of *Time* or haunting *Oprah, Letterman,* or even TED talks. Yet for more than two decades they have been stirring a heated argument among geneticists and other evolutionary theorists. As evidence of the power of rapid gene expression and other complex genomic dynamics mounts, these questions might (or might not, for pesky reasons we'll get to) begin to change not only mainstream evolutionary theory but our more everyday understanding of evolution.

Twenty years ago, phase changes such as those that turn grasshopper to locust were relatively unknown, and, outside of botany anyway, rarely viewed as changes in gene expression. Now, notes Mary Jane West-Eberhard, a wasp researcher at the Smithsonian Tropical Research Institute in Panama, sharp phenotype changes due to gene expression are "everywhere." They show up in gene-expression studies of plants, microbes, fish, wasps, bees, birds, and even people. The genome is continually surprising biologists with how fast and fluidly it can change gene expression—and thus phenotype.

These discoveries closely follow the recognition, during the 1980s, that gene-expression changes during very early development—such as in embryos or sprouting plant seeds—help to create differences between species. At around the same time, genome sequencing began to reveal the startling overlaps mentioned above between the genomes of starkly different creatures. (To repeat: you are 80 percent cow.)

Gregory Wray, a biologist at Duke University in North Carolina who studies fruit flies, sees this flexibility of genomic interpretation as a short path to adaptive flexibility. When one game plan written in the book can't provide enough flexibility, fast changes in gene expression—a change in the book's reading—can provide another plan that better matches the prevailing environment.

"Different groups of animals succeed for different reasons," says Wray. "Primates, including humans, have succeeded because they're especially flexible. You could even say flexibility is the essence of being a primate."

According to Wray, West-Eberhard, and many others, this recognition of gene expression's power, along with other dynamics and processes unanticipated by mainstream genetic theory through the middle of last century, requires that we rethink and expand the way we view genes and evolution. For a century, the primary account of evolution has emphasized the gene's role as architect: a gene (or gene variant) creates a trait that either proves advantageous or not, and is thus selected for, changing a species for the better, or not. Thus, a genetic blueprint creates traits and drives evolution.

This gene-centric view, as it is known, is the one you learnt in high school. It's the one you hear or read of in almost every popular account of how genes create traits and drive evolution. It comes from Gregor Mendel and the work he did with peas in the 1860s. Since then, and especially over the past fifty years, this notion has assumed the weight, solidity, and rootedness of an immovable object.

But a number of biologists argue that we need to replace this gene-centric view with one that more heavily emphasizes the role of more fluid, environmentally dependent factors such as gene expression and intragenome complexity—that we need to see the gene less as an architect and more as a member of a collaborative remodeling and maintenance crew.

They ask for something like the rejection a century ago of the Victorian-era "Great Man" model of history. This revolt among historians recast leaders not as masters of history, as Tolstoy put it, but as servants. Thus the Russian Revolution exploded not because Marx and Lenin were so clever, but because fed-up peasants created an impatience and an agenda that Marx articulated and Lenin ultimately hijacked. Likewise, D-Day succeeded not because Eisenhower was brilliant but because US and British soldiers repeatedly improvised their way out of disastrously fluid situations. Wray, West-Eberhard, and company want to depose genes likewise. They want to cast genes not as the instigators of change, but as agents that institutionalize change rising from more dispersed and fluid forces.

This matters like hell to people like West-Eberhard and Wray. Need it concern the rest of us?

It should. We are rapidly entering a genomic age. A couple of years ago, for instance, I became one of what is now almost a half million 23andMe customers, paying the genetic-profiling company to identify hundreds of genetic variants that I carry. I now know "genes of interest" that reveal my ancestry and help determine my health. Do I know how to make sense of them? Do they even make sense? Sometimes; sometimes not. They tell me, for instance, that I'm slightly more likely than most to develop Alzheimer's disease, which allows me to manage my health accordingly. But those genes also tell me I should expect to be short and bald, when in fact I'm 6'3" with a good head of hair.

Soon, it will be practical to buy my entire genome. Will it tell me more? Will it make sense? Millions of people will face this puzzle. Along with our doctors, we'll draw on this information to decide everything from what drugs to take to whether to have kids, including kids a few days past conception—a true make-or-break decision.

Yet we enter this genomic age with a view of genetics that, were we to apply it, say, to basketball, would reduce that complicated team sport to a game of one-on-one. A view like that can be worse than no view. It tempts you to think you understand the game when you don't. We need something more complex.

"And it's not as if people can't handle things more complex," says Wray. "Educated people handle ideas more complex than this all the time. We have a more complicated understanding of football than we do genetics and evolution. Nobody thinks just the quarterback wins the game.

"We're stuck in an outmoded way of thinking that should have fallen long ago."

. . .

This outmoded thinking grew from seeds planted 150 years ago by Gregor Mendel, the monk who studied peas. Mendel spent seven years breeding peas in a five-acre monastery garden in the town of Brno, now part of the Czech Republic. He crossed plants bearing wrinkled peas with those bearing smooth peas, producing 29,000 plants altogether. When he was done and he had run the numbers, he had exposed the gene.

Mendel didn't expose the physical gene, of course (that would come a century later), but the conceptual gene. And this conceptual gene, revealed in the tables and calculations of this math-friendly monk, seemed an agent of mathematical neatness. Mendel's thousands of crossings showed that the traits he studied—smooth skin versus wrinkled, for instance, or purple flower versus white—appeared or disappeared in consistent ratios dictated by clear mathematical formulas. Inheritance appeared to work like algebra. Anything so math friendly had to be driven by discrete integers.

It was beautiful work. Yet when Mendel first published his findings in 1866, just seven years after Charles Darwin's *On the Origin of Species,* no one noticed. Starting in 1900, however, biologists rediscovering his work began to see that these units of heredity he'd discovered—dubbed *genes* in 1909—filled a crucial gap in Darwin's theory of evolution. This recognition was the *Holy Shit!* moment that launched genetics' *Holy Shit!* century. It seemed to explain *everything.* And it saved Darwin.

Darwin had legitimized evolution by proposing for it a viable mechanism—natural selection, in which organisms with the most favorable traits survive and multiply at higher rates than do others. But he could not explain what created or altered traits.

Mendel could. Genes created traits, and both would spread through a population if a gene created a trait that survived selection.

That much was clear by 1935. Naturally, some kinks remained, but more math-friendly biologists soon straightened those out. This took most of the middle part of the twentieth century. Biologists now call this decades-long project the modern evolutionary synthesis. And it was all about math.

The first vital calculations were run in the 1930s, when Ronald Fisher, J.B.S. Haldane, and Sewall Wright, two Brits and an American working more or less separately, worked out how Mendel's rather binary genetic model could create not just binary differences such as smooth versus wrinkled peas but the gradual evolutionary change of the sort that Darwin described. Fisher, Haldane, and Wright, working the complicated math of how multiple genes interacted through time in a large population, showed that significant evolutionary change often revealed itself as many small changes yielded a large effect, just as a series of small nested equations within a long algebra equation could.

The second kink was tougher. If organisms prospered by out-competing others, why did humans and some other animals help one another? This might seem a non-mathy problem. Yet in the 1960s, British biologist William Hamilton and American geneticist George Price, who was working in London at the time, solved it too with math, devising formulas quantifying precisely how altruism could be selected for. Some animals act generously, they explained, because doing so can aid others, such as their children, parents, siblings, cousins, grandchildren, or tribal mates, who share or might share some of their genes. The closer the kin, the kinder the behavior. Thus, as Haldane once allegedly quipped, "I would lay down my life for two brothers or eight cousins."

Thus math reconciled Mendel and Darwin and made modern genetics and evolutionary theory a coherent whole. Watson and Crick's 1953 discovery of the structure of DNA simply iced the cake: now we knew the structure that performed the math.

Finally, also in the 1960s, Hamilton and American George Williams upped the ante on the gene's primacy. With fancy math, they argued that we should view any organism, including any human, as merely a sort of courier for genes and their traits. This flipped the usual thinking. It made the gene vital and the organism expendable. Our genes did not exist for us. We existed for them. We served only to carry these chemical codes forward through time, like those messengers in old sword-and-sandal war movies who run nonstop for days to deliver data and then drop dead. A radical idea. Yet it merely extended the logic of kin selection, in which any gene-courier—say, a mom watching her children's canoe overturn—would risk her life to let her kin carry forth her DNA.

This notion of the gene as the unit selected, and the organism as a kludged-up cart for carrying it through time, placed the gene smack at the center of things. It granted the gene something like agency.

At first, not even many academics paid this any heed. This might be partly because people resist seeing themselves as donkey carts. Another reason was that neither Hamilton nor Williams were masterly communicators.

But fifteen years after Hamilton and Williams kited this idea, it was embraced and polished into gleaming form by one of the best communicators science has ever produced: the biologist Richard Dawkins. In his magnificent book *The Selfish Gene* (1976), Dawkins gathered all the threads of the modern synthesis—Mendel, Fisher, Haldane, Wright, Watson, Crick, Hamilton, and Williams—into a single shimmering magic carpet.

These days, Dawkins makes the news so often for things like pointing out that a single college in Cambridge has won more Nobel Prizes than the entire Muslim world, that some might wonder how he ever became so celebrated. *The Selfish Gene* is how. To read *The Selfish Gene* is to be amazed, entertained, transported. For instance, when Dawkins describes how life might have begun—how a randomly generated strand of chemicals pulled from the ether could happen to become a "replicator," a little machine that starts to build other strands like itself, and then generates organisms to carry it—he creates one of the most thrilling stretches of explanatory writing ever penned. It's breathtaking.

Dawkins assembles genetics' dry materials and abstract math into a rich but orderly landscape through which he guides you with grace, charm, urbanity, and humor. He replicates in prose the process he describes. He gives agency to chemical chains, logic to confounding behavior. He takes an impossibly complex idea and makes it almost impossible to misunderstand. He reveals the gene as not just the center of the cell but the center of all life, agency, and behavior. By the time you've finished his book, or well before that, Dawkins has made of the tiny gene—this replicator, this strip of chemicals little more than an abstraction—a huge, relentlessly turning gearwheel of steel, its teeth driving smaller cogs to make all of life happen.

It's a gorgeous story. Along with its beauty and other advantageous traits, it is amenable to math and, at its core, wonderfully simple. It has inspired countless biologists and geneticists to plumb the gene's wonders and do brilliant work. Unfortunately, say Wray, West-Eberhard, and many others, the selfish-gene story is so focused on the gene's singular role in natural selection that in an age when it's ever-more clear that evolution works in ways far more clever and complex than we realize, the selfish-gene model increasingly impoverishes both scientific and popular views of genetics and evolution. As both conceptual framework and metaphor, the selfish-gene has helped us see the gene as it revealed itself over the twentieth century. But as a new age and new tools reveal a more complicated genome, the selfish gene is blinding us.

. . .

For over two decades, Wray, West-Eberhard, and other evolutionary theorists—such as Massimo Pigliucci, professor of philosophy at the City University of New York; Eva Jablonka, a geneticist and historian of science at Tel Aviv University, London; Stuart Kauffman, professor of biochemistry and mathematics at the University of Vermont; Stuart A Newman, professor of cell biology and anatomy at the New York Medical College; and the late Stephen Jay Gould, to name a few—have been calling for an "extended modern synthesis" to replace the gene-centric view of evolution with something richer. They do so even though they agree with most of what Dawkins says a gene does. They agree, in essence, that the gene is a big cog, but would argue that the biggest cog doesn't necessarily always drive the other cogs. In many cases, the other cogs drive the gene. The gene, in short, just happens to be the biggest, most obvious part of the trait-making inheritance and evolutionary machine. But not the driver.

Another way to put it: Mendel stumbled over the wrong chunk of gold.

Mendel ran experiments that happened to reveal strong single-gene dynamics whose effects—flower color, skin texture—can seem far more significant than they really are. Many plant experiments since then, for instance, have shown that environmental factors such as temperature changes can spur gene-expression changes that alter a plant far more than Mendel's gene variants do. As with grasshoppers, a new environment can quickly turn a plant into something almost unrecognizable from its original form. If Mendel had owned an RNA-sequencing machine and was in the habit of tracking gene expression changes, he might have spotted these. But sequencers didn't exist, so he crossed plants instead, and saw just one particularly obvious way that an organism can change.

The gene-centric view is thus "an artifact of history," says Michael Eisen, an evolutionary biologist who researches fruit flies at the University of California, Berkeley. "It rose simply because it was easier to identify individual genes as something that shaped evolution. But that's about opportunity and convenience rather than accuracy. People confuse the fact that we can more easily study it with the idea that it's more important."

The gene's power to create traits, says Eisen, is just one of many evolutionary mechanisms. "Evolution is not *even* that simple. Anyone who's worked on systems sees that natural selection takes advantage of the most bizarre aspects of biology. When something has so many parts, evolution will act on all of them. It's not that genes don't sometimes drive evolutionary change. It's that this mutational model—a gene changes, therefore the organism changes—is just one way to get the job done. Other ways may actually do more."

Like what other ways? What significant and plausible evolutionary dynamics stand in tension with a single-gene-centered model? What gets obscured by the insistence that a "selfish gene," a coherent, solitary replicator, is the irreducible and ever-present driver of evolution?

A shortlist of such dynamics would include some of the evolutionary dynamics being proposed by anthropologists, such as cultural transmission of knowledge and behavior, that allow social species ranging from bees to humans to adapt to changing environments without genetic alterations; and culture-gene evolution, a related idea, in which culture is not the "handmaiden" of genes, but another source of transmissible adaptive information whose elements coevolve with genes, each affecting the other.

Also in tension with the selfish-gene model are epigenetic changes suggested by recent research, such as methylation and other alterations to chemical wrappings around DNA, that can modulate DNA's expression without changing its sequence. Such epigenetic changes may provide a way to pass heritable traits down through at least a few generations without changing any actual genes. To be sure, this research is still unproven as a significant evolutionary force. But while it is clearly important enough to pursue, many defenders of the selfish-gene model dismiss it out of hand.

Finally, the selfish-gene model is in tension with various "interesting evolutionary phenomena," as Gregory Wray puts it in *Evolution: The Extended Synthesis*, "that are apparent only at the scale of hundreds or thousands of genes"—a scale only made viewable during the past decade or so, as we've learnt to rapidly sequence entire genomes.

Of these genomic dynamics, perhaps the most challenging to the selfish-gene story are epistatic or gene-gene interactions. Epistasis refers to the fact that the presence of some genes (or their variants) can have profound and unpredictable influences on the activity and effects generated by other genes. To put it another way, a gene's effect can vary wildly depending on which combination of other genes it finds itself with. (Think Jerry Garcia playing with different musical partners.)

Epistasis is hardly a new concept. In fact, geneticists have been arguing about its importance ever since R.A. Fisher and Sewall Wright bickered about it in the 1920s. Dawkins acknowledges a role for gene-gene interactions in *The Selfish Gene,* noting that "the effect of any one gene depends on interaction with many others." But research since then show that these interactions take place in nonlinear, nonadditive ways of a complexity impossible to understand at the time Dawkins wrote his book. Casey Greene and Jason Moore of Dartmouth, for instance, recently found that in some cases epistatic interactions seem to warp conventional gene-trait relationships so profoundly that they can often negate the gene as a trait's reliable carrier.

This is not merely a matter of one gene muffling or amplifying another, though both these things happen. And it's not a matter of additive effects, such as four "tall" genes making you taller than would two. Rather, these multigene epistatic interactions can create endless possible combinations of mutual influence in which any given gene's contribution seems to rise less from its inherent trait-making power than from what company that gene finds itself keeping. To draw on P.Z. Myers's apt analogy, epistasis means that single genes often carry little more inherent significance than individual playing cards do in poker. In a poker hand, the significance and effect of a two of hearts—its "trait"—depend so heavily on the other cards you're holding that it's almost meaningless to say the card has any replicable power on its own. It's replicable in that it's a two of hearts every time it's dealt. But it can deliver the same effect in subsequent generations only if it's dealt not just into the exact same handful of cards, but into a round in which all the other players at the table also hold the same cards as before— and happen to bet, hold, and fold in exactly the same way. Not something to count on.

And a two of hearts is a far more coherent thing than is a gene. One of the peskiest problems of leaning too heavily on a gene-centric model these days is that the definition of the word *gene* gets ever-more various and slippery.

Even as a technical term, the word carries at least a half dozen meanings, and more are added as science finds new tools for exploring the genome. This alone makes it either a poor candidate for a popular meme—or, if you value flexibility over exactitude, perhaps a perfect one, since its meaning can be defended or reshaped or expanded to suit the occasion. If you expand the meaning to be "the thing essential to all true heredity and selection," you can then give the gene primary credit for any discovered or proposed evolutionary force in which the gene seems to be involved—and reject outright any proposed evolutionary force that doesn't seem to involve genes.

But the gene's definition is not just semantically vague. As geneticists explore the genome's previously uncharted stretches, they're finding that a lot of the work conventionally attributed to "genes" (in the sense of consistent, reasonably well-defined clusters of DNA) appears to be done instead by networks of genes and strange DNA elements that doubly defy the selfish-gene model.

These regulatory networks challenge the selfish-gene model first because they include DNA elements not conventionally defined as genes. More important, some researchers believe these networks challenge the selfish-gene model because they often seem to behave not like selfish entities balancing their separate agendas, in selfish-gene style, but like managerial teams regulating the behavior of individual genes for the interest of the organism. The chromosome's three-dimensional nature brings those regulatory chunks into contact with individual genes in highly unpredictable ways. With each gene "surrounded by an ocean" of such regulatory elements, as molecular biophysicist Joe Dekker told *WIRED*, each gene "can touch and interact with a whole collection of them." Yale geneticist Mark Gerstein found that the genes in these networks sometimes seem to get selected for even if they don't have important effects on their own. In other cases they seem to have effects but be exempt from selection pressure.

These regulatory elements now appear to grossly outnumber the actual genes, possibly by as much as fifty to one. As Yale geneticist Mark Gerstein politely notes, the complexity of these regulatory networks, along with their ad hoc management-team nature, raise the question of what's being selected: individual genes, as the selfish-gene model proposes, or the management team, by some process still hidden amid all this complexity. Others, such as Cold Spring Harbor geneticist Thomas Gingeras, question outright whether the transcript (the marching orders a gene issues to begin gene expression) should replace the gene as the genome's functional unit. These issues are not merely academic; resolving them could help solve mysteries about cancer and other diseases.

Such dynamics have emerged only in the last decade or so, as researchers have been able to examine the genome more closely. Yet even though we so far "have only a dim idea of how all this works," as Gregory Wray wrote in 2010, it "is clear . . . that these kinds of assumption-violating exceptions are not rare."

Wray's language here is crucial: he's not saying these findings refute the details of the gene-centric model. He's saying they violate the model's assumptions.

And this is the crux of this entire dispute: The point is not whether the findings of a genomic age or of anthropology refute the selfish-gene model, invalidate its theoretical details, or debunk the modern synthesis. Mostly they don't. The selfish-gene model is roomy enough to host many of these findings. It has shown an uncanny ability to do so. But as time passes it does so ever more uncomfortably, for both host and guests. Some findings or ideas must be almost forced in. Others get prematurely locked out.

The selfish-gene model and metaphor can probably be stretched even more to account for some of these things. But in an age when assumption-violating ideas from genomic studies, anthropology, and other fields are flourishing, does the selfish gene story remain the best way to account for them? Does it make sense to attach these proliferating findings and ideas on to the selfish-gene story as appendices? Or is it time to find another story? It may be that the gene is always a player. But it is rarely the only player. And—may I speak metaphorically?—it may (or may not) be that the gene always behaves as if it were selfish. But that doesn't mean it always gets its way.

· · ·

One of the assumption-violating exceptions Wray refers to is gene expression's breadth of power. In the social wasps that Mary Jane West-Eberhard has been studying in Panama since 1979, many of the most important distinctions among a colony's individuals rise not from differences in their genomes, which vary little, but from the plasticity born of gene expression. This starts with the queen, who is genetically identical to her thousands of sisters yet whose gene expression makes her not only larger, but singles her out as the colony's reproductive unit. Likewise with most honeybees. In social honeybees, the differences between workers, guards, and scouts all arise from gene expression, not gene sequence. Individual bees morph from one form to another—worker to guard to scout—by gene expression alone, depending on the needs of the hive.

As described above, the questionable coherence of genes seems to apply especially to gene regulation—as do epistatic networks that further undermine the gene's primacy. So while it's clear that DNA plays a key role in regulating gene expression, it is not clear that all these "regulatory genes" are the selfish genes of the Dawkins model.

This is but one reason why West-Eberhard, among others, has been long trying to cure the "cyclic amnesia" that she says has ignored 150 years of evidence that the gene's centrality is overplayed. West-Eberhard is a particularly articulate advocate. Yet she's frustrated at how little she's been able to change things.

As a David to Dawkins's Goliath, West-Eberhard faces distinct challenges. For starters, she's a she while Dawkins is a he, which should not matter but does. And while Dawkins holds forth from Oxford, one of the most prestigious universities on earth, and deploys from London an entire foundation in his name, West-Eberhard studies and writes from a remote outpost in Central America. Dawkins commands locust-sized audiences any time he speaks and probably turns down enough speaking engagements to fill five calendars; West-Eberhard speaks mainly to insect-crazed colleagues at small conferences. Dawkins wrote a delicious three-hundred-page book that has sold tens of millions of copies; West-Eberhard has written a bunch of fine obscure papers and an eight-hundred-page tome, *Developmental Plasticity and Evolution* (2003), which, though not without its sweet parts, is generally consumed as a meal of obligation.

She does have her pithy moments. There are times, she says, when "the gene does not lead. It follows."

Massimo Piglucci and Gerd Muller use the same language in *Evolution: The Extended Synthesis.* By "the gene follows," they mean that in complex organisms particularly, dynamics other than gene alterations, ranging from gene expression to complex gene regulation to developmental pathways formed by culture, can create heritable adaptations that either remain on their own or later become "fixed" or locked in by genes.

One way in which the gene follows is through genetic assimilation—a clunky term for a graceful process. This can look Lamarckian, but it is not. It's the development of a heritable change through flexible gene-expression responses that later get "fixed," or locked in, by a change in genotype. . . . This isn't the gene-centric world in which genotype creates phenotype. It's a phenotype accommodating a new genotype by making it valuable.

Genetic assimilation was recognized as a possibility in the 1940s, but as Massimo Pigliucci and Courtney Murren put it, it was "attacked as of minor importance during the 'hardening' of the neo-Darwinian synthesis and . . . relegated to a secondary role for decades." Interest has surged lately as gene expression becomes more apparent, and biologists are starting to spot the process in the field. No one proposes that genetic assimilation happens all the time or even commonly, or that it widely replaces conventional gene-driven evolution. But its existence suggests how gene expression's fluidity can combine with conventional genetic dynamics to broaden evolution's reach.

Gene Robinson, an entomologist who studies honeybees at the University of Illinois, says genetic assimilation could well have helped to create African honeybees, the "killer bee" subspecies that is genetically distinct from the sweeter European honeybees that most beekeepers keep. Honeybee hives in certain parts of Africa, he says, were and are raided by predators more often than hives elsewhere, so their inhabitants had to react more sharply to attacks. This encouraged gene-expression changes that made the African bees respond more aggressively to threat. When new genes showed up that reinforced this aggression, those genes would have been selected for and spread through the population. This, Robinson says, is quite likely how African bees became genetically distinct from their European honeybee cousins. And they'd have been led there not by a gene, but by gene expression.

. . .

After several weeks of reading and talking to this phenotypic plasticity crowd, I phoned Richard Dawkins to see what he thought of all this. Did genes follow rather than lead? I asked him specifically about whether processes such as gene assimilation might lead instead. He said that genetic assimilation doesn't really change anything, because since the gene ends up locking in the change and carrying it forward, it all comes back to the gene anyway.

"This doesn't modify the gene-centric model at all," he said. "The gene-centric model is all about the gene being the unit in the hierarchy of life that is selected. That remains the gene."

"He's backfilling," said West-Eberhard. "He and others have long been arguing for the primacy of an individual gene that creates a trait that either survives or doesn't."

Yet West-Eberhard understands why many biologists stick to the gene-centric model. "It makes it easier to explain evolution," she says. "I've seen people who work in gene expression who understand all of this. But when they get asked about evolution, they go straight to Mendel. Because people understand it more easily." It's easy to see why: even though life is a zillion bits of biology repeatedly rearranging themselves in a webwork of constantly modulated feedback loops, the selfish-gene model offers a step-by-step account as neat as a three-step flow chart. Gene, trait, phenotype, done.

In other words, the gene-centric model survives because simplicity is a hugely advantageous trait for an idea to possess. People will select a simple idea over a complex idea almost every time. This holds especially in a hostile environment, like, say, a skeptical crowd. For example, Sean B. Carroll, professor of molecular biology and genetics at the University of Wisconsin, spends much of his time studying gene expression, but usually uses gene-centric explanations, because when talking to the public, he finds a simple story is a damned good thing to have.

Which drives West-Eberhard nuts.

"Dawkins understands very well that gene expression is powerful," she says. "He sees things are more complex than a selfish gene. He could turn on its head the whole language."

Yet Dawkins, and with him much of pop science, sticks to the selfish gene. The gene explains all. So far it has worked. The extended synthesis crowd has published scores of papers, quite a few books, and held meetings galore. They have changed the way many biologists think about evolution. But they have scarcely touched the public's understanding. And they have not found a way to displace a meme so powerful as the selfish gene.

This meme, methinks, forms the true bone of contention and the true obstacle to progress. It's one of the odd beauties of this whole mess that Dawkins himself coined the term *meme,* and did so in *The Selfish Gene.* He defined it as a big idea that competes for dominance in a tough environment—an idea that, like a catchy tune or a good joke, "propagates itself by leaping from brain to brain."

The selfish-gene meme has done just that. It has made of evolutionary theory a vehicle for its replication. The selfish gene has become a selfish meme. If you're West-Eberhard or of like mind, what are you to replace it with? The slave-ish gene? Not likely to leap from brain to brain. The cooperative gene? Dawkins himself considered this but rejected it and I agree that it lacks sufficient bling. And as West-Eberhard notes, any phrase with "gene" in it still encourages a focus on single genes. And "evolution is not about single genes," she says. "It's about genes working together."

Perhaps better, then, to speak not of genes but the genome—all your genes together. And not the genome as a unitary actor, but the genome in conversation with itself, with other genomes, and with the outside environment. If grasshoppers becoming locusts, sweet bees becoming killers, and genetic assimilation are to be believed it's those conversations that define the organism and drive the evolution of new traits and species. It's not a selfish gene or a solitary genome. It's a social genome.

What would Mendel think of that? Let's play this out.

Mendel actually studied bees as a boy, and he studied them again for a couple years after he finished his pea-plant studies. In crossbreeding two species at the monastery, he accidentally created a strain of bees so vicious that he couldn't work with them. If he'd had an RNA sequencer, he, like Gene Robinson, could have studied how much of the bees' aggression rose from changes in the genetic code or how much rose from gene expression in response to the environment. If he had, the father of genetics might have seen right then that traits change and species evolve not just when genes change, when a creature and its genome and hive mates respond to an environment. He might have discovered not just genes, but genetic assimilation. Not the selfish gene, but the social genome.

Alas, no such equipment existed, and Mendel worked in a monastery in the middle of town. His vicious bees promised not a research opportunity but trouble. So he killed them. He would found genetics not through a complex story told by morphing bees, but through a simple tale told by one pea wrinkled and one pea smooth.

52. Toward Race Impact Assessments

Osagie K. Obasogie

THE PERSISTENCE OF TYPOLOGICAL THINKING ABOUT RACE IN SCIENCE

Race-based medicines, genetic ancestry tests, and DNA forensics can each potentially benefit racial minorities and society in general by providing life-saving medicines targeted for vulnerable and underserved populations, increasing individuals' knowledge about their ancestry, and offering tools to law enforcement to help solve crimes. However, these technologies are also united by a tendency to promote typological perspectives on race that are reminiscent of the nineteenth and early twentieth centuries when "pure" or "real" races were thought to independently exist and each race was thought to share a set of unique biological traits that could be identified and measured.[1] This contrasts with mainstream scientific perspectives that view the distribution of human traits on a continuum that does not have the discrete breaks that typologists find indicative of biological race.[2]

Despite this tension, biological race continues to be salient in lay and scientific discourses. Increasing public exposure to race and genetics research is shaping lay opinions about the relevance of genes to racial disparities in social and health outcomes. At the same time, scientists continue to espouse typological approaches that give coherence to biological

Original publication by Osagie Obasogie. "Toward Race Impact Assessments." Excerpt from "The Return of Biological Race? Regulating Innovations in Race and Genetics Research Through Race Impact Assessments." *Southern California Interdisciplinary Law Journal* 22, no. 1 (Fall 2012). Reprinted by permission of Osagie Obasogie. Due to space limitations, portions of this chapter have been deleted or edited from the original.

understandings of racial difference. Sociologist Ann Morning, who has conducted research on scientists' perspectives on race, observes that:

> Social and biological scientists hold a wide range of beliefs about the nature of racial difference; contrary to some scholars' expectations, they are far from any consensus, either within or between disciplines. . . . The essentialist proposition that races are biologically grounded entities remains a compelling view for many contemporary scientists.[3]

This all leads to an important question: Given law's past complicity in furthering racial subordination through promoting biological race, what normative role should government take in regulating new biotechnologies that advance biological understandings of racial difference? This is a difficult question because the social and scientific contexts have changed between past articulations of biological race and today's innovations, but the potential risks to racial minorities remain quite similar. While overly strict regulations might unduly prevent access to life-saving or life-enhancing technologies, overly permissive approaches may lead to new forms of racial subordination and promote an impoverished understanding of race among the public.

REGULATORY GULF: EXPANDING USES, DIMINISHING OVERSIGHT

Biotechnologies that implicate race are expanding in their use and development. For example, biologists Sarah Tate and David Goldstein observed in a 2004 *Nature Genetics* article that, while controversial, "At least 29 medicines (or combination of medicines) have been claimed, in peer-reviewed scientific or medical journals, to have differences in either safety or, more commonly, efficacy among racial or ethnic groups."[4] This suggests that more race-based medicines such as BiDil are in development and may very well be on their way. Additionally, genetic ancestry tests are also becoming increasingly popular. Bolnick and several colleagues note that "at least two dozen companies now market 'genetic ancestry tests,' . . . [and that m]ore than 460,000 people have purchased these tests over the past 6 years, and public interest is still skyrocketing."[5] Most recently, scholar Henry Louis Gates Jr. and others joined 23andMe as advisors to the direct-to-consumer genetics company's "Roots into the Future" program, which hopes to attract ten thousand African Americans to its ancestry-testing services.

New applications of DNA forensics are probably expanding the fastest. For example, Murphy notes that while "it took Virginia nearly eight years, from 1993 to 2001, to reach its first 1,000 'cold hits' [identifying suspects by running unknown samples left at crime scenes across DNA databases], the state reached its second 1,000 in a matter of eighteen months. Since 2001, the laboratory has averaged at least one 'cold hit' a day, and as of July 2002, that figure had doubled to two and one half hits a day."[6]

Applications such as molecular photofitting [using DNA left at a crime scene to create a visual depiction of a suspect's face] are being developed to achieve higher levels of sophistication, while practices like familial testing [creating a profile from crime scene evidence and running it through a DNA database to find a close match that might point to a known relative and therefore identify the main suspect] are becoming increasingly commonplace in criminal investigations.[7] For example, the California Attorney General's office has issued guidelines for familial testing for its state database,[8] which will likely accelerate the use of this technology by state and local law enforcement.[9]

At the same time that these technologies are expanding, regulatory oversight of their scientific rigor and public impact remains inadequate. The FDA [Food and Drug Administration] does not subject new drug applications seeking race-specific labeling to any other standard outside the agency's traditional emphasis on safety and efficacy.[10] Direct-to-consumer genetic tests have been criticized by some federal agencies including the FTC [Federal Trade Commission],[11] but the specific issues related to race and genetic ancestry have not been a significant part of the conversation. Genetic ancestry tests fall outside of the FDA's regulatory authority because "a genetic test is only subject to FDA oversight if it is a medical device . . . whereas a test to determine ancestry is not a device."[12] Moreover, the FBI, in coordination with state and local law enforcement agencies, continues to expand DNA forensics into questionable areas such as familial testing and molecular photofitting.

These applications can significantly impact minority communities, as well as alter public understandings of race.[13] Therefore, a more robust regulatory response is needed to ensure that these innovations do not legitimize biological understandings of race in a manner that supersedes their potential benefits. . . .

THE NEED FOR ADMINISTRATIVE AGENCY IMPACT ASSESSMENTS ON RACE AND HUMAN BIOTECHNOLOGY

Each application discussed in this article falls under the administrative authority of an existing government agency that can set standards for how the public engages these innovations: the FDA for race-based medicines, the

FTC for genetic ancestry tests, and the FBI for DNA forensics. As discussed earlier, existing regulatory approaches to these applications by these agencies are thin; race simply is not taken seriously as a regulatory matter despite the potential risks that these applications portend for reinventing biological race in a manner that may disadvantage minorities.

To avoid these risks and to give minorities access to potential benefits that may stem from these technologies, I propose race impact assessments as a new tool for administrative agencies that are responsible for overseeing any new biotechnology that implicitly or explicitly makes a claim about the biological significance of social categories of race, or that may disproportionately affect minority communities. Generally, impact assessments are evaluative mechanisms used by government agencies to analyze the risks and benefits of new proposals so as to promote individual and social well-being. Most notably, environmental impact assessments have played a significant role in making sure that government agencies consider the potential consequences that a new project or initiative might have on the environment before moving forward. The National Environmental Policy Act (NEPA), signed into law in 1970,[14] requires federal agencies to determine whether certain proposed actions—road construction, building construction, etc.—might have an adverse effect on the human environment.[15] NEPA's most significant legal requirement is the Environmental Impact Statement (EIS).[16] Generally, agencies will first conduct an initial environmental assessment.[17] This initial assessment determines if a proposed action might have an adverse effect on the human environment that requires an EIS. The more rigorous EIS process seeks to flesh out the potential harms of a proposed action and determine if viable and less disruptive alternatives exist.[18] According to 40 C.F.R § 1502.1, the purpose of an EIS is to "serve as an action-forcing device to insure that the policies and goals defined in the Act are infused into the ongoing programs and actions of the Federal Government."[19] Brian Cole and his co-authors note that "the authors of NEPA recognized . . . that [the] problems in one sector are shaped to a large extent by actions in other sectors[,] . . . [whereby] the assessment and consideration of environmental impacts has become a routine part of decision making in federal, state, and local agencies."[20] Thus, the major achievement of environmental impact assessments has been to change the culture of administrative agencies by raising awareness and improving sensitivity to the way that federal actions can damage the environment and to the crucial role that regulatory agencies can play in mitigating these harms.

Much of this cultural change has occurred through the interdisciplinary and cooperative nature of environmental impact assessments as dictated by federal law.[21] By simultaneously engaging in prospective assessment of potential impacts and making these findings publicly available for comment and feedback, environmental impact statements have been able to institutionalize environmental concerns and an ethos of public engagement into regulatory agencies' organizational behavior. In doing this, NEPA and its environmental impact statements have led to better "integration of environmental goals into agency decision making, improved planning, and transparency and public involvement for improved agency decision making."[22] These successes have led to impact assessment proposals in other areas of federal oversight. For example, health impact assessments (HIAs) have been proposed to identify activities and policies "likely to have major impacts on the health of a population in order to reduce the harmful effects on health and to increase beneficial effects."[23] HIAs evaluate the potential health impacts of a proposal based on "a broad model of health, which proposes that economic, political, social, psychological, and environmental factors determine population health," and take into consideration the "opinions and expectations of those who may be affected by a proposed policy."[24] Scholars have proposed other types of impact assessments that draw on these themes. For example, social impact assessments have been developed as a way to "analy[ze] . . . and manag[e] the intended and unintended consequences on the human environment of interventions . . . and social change processes so as to create a more sustainable biophysical and human environment."[25] Similarly, human rights impact assessments have been suggested to "help evaluate the effects of public health policies on human rights and dignity."[26]

Impact assessments of this nature share at least three relevant characteristics that are informative for developing race impact assessments. First, impact assessments are evidence-based; data collection is central to the regulatory decision-making process. No one type of data is privileged; impact assessments can be quantitative or qualitative. These assessments place a premium on engaging with, and thinking through, various policy proposals' real-world implications. As a second related trait, impact assessments are multidisciplinary. Just as no one type of data is privileged, neither is any one disciplinary approach. To be sure, the strength of impact assessments stems from their use of multiple methods and multiple disciplinary perspectives so as to provide a holistic analysis of the many ways in which health, society, human rights, or any other issue may be affected by a particular proposal. Third, impact assessments are characterized by involving multiple stakeholders. While many of the issues analyzed by impact assessments are highly technical and deeply embedded in cutting-edge science, impact assessments are used to bring together a wide range of people—experts, nonexperts, community members, and others—in a deliberative and collaborative effort to ensure that decision makers are informed of all perspectives.

These fundamental characteristics of impact assessments produce at least three significant benefits. First, impact assessments help root out facially innocuous practices that may have harmful effects. By being sensitive to topics such as the environment, health, or human rights and how federal decision making can affect them, policy makers can anticipate and mitigate unintended harms—especially those affecting vulnerable populations. A second benefit is that impact assessments increase cooperation and deliberation between government agencies, experts, and the public. The enhanced contact and communication between these various stakeholders encourages a more deliberative democracy by creating a process involving multiple levels of engagement and accountability. Third, the collaborative effort facilitated by impact assessments encourages multiple government agencies to engage with one another about their shared responsibilities. This decreases the likelihood of important issues falling in between regulatory gaps where agencies can end up pointing the finger at each other.

These traits and benefits suggest that the implementation of race impact assessments would significantly assist administrative agencies in predicting the risks and benefits of biotechnologies that implicate race so as to mitigate the former and promote the latter. Not only do race impact assessments provide an opportunity for a broad appraisal of the scientific claims made by race-based medicines, genetic ancestry tests, and DNA forensics, but they also provide a forum where multiple stakeholders—such as government officials, scientists, constituency-based groups, and others—can exchange ideas. These stakeholders will be able to provide guidance on how the public can gain access to biotechnologies' potential benefits without unduly subjecting racial minorities to the risks associated with government reasserting questionable linkages between race and biology.

Impact assessments are not entirely unproblematic. For example, the assessment process can be quite lengthy and take several years. This is particularly troublesome with regard to potentially lifesaving medicines or law enforcement practices that can solve open cases and prevent future crimes. Second, impact assessments are resource-intensive and can be costly. Who is going to pay for these assessments and who has time outside of their regular professional and daily obligations to participate in another round of bureaucratic fact-finding? In addition to these hurdles, there is also a concern that, without any substantive or normative claims supporting them, impact assessments can become a mere procedural tool that may not be able to create the change they seek. The impact assessment process may be vulnerable to co-optation by contrary interests that may work against the very concerns giving rise to the assessment process itself and further legitimize questionable and unquestioned practices.

These are surely important concerns. But they are not insurmountable, given the remarkable stakes at hand. The unchecked proliferation of biological race has been at the center of some of the most brutal acts in human history. While those who promote new biotechnologies that implicate race often have laudable objectives, it is important to remain aware of the possible dangers. Given the government's historical complicity in promoting biological race in a manner that harmed the most vulnerable members of society,[27] it has a moral and ethical responsibility to support race impact assessments to atone for past wrongs and to promote a future where minorities can partake in the benefits of scientific innovation without remaining perpetually vulnerable to its risks. Moreover, while the risk of co-optation is real, improving and diversifying deliberations while making the process more transparent—practices embedded in the impact assessment process— can be effective checks.

MODELING RACE IMPACT ASSESSMENTS

. . . It would be premature to propose a full race impact assessment model at this point; successful impact assessments need to develop out of robust empirical examinations of each agency's organizational design and culture in relation to current decision-making processes on race-related innovations—information that is not yet fully available. Moreover, mature impact assessments require the collaboration of experts across multiple fields and affected stakeholders to create model tools that balance new technologies' potential benefits with the potential risks of reifying social categories of race as biologically significant lines of human difference. However, it may be productive to sketch the next steps that need to be taken to move this conversation forward while also broadly mapping the ways in which race impact assessments might be integrated into federal agencies as part of their review of new technologies that implicate race.

As an initial matter, it is important to note that, before any agency moves forward with an assessment of this nature, it would need some type of statutory authority—like NEPA—from which to proceed. NEPA provides an excellent model that Congress can mimic to charge federal agencies to engage in holistic assessments of projects, innovations, or proposals that fall under their administrative authority that may potentially harm race relations by promoting biological understandings of racial difference and disparities. Like the environment, race can be seen as a shared and connected ecosystem that requires federal protection for the benefit of human health and social relations. Moreover, just as NEPA established a Council on Environmental Quality (CEQ) within the executive branch to assist the president in overseeing NEPA's implementation across federal agencies,[28] so too could a similar council within the executive branch play a key role regarding racial issues. This council could monitor the diverse ways in which race may be implicated in administrative agency decision making—from new drug applications to innovative forensic technologies—and encourage consistency across federal administrative agencies. . . .

This article reserves a more detailed articulation of the substance and integration of race impact assessments into administrative agencies until future research can lay the empirical groundwork [for gaining a particular understanding of how race has been treated within each agency and available mechanisms for oversight]. But for now, it is useful to briefly sketch the work that race impact assessments might do. Take, for example, a hypothetical new drug not unlike BiDil, where the manufacturer seeks a race-specific indication for Latinos with renal disease based on impressive clinical trial results with participants that self-identify as Latino. An FDA-led race impact assessment might begin by identifying relevant stakeholders to participate in not only analyzing the claims' scientific merit and clinical trial results, but also the impact that this particular race-specific indication might have on racial minorities. This would be balanced with the potential benefit produced by the race-specific indication, such as the ability to identify and treat more Latinos suffering from this condition or to increase compliance within this group. Multiple methods would be used to evaluate the evidence—both the statistical assessment of clinical trial data and the qualitative assessment of constituent perspectives. A final report would be presented to FDA officials to aid their determination of whether a race-specific indication is warranted for the drug.

The FTC might engage in a similar process to assess products claiming to use genetic technologies to determine individuals' racial backgrounds. A diverse committee of experts—from population geneticists to legal scholars and philosophers—in addition to laypersons would examine the claims in relation to the strength and limitations of the company's methods and data. The committee would also collect and assess qualitative data from affected stakeholders to analyze the ways this technology might affect certain communities and how the public perceives the biological relevance of race. The committee would then provide a report to FTC officials to inform their decision about how to oversee the sale and marketing of these products.

The FBI could also use race impact assessments to examine the implications of emerging forensic techniques that may disproportionately affect minority communities. Not only would an external, quantitatively driven scientific evaluation of techniques help assess whether the race-based or race-impacting methods are valid, but a qualitative assessment of stakeholder sentiments might also help the FBI develop procedures that both assist them in law enforcement and respect community concerns.

CONCLUSION

When the first draft of the human genome was completed in 2000 and showed that all humans are 99.9 percent similar at the molecular level,[29] scientists, politicians, and the media rejoiced in declaring that there are no biological differences between racial groups, not unlike the way biological race was publicly discredited after World War II.[30] President Bill Clinton summed up the sentiment in a statement made from the East Room of the White House when he pronounced that "modern science has confirmed what we first learned from ancient fates. The most important fact of life on this Earth is our common humanity."[31]

However, just as biological race remained a salient, if not prominent, variable in scientific research after the 1950 UNESCO statement publicly declared its death,[32] so too has it remained a powerful lens through which we understand human difference in the genomic era. Rather than focusing on our shared humanity, researchers have focused intensely on the less than 1 percent of genetic variation thought to explain racial difference and disparities.

There may very well be important innovations emerging from this renewed focus on biological race. But, given the horrific track record that we have with using science to measure and define racial difference, the government needs to take strong steps to ensure that these technologies are used responsibly. Administrative agency race impact assessments are an important first step to providing a democratic, deliberative, and collaborative space to collect and analyze the data necessary to inform decision makers on how to sensibly regulate these new innovations.

NOTES

1. Ernst Mayr, "Typological versus Population Thinking," in *Conceptual Issues in Evolutionary Biology*, ed. Elliott Sober, 3rd ed. (Cambridge, MA: MIT Press, 2006), 325, 327. . . .

2. See id., 326–28.

3. Ann Morning, *The Nature of Race: How Scientists Think and Teach about Human Difference* (Berkeley: University of California Press, 2011), 221.

4. Sarah K. Tate and David B. Goldstein, "Will Tomorrow's Medicines Work for Everyone?," supplement, *Nature Genetics* 36 (2004), S34. . . .

5. Deborah A. Bolnick et al., "The Science and Business of Genetic Ancestry Testing," *Science* 318 (2007), 399.

6. Erin Murphy, "The New Forensics: Criminal Justice, False Certainty, and the Second Generation of Scientific Evidence," *California Law Review* 95 (2007), 740.

7. See Osagie K. Obasogie, "The Return of Biological Race? Regulating Innovations in Race and Genetics Research through Race Impact Assessments," *Southern California Interdisciplinary Law Journal* 22, no. 1 (fall 2012), part III.C.2.b.

8. See information bulletin from Lance Gima, chief, Bureau of Forensic Services, and Edmund G. Brown Jr., California attorney general, to all California Law Enforcement Agencies and District Attorney's Offices, *DNA Partial Match (Crime Scene DNA Profile to Offender) Policy* (2008), http://ag.ca.gov/cms_attachments /press/pdfs/n1548_08-bfs-01.pdf.

9. See Maura Dolan, "State to Double Crime Searches Using Family DNA," *Los Angeles Times*, May 9, 2011, http://articles.latimes.com/2011/may/09/local/la-me-familial-dna-20110509.

10. See David E. Winickoff and Osagie K. Obasogie, "Race Specific Drugs: Regulatory Trends and Public Policy," *Trends in Pharmacological Science* 29 (2008), 278. . . .

11. See Federal Trade Commission, *At Home Genetic Tests: A Healthy Dose of Skepticism May Be the Best Prescription* (July 2006), www.ftc.gov/bcp/edu/pubs /consumer/health/hea02.shtm. . . .

12. Jeffrey Shuren, director, Center for Devices and Radiological Health, F.D.A., "Statement before the U.S. House of Representatives Subcommittee on Oversight and Investigations Committee on Energy and Commerce: Direct-to-Consumer Genetic Testing and the Consequences to the Public" (July 22, 2010), www.fda.gov /NewsEvents/Testimony/ucm219925.htm.

13. See Obasogie, supra note 7, part III.C.2.b.

14. National Environmental Policy Act of 1969, Pub. L. No. 91–190, 83 Stat. 852 (codified as amended at 42 U.S.C. §§ 4321, 4331–35, 4341–47 [2012]).

15. NEPA requires agencies to "include in every recommendation or report on proposals for legislation and other major federal actions significantly affecting the quality of the human environment" an environmental impact statement. 42 U.S.C. § 4332(C). These statutory requirements have been defined within the Code of Federal Regulations, 40 C.F.R. § 1502.3.

16. 42 U.S.C. § 4332(C)(i).

17. 40 C.F.R. § 1508.9.

18. 42 U.S.C. § 4332(C)(i)–(iv).

19. 40 C.F.R. § 1502.1.

20. Brian L. Cole et al., "Prospects for Health Impact Assessment in the United States: New and Improved Environmental Impact Assessment or Something Different?," *Journal of Health Politics, Policy and Law* 29 (2004), 1153, 1157.

21. 40 C.F.R. § 1502.6 (2012). . . .

22. Cole et al., supra note 20, 1167–68.

23. Northern and Yorkshire Public Health Observatory, "An Overview of Health Impact Assessment" (2001), 1, http://dro.dur.ac.uk/5613/1/5613.pdf.

24. Id.

25. Frank Vanclay, "Social Impact Assessment" (working paper, World Commission on Dams, 2000), 2, www.scribd.com/doc/81829611/Social-Impact-Assessment-Vanclay.

26. Lawrence Gostin and Jonathan M. Mann, "Towards the Development of a Human Rights Impact Assessment for the Formulation and Evaluation of Public Health Policies," *Health and Human Rights* 1 (1994), 59, 60.

27. See Obasogie, supra note 7, part II.

28. 42 U.S.C. §§ 4342–4347 (2012).

29. See Nicholas Wade, "Reading the Book of Life: Now, the Hard Part: Putting the Genome to Work," *New York Times*, June 27, 2000, F1. . . .

30. See Obasogie, supra note 7, part II.C.

31. Bill Clinton, president of the United States, Tony Blair, prime minister of England, Dr. Francis Collins, director of the National Human Genome Research Institute, and Dr. Craig Venter, president and chief science officer, Celera Genomics Corp., "Remarks on the Completion of the First Survey of the Entire Human Genome Project" (June 26, 2000), www.ornl.gov/sci/techresources/Human_Genome /project/clinton2.shtml.

32. See Obasogie, supra note 7, part II.C.

53. Human Genetic Engineering Demands More Than a Moratorium

Sheila Jasanoff, J. Benjamin Hurlbut,
and Krishanu Saha

On April 3, 2015, a group of prominent biologists and ethicists writing in *Science* called for a moratorium on germline gene engineering; modifications to the human genome that will be passed on to future generations. The moratorium would apply to a technology called CRISPR/Cas9, which enables the removal of undesirable genes, insertion of desirable ones, and the broad recoding of nearly any DNA sequence.

Such modifications could affect every cell in an adult human being, including germ cells, and therefore be passed down through the generations. Many organisms across the range of biological complexity have already been edited in this way to generate designer bacteria, plants, and primates. There is little reason to believe the same could not be done with human eggs, sperm, and embryos. Now that the technology to engineer human germlines is here, the advocates for a moratorium declared, it is time to chart a prudent path forward. They recommend four actions: a hold on clinical applications; creation of expert forums; transparent research; and a globally representative group to recommend policy approaches.

If these recommendations seem familiar, it is because this is not the first time science leaders have responded to a similar problem. In calling for a moratorium on germline modification, the group invoked a famous precedent: the 1975 meeting at Asilomar, California, on recombinant DNA. Two years before that meeting, scientists declared a voluntary moratorium on experiments that they worried might endanger human health and the environment. The moratorium allowed for a period of reflection to ensure

Original publication by Sheila Jasanoff, J. Benjamin Hurlbut and Krishanu Saha. "Human Genetic Engineering Demands More Than a Moratorium." *The Guardian.* April 7, 2015. Reprinted by permission of *The Guardian.*

that scientific progress would proceed without putting society's well being at risk.

Asilomar is remembered as a great success because it defused public anxiety and opened up the market for biotechnology. It is frequently cited when people are uncertain about what is at stake in emerging scientific domains. Asilomar offers an easy recipe for public policy: a research moratorium followed by an expert assessment of which risks are acceptable and which warrant regulation. It is a tonic to cure public anxiety and create safe spaces for science.

But how good is the Asilomar model for governing controversial biotechnological advances? The answers are ambiguous at best. The Asilomar meeting achieved agreement in part by bracketing off three serious concerns: environmental release of engineered organisms; biosecurity; and ethical and social aspects of human genetic engineering. Decades later, these are precisely the issues we are still wrestling with in the public domain.

The molecular biologists at Asilomar sidestepped ecological concerns by prohibiting the release of genetically engineered organisms into the environment. A few years later, scientists on the US Recombinant DNA Advisory Committee unilaterally tried to end the prohibition on release of genetically modified (GM) organisms into the environment without asking for an environmental impact assessment. They judged the issues to be purely technical, to be resolved by scientific expertise alone. But history suggests their confidence was misplaced. Persistent controversy over GM crops and foods, longstanding in Europe and on the rise in the United States, indicates that legitimate public concerns about the benefits of these technologies could not simply be wished away.

Biosecurity came back to haunt us in 2012 when a Dutch researcher at Erasmus University used research funding from the US National Institutes of Health to create an H5N1 flu virus that could cause a pandemic. He sought to publish the results, raising sudden concern that this knowledge could be used as a tool of bioterrorism. Yet multiple rounds of scientific peer review had essentially failed to question whether the research itself was beneficial or appropriate.

Ethical questions, too, continue to swirl around genetic engineering and embryo research. [As of April 2015,] [r]umors that genes of human embryos have already been edited resurrect anxieties that the Asilomar scientists neither defused nor eliminated.

The experts calling for a moratorium on human germline editing assert that more research must be done before we can judge the ethical propriety of genetically modifying children. Besides, they argue, global publics must be educated by experts before an informed dialogue can take place. But the problem is not simply a lack of technical knowledge. The answer to how we should act does not lie in the technological details of CRISPR. It is our responsibility to decide, as parents and citizens, whether our current genetic preferences should be edited, for all time, into our children and our children's children.

A moratorium without provisions for ongoing public deliberation narrows our understanding of risks and bypasses democracy. Regrettably, we have not yet developed the habits of deliberation that could guide research agendas before technological innovation renders neglected ethical questions immediate and urgent. Even in technologically advanced societies, we tend to defer to expert judgments about which risks are reasonable to worry about, and which are not. This is a democratic deficit. It inhibits our capacity to participate thoughtfully in imagining the futures we want and governing technological change accordingly.

An effective moratorium must be grounded in the principle that the power to modify the human genome demands serious engagement not only from scientists and ethicists but from all citizens. We need a more complex architecture for public deliberation, built on the recognition that we, as citizens, have a duty to participate in shaping our biotechnological futures, just as governments have a duty to empower us to participate in that process. Decisions such as whether or not to edit human genes should not be left to elite and invisible experts, whether in universities, ad hoc commissions, or parliamentary advisory committees. Nor should public deliberation be temporally limited by the span of a moratorium or narrowed to topics that experts deem reasonable to debate.

Education has a vital role to play in remedying the democratic deficit, but what citizens need is not simply more STEM (science, technology, engineering, and mathematics) courses. Knowing science does not teach us how to live well with its power. Our universities need to devote more resources to teaching the relationship between science, technology, and society so as to produce the citizens, the concepts, and the conversations capable of guiding our common future.

Prudence demands that we marshal the full force of democracy to imagine the lives we want. Otherwise we will find ourselves governed by technologies whose implications we did not foresee and whose development we chose to neglect.

54. "Moral Questions of an Altogether Different Kind"

Progressive Politics in the Biotech Age

Marcy Darnovsky

Charles Darwin's theory of evolution was far more than an advance in scientific understanding. It famously upended traditional ways of thinking about the origins of life and the place of humanity in history and the cosmos. Especially in the United States, some religious believers came to see the theory as a challenge to basic tenets of their faith, which opened up societal and political rifts that remain gaping today. But recent developments in human biotechnology have the potential to surpass the political and cultural intensity triggered by those earlier debates.

The study of evolution is at heart an explanatory endeavor. Human biotechnologies—applications of reproductive, genetic, and biomedical science and related emerging fields including nanotechnology and synthetic biology—offer tools as well as explanations. They promise, and sometimes deliver, results that in ancient times were thought to be achievable only by divine intervention: they heal the sick, let the crippled walk again, and give children to the barren. They act directly on our bodies, behaviors, and minds and alter the way we understand ourselves in the process.

Many biotechnology products and practices enjoy and deserve widespread support across the political spectrum. Among these are new kinds of drugs and other medical treatments; innovative diagnostic tests that can identify disease while it is more amenable to cure; forensic DNA analyses that help identify the guilty and exonerate the innocent; and more. All can be misused, to be sure, and many need to be regulated. But we do not

Original publication by Marcy Darnovsky. "'Moral Questions of an Altogether Different Kind': Progressive Politics in the Biotech Age." *Harvard Law & Policy Review* 4, no. 1 (January 2010): 99–119. Reprinted by permission of Marcy Darnovsky. Due to space limitations, portions of this chapter have been deleted or edited from the original.

demand and should not expect perfection from any of our tools, including those based in biotechnology.

Nevertheless, some applications of these emerging technologies pose challenges to core progressive and liberal values. Consider a few current practices: Poor villagers in developing countries sell their kidneys and rent their wombs to global elites. Clinical researchers scour underdeveloped regions for human subjects; at home, they look to immigrants, prisoners, and people with no access to health care. Fertility clinics and brokers offer as much as a hundred thousand dollars to the best and brightest on America's college campuses in exchange for their eggs. Drugs and gene tests based on molecular differences between populations threaten to revive discredited notions about biology and race.[1] Biotech scientists involve themselves in profit-making ventures, while continuing to receive large amounts of public funding.

Moreover, these technologies are taking us into uncharted moral and political waters. The recent controversy over embryonic stem cell research may be merely an early warning of biopolitical storms already on the horizon. Biotechnology-based products and procedures now under development will pose social and ethical questions unprecedented in human history. Some are close at hand. When and how should children conceived with high-tech assistance learn that they have two or even three biological mothers? Should researchers transfer human genes or brain cells into non-human animals? Should they, as has recently been proposed, attempt to use cloning techniques to resurrect a Neanderthal?

Most socially consequential is the prospect of manipulating the traits of future children and generations. Some enthusiasts advocate the development of designer baby technologies that would give parents the option to engineer their children's appearance and talents and would allow the current generation to take control of human evolution. Undeterred by the eugenic abuses of the twentieth century, these enthusiasts eagerly anticipate the emergence of a consumer-based eugenics in the twenty-first century. They view human improvement as better realized by biological enhancement than by social change and are strategizing about how best to encourage the emergence of what they call "transhumans" or "posthumans."

Taken together, these visions constitute sociopolitical narratives whose effects on politics and culture are being felt even in advance of their technical feasibility. . . . As we face these unprecedented biotechnological powers that pose novel moral questions and cause an unavoidable political engagement with spiritual questions, where can liberals and progressives look for guidance? How can we prevent harmful uses of human biotechnologies while preserving our commitment to science as a reliable method for producing shared knowledge? How can we open the public sphere to concerns about the ways in which certain applications of human biotechnology could undermine the common good and moral values, yet protect the public sphere from the narrow-mindedness and coercion that liberal tolerance is meant to avoid?

Unfortunately, two currents in recent liberal and progressive thought leave us ill-equipped for these challenges. One is a tendency to embrace technological and scientific developments without adequate attention to the risks they pose and the deep impact that they can have on our politics and culture. The other is a reluctance to directly address moral controversies, especially when strongly held religious beliefs are in play. These currents converge in the quarrel between science and religion that has dogged the United States over the past century.

This essay takes issue with both these currents. It argues that human biotechnologies, like other powerful technological innovations, should be subject to democratic governance and shaped by public policies. And it asserts that progressives and liberals must seriously engage the moral controversies and qualms that human biotechnologies raise.

MORAL VALUES IN AMERICAN POLITICS

From the 1960s until the past few years, most liberals and progressives treated morality and religion in politics quite gingerly.[2] The dominant left and center view during these decades was liberal in the classical sense: it saw religious freedom and religious tolerance as important, but considered religious commitment a matter of personal belief and individual choice. Moral and ethical values were seen as private concerns, unwelcome in the public sphere, and certainly not matters on which government policy should take sides.

This version of liberalism is meant to protect individual liberties and to guard against religious intolerance—critically important goals. But its shortcomings are also significant: it discourages efforts to address the moral dimensions of public policy and dampens public deliberations about shared values and the common good. When classical liberalism tilts toward a libertarian prioritization of personal liberty, it tends to shortchange social justice and solidarity; when joined with market liberalism, it provides few conceptual resources for resisting incursions of commercial dynamics into ever larger areas of both public and private life. These deficits have been apparent in many liberals' and progressives' recent encounters with the politics of human biotechnology.

Both pragmatism and principle counsel that progressives and liberals should more robustly engage the moral values and spiritual questions that human biotechnologies put on the political agenda, and should consider the role of religious traditions in grappling with these issues. The pragmatic argument is that many Americans care deeply about the moral and spiritual aspects of political issues and want public policies and electoral campaigns to address them. Unless we secular liberals and progressives work to lessen the suspicion and antagonism that often divide us from people of faith, we will continue to cede crucial ground to Republicans and religious fundamentalists.

This perspective gathered steam in the aftermath of the 2004 presidential election. Politics and religion writer Amy Sullivan put the pragmatic case bluntly: "Trying to understand American politics without looking at religion would be like trying to understand the politics of the Middle East without paying attention to oil."[3] And in a symposium of political writers pondering "the question of why the Democratic Party—which has now lost five of the past seven presidential elections and solidified its minority status in Congress—keeps losing elections,"[4] Robert Reich wrote:

> I'm not saying Democrats have to adopt my particular moral positions. But unless or until Democrats return to larger questions of public morality, they won't inspire the American public. Plans and policies are important, of course. But there's no substitute for offering a vision of what we can become as a nation—and giving citizens the faith we can get there.[5]

The principled argument for cultivating the habit of moral inquiry is that doing so enriches the moral imagination and social vision of progressives and liberals. Such a project is likely to involve inquiry into religious traditions, if only because individuals and societies so often derive their moral understandings from them. And while the "new atheists" ferociously denounce religion (the subtitle of Christopher Hitchens' 2007 book, for example, is *How Religion Poisons Everything*[6]), they do not—and cannot—deny the salience of vital moral questions. The history of moral thought is inextricably interwoven with the history of religion, and progressives and liberals can learn from that long tradition even while we reject dogmatism and intolerance. . . .

The deep and damaging political polarization of recent decades has made thoughtful deliberation about matters of meaning and morality difficult. The impulse to bracket them—to shunt them into private life or avoid them altogether—is strong. But the price of doing so is far too high.

LESSONS OF THE STEM CELL WARS

How have the shifting progressive and liberal sensibilities about "addressing issues in moral terms"[7] played out in the biopolitical realm? The Bush-era battles over embryonic stem cell research clearly demonstrated the need for a revised progressive and liberal approach to the politics of human biotechnology, and the Obama administration has crafted one.

The stem cell wars generated polarized stereotypes of progressives as "pro-science" and religious conservatives as "anti-science." In fact, religious conservative opposition to embryonic stem cell research was (at least in many cases) based not on antipathy to science, but on a moral conviction that human embryos should not be destroyed even in a worthy endeavor such as medical research. In reaction, many liberals and progressives came to equate moral concerns about human biotechnology solely with this theological commitment to the personhood of human embryos.

The rancor and polarization that developed during the Bush years kept the status of embryos in the spotlight and consigned other moral and political issues raised by stem cell research to the shadows. Concerns about stem cell research that had nothing to do with the personhood of embryos were typically ignored. But a number of progressives did raise such concerns. Women's health advocates questioned the subset of stem cell research that involves a cloning technique because it requires large numbers of human eggs, the acquisition of which poses risks to the young women who provide them. Some progressives who support embryonic stem cell research (including my own organization, the Center for Genetics and Society) nonetheless opposed the 2004 California voter initiative earmarking three billion dollars for the endeavor because it enshrines conflicts of interest in the state agency it established and violates basic precepts of democratic accountability and governance. The hyperbolic exaggerations about the imminence of medical breakthroughs, in which so many stem cell research supporters indulged, were a deep disservice both to political and scientific integrity.

The partisan stem cell divide also obscured important political complexities. Notable conservatives such as Nancy Reagan and Orrin Hatch supported embryonic stem cell research, as did numerous liberal and moderate religious bodies. Similarly, the larger politics of human biotechnology confound conventional ideological alignments. . . .

The Obama administration has now significantly cooled the stem cell wars. The president lifted the Bush administration's restrictions on federal funding of research that uses stem cells derived from embryos created but not needed for fertility treatments.[8] He also directed the National Institutes of Health to draw up research guidelines—"strict guidelines," he said, "which we will rigorously enforce, because we cannot ever tolerate misuse or abuse."[9] Firmly rejecting the view that early-stage embryos have the rights of personhood, President Obama went on to affirm the importance of applying moral judgments to scientific work. And he articulated a clear position on human reproductive cloning: "It is dangerous, profoundly wrong, and has no place in our society, or any society."[10]

This statement is significant because it goes beyond the narrow safety and procedural considerations that many researchers and bioethicists cite as the only legitimate reasons to forgo efforts to clone human beings and to pursue other widely opposed forms of human genetic manipulation. Instead, while recognizing that science and technology are uniquely positioned to tell us what can be done, it asserts that social values, mediated and expressed through democratically accountable institutions, must determine what should be done.

Now that the new administration has lowered the temperature on the stem cell issue, how should liberals and progressives think about the difficult questions involved in assessing and regulating human biotechnologies? As a starting point, we should take stock of some recent missteps, in the hope that we can do better on the biotech-related challenges to come. The sections that follow consider the tendency of liberals and progressives to avoid the substantive moral dilemmas posed by biotechnological innovations; to elevate individual liberty over social justice in considerations of biotech issues; to downplay the excess of market dynamics that liberals appropriately challenge in other spheres; and to champion science—especially biotech science—as unquestionably beneficial and properly protected from public policy and oversight.

BIOETHICS: THINNING THE DEBATE, NARROWING THE QUESTIONS

Bioethics emerged in the mid-twentieth century out of concerns about the ethical and philosophical implications of then-new procedures and technologies, including end-of-life decision making, organ donation, human experimentation, and genetic engineering. In the field's early decades, bioethicists often grappled with the broad social consequences of new developments in the life sciences. Later, much of their attention shifted to important but narrower topics involving individuals' relationships with research and medicine, such as informed consent, patient safety, and rules for the conduct of research.

This narrowing has been especially unfortunate because bioethics is widely seen as the most appropriate locus of deliberation about emerging biotechnologies. Though bioethicists typically present themselves as experts rather than as spokespeople, their work and their words can too easily function as a substitute for democratic participation in biotechnology-related policy matters. . . .

The 2005 guidelines for stem cell research issued by the National Academies of Sciences (NAS) are an example of this trend.[11] One section addresses concerns raised by chimeras that are created by adding human stem cells to animals or animal embryos.[12] Most scientists consider such chimeras to be legitimate research tools. Yet many Americans—scientists and lay people alike—are uneasy about research that may produce higher animals with human traits. Just how far should researchers go in their work to humanize animals? What if these experimental creatures begin to display some degree of human consciousness?

The NAS guidelines acknowledge the possibility that "cell transfer [could] result in the animal's acquiring characteristics that are valued as distinctly human"[13] and concur that the "idea that human neuronal cells might participate in 'higher order' brain functions in a nonhuman animal, however unlikely that may be, raises concerns that need to be considered."[14] But rather than actually moving to such a consideration, the guidelines merely advise that reviewers ask the question, "If visible human-like characteristics might arise, have all those involved in these experiments, including animal care staff, been informed and educated about this?"[15]

This is at best an end run around rocky ethical and moral terrain. And while it may be understandable for research guidelines to avoid what could well become a political quagmire, most liberals and progressives have also declined to address these concerns in other spheres. Instead, when George W. Bush proposed a ban on "human-animal hybrids" in his 2006 State of the Union address,[16] liberal bloggers and bioethicists took the opportunity to ridicule him.

True, Bush's way of addressing the issue was confused and misleading, and his choice of the State of the Union as a venue was strange. The scathing liberal and progressive responses to Bush's comment hit these easy targets dead center. But they missed the mark by failing to acknowledge that many Americans have reasonable concerns about lab-made animals in which "visible human-like characteristics might arise."[17] Some took the view that the only concerns that exist are those based in religious belief, and that any concern connected to religion should be dismissed. In short, their attitude seemed to be that only the scorn-worthy could possibly imagine that there is anything to worry about. As liberal science blogger PZ Myers put it, "Once again, the ignorance and the bigotry of the religious right wins out over reason and humanitarianism."[18]

Mockery is thus one response to moral concerns about biotechnology. Another is to retreat to procedural recommendations, such as additional training for animal caretakers. A third response, expressed by bioethicist Paul Root Wolpe . . . acknowledges that "substantive rationality" has a place in liberal societies—and that this place is outside of policy bodies such as bioethics commissions:

> Our society is designed to have its discussion of ultimate ends as part of
> civil society—in its newspapers, from its pulpits, and around
> watercoolers. Presidential commissions and the resulting legislative
> recommendations such commissions make are probably better off
> staying away from discussions of ultimate ends.[19]

This is, in fact, the classic liberal view. Many progressives also adhere to it, as do many government institutions, including the federal agency that most directly deals with human biotechnologies, the Food and Drug Administration (FDA). The FDA assesses the safety and efficacy of the products it regulates; it does not consider social consequences or moral questions.

Outside the United States, by contrast, such concerns are widely considered a legitimate part of public debate and public policy. This is the case in many European countries, most of whose populations are far more secular than Americans. In 1998, for example, the European Parliament and the Council of the European Union issued a directive forbidding patents on inventions that contravene what they term "public morality."[20] Similar provisions exist in the patent laws of many countries. Even the Agreement on Trade-Related Aspects of Intellectual Property Rights (TRIPS) of the World Trade Organization, the primary international agreement governing intellectual property, permits denials if a patent would be contrary to "*ordre public,* or morality."[21]

Unlike the United States, Europe has explicitly prohibited the human biotechnology practices that generate the most concern, including reproductive human cloning, inheritable genetic modification, and sex selection for nonmedical purposes. These prohibitions are justified by an appeal to "human rights and fundamental freedoms" in the Council of Europe's 1998 Convention on Human Rights and Biomedicine.[22] Canada has adopted similar provisions in its Assisted Human Reproduction Act (AHRA) of 2004.[23] The AHRA includes an explicit "declaration of principles" that mentions a number of values unlikely to be specifically cited in US policy, including "dignity and rights in the use of these technologies and in related research."[24] In addition, the AHRA says that "trade in the reproductive capabilities of women and men and the exploitation of children, women, and men for commercial ends raise health and ethical concerns that justify their prohibition."[25]

Other biotech-related policy documents also include explicitly moral language. The recently drafted *Declaration of Istanbul on Organ Trafficking and Transplant Tourism*, for example, says that "organ trafficking and transplant tourism violate the principles of equity, justice, and respect for human dignity and should be prohibited."[26] In comparison, the US National Organ Transplant Act of 1984, which also prohibits people from selling their organs, does not include any comparable language.[27]

Not so long ago, however, many American liberals and progressives were comfortable with moral appeals. Eleanor Roosevelt, for example, was chair of the United Nations Commission that drafted the *Universal Declaration of Human Rights*, which asserts the legitimacy of "meeting the just requirements of morality, public order and the general welfare in a democratic society."[28]

DELIVERING A BABY: JUSTICE AND LIBERTY, MORALITY, AND MARKETS

Most progressives and liberals take social justice as a touchstone commitment but consider individual autonomy to be decisive in matters that bear on reproduction. The battle for abortion rights has become all but synonymous with the term "choice," and arguments that draw on other values have long been more muted. For example, abortion rights advocates seldom ground their case in a social vision—in the argument, for instance, that women's freedom to decide whether and when to bear children is necessary to the kind of society we want to build.

During the Bush years, some reproductive rights advocates extended their understanding of "choice" to encompass support for even the most extreme reproductive technologies, including reproductive cloning. In this view, the right to decide whether and when to bear a child is conflated with a right to determine the precise traits of a child that one will bear.[29] In fact, in 2002, Planned Parenthood Federation of America came close to officially endorsing human reproductive cloning as an extension of women's reproductive freedom.[30] Fortunately, Planned Parenthood instead affirmed that strong support for abortion rights in no way precludes clear opposition to the creation of cloned children.

In recent years Planned Parenthood and other reproductive rights groups, partly in response to the reproductive justice approach developed by women of color organizations, have begun to appreciate the limitations of the "choice" framework, including its inadequacy for grappling with the issues that reproductive and genetic innovations raise. . . .

In some cases, [these] technologies are driven as much by the market as by technical developments. Consider, for example, the booming practice of pregnancy outsourcing, now nearly a half-billion-dollar-a-year business in India.[31] Fertility clinics there recruit poor rural women to serve as surrogates, housing them in dormitories during their pregnancies. The clinics closely monitor the surrogates' diets and sexual contacts and require them to abide by other behavioral restrictions and medical stipulations—some insist on caesarean section births—enumerated in contracts that the surrogates sometimes cannot read. The clinics' clients are far more affluent, though not necessarily wealthy by developed world standards. Surrogacy costs in India are about one-fifth of US rates,[32] putting it within reach for middle-class Americans, Europeans, and others.

Many commissioning parents turn to India because commercial surrogacy is illegal where they live. This is the case in the United Kingdom, Canada, many European nations, and some American states. India is also attractive because surrogates recruited there are highly unlikely to make a serious bid to keep the baby, as has happened in several high-profile cases in the United States.[33]

The payment that Indian surrogates receive for their services is typically a windfall for them. Reporters who have traveled to India to report on the "rent-a-womb" phenomenon have no trouble finding surrogates who receive more money for one pregnancy than they could make in ten or more years of other work. One told *Marie Claire* reporter Abigail Haworth that helping her husband in his scrap-metal business typically nets the two of them $1.20 to $1.45 a day, while a successful surrogate pregnancy would bring in $5,500. That, she said, would "give my children a future."[34]

How should we evaluate this new global industry? Some observers—including some feminists and social liberals and progressives, as well as market liberals and economic conservatives—argue that women must be free to sell their eggs and rent their wombs; that these practices are no different from other forms of wage labor, and should be treated as private matters. Articles about commercial surrogacy in women's magazines are often written as heartwarmers with fairy-tale endings. They almost always take the point of view of the contracting parents and almost never interrogate their desire for a genetically related child. In a 2007 segment of the *Oprah Winfrey Show* titled "Wombs for Rent," the media megastar told her eight million viewers that surrogacy outsourcing is not exploitation; rather, it is a beautiful example of women helping women and a "confirmation of how close our countries can really be."[35] Others find "reproductive tourism" deeply troubling, and argue that we should not extend the ethos and dynamics of the marketplace to pregnancy and childbearing. . . .

In India, commercial surrogacy exemplifies—as it intensifies—the social inequalities that characterize the global economy. Wherever it takes place, it turns gestation and childbirth into a market transaction. It is difficult to see how support for this new form of exploitation can coexist with political commitments to restraining market excesses and expanding social justice.

TECHNO-SKEPTICISM AND TECHNO-TRIUMPHALISM

. . . In 1975, space technology enthusiasts formed the L5 Society to promote their vision of space colonies. Its sensibility was libertarian; many in its orbit, so to speak, went on to advocate a variety of techno-utopian fantasies. Its antipathy toward social solidarity and its hostility to religious tradition are embedded in its slogan, "The meek shall inherit the earth. The strong and the wise keep moving."[36]

Such assertions of technological triumph and trivializations of spirituality were echoed a quarter-century later by James Watson, the Nobel Laureate who helped to deduce the helical structure of DNA. In 1989, Watson told *Time* magazine, "We used to think our fate was in our stars. Now we know, in large measure, our fate is in our genes."[37] In a 2000 address to the British Parliamentary and Scientific Committee, he asked, "If scientists don't play god, who will?"[38]

Watson is also an advocate of a new eugenics enabled by advances in genetic and reproductive technologies. At a 1998 University of California, Los Angeles (UCLA) conference called "Engineering the Human Germline," organized by a small but disturbing number of influential scientists who share these views, Watson told the audience:

> I can't indicate how silly I think it is [the sanctity of the gene pool]. I mean, we have great respect for the human species. . . . But saying we're sacred and should not be changed? Evolution can be just damn cruel, and to say that we've got a perfect genome and there's some sanctity? I'd like to know where that idea comes from, because it's utter silliness.

We should treat other people in a way that maximizes the common good of the human species. That's about all we can do.[39]

The UCLA conference was attended by a thousand people and covered on the front pages of the *New York Times* and the *Washington Post*.[40] Among the speakers was Princeton University biologist Lee Silver, notorious for predicting that emerging technologies will result in the division of humanity into separate castes (and eventually separate species): the ruling "GenRich" and their inferiors, the "Naturals."[41] At the UCLA event, Silver was thrilled that "we now have the power to seize control of our evolutionary destiny."[42] With similar enthusiasm, conference organizer Gregory Stock wrote the following year that inheritable genetic modification "will force us to re-examine even the very notion of what it means to be human [as] we become subject to the same process of conscious design that has so dramatically altered the world around us. . . . Through this technology, we will seize control of our own evolution."

The UCLA conference was not strictly an academic affair. Its goal, Stock explained to *Nature Biotechnology*, was to make inheritable genetic engineering "acceptable" to the public.[43] Such techno-enthusiasm is extreme, but these advocates are hardly fringe figures. Perhaps because several of them have high public profiles and have won prestigious scientific awards, few scientists or other public intellectuals challenged their open embrace of technologies that are likely to create entirely new forms of human inequality. For the most part, liberals and progressives were also silent.

SCIENCE'S WHITE KNIGHTS?

By the 1990s and early 2000s, critical thinking about powerful new technologies had dimmed in progressive circles. This trend accelerated in the confrontation with the George W. Bush administration's pattern of suppressing scientific evidence, distorting research findings, and disregarding experts whenever their advice was politically inconvenient.[44]

Progressives were clearly correct to oppose this egregious approach to science policy. But they took a wrong turn when they interpreted what Chris Mooney called the "Republican War on Science" as primarily a religious crusade rather than one also significantly driven by corporate interests.[45] They also erred when they fought back by emptying their approach to science and technology policy of moral content. . . .

During the Bush years, liberals' and progressives' unqualified defense of science led to additional unintended consequences. They too often discounted the importance of public regulation and oversight; overlooked conflicts of interest and corporate encroachments; and portrayed science as an endeavor above and outside social values and power dynamics, a protected zone from which political "interference" should be excluded.[46]

In some quarters, this attitude has persisted in the post-Bush era. A case in point concerns the issue of paying young women to provide eggs for cloning-based stem cell research.[47] Egg extraction has been common in the fertility field for several decades. While it poses both short-term and long-term health risks to women who undergo it, there have been very few studies to determine the frequency and seriousness of the risks and adverse reactions.[48]

In the fertility context, the benefit of egg extraction is clear: a significant fraction of such procedures results in a baby being born. Cloning-based stem cell research, by contrast, is a speculative endeavor that has not yet produced any stem cell lines. For years, some researchers and advocates made fanciful claims about the imminence and importance of cloning-based stem cell work. In the 2004 presidential elections, it became a Democratic cause célèbre. At the party's national convention that year, Ron Reagan, Jr. asserted that cloning would soon provide each of us with our own "personal biological repair kit standing by at the hospital."[49]

Since that time, the prospects for cloning-based stem cell research have diminished considerably.[50] In late 2005, a major scientific scandal erupted when celebrity Korean researcher Hwang Woo Suk, the only scientist ever to claim to have produced stem cells from cloned embryos, was found to have fabricated his data.[51] Since 2007, after the discovery of methods to "reprogram" ordinary body cells into fully potent stem cell lines, the case for cloning as the path to disease-specific and patient-specific cell lines has further eroded. Many scientists, including Ian Wilmut, famous for his role in cloning Dolly the sheep in 1996, have abandoned the work.[52]

Even before the Hwang scandal and the emergence of cell reprogramming, paying women to provide eggs for stem cell research was opposed in some mainstream scientific circles. A number of prestigious scientific bodies, including the National Academy of Sciences and the California Institute for Regenerative Medicine, ruled that women could be reimbursed for expenses they incur in the course of egg-harvesting procedures, but not beyond that.[53] A number of industrialized countries have also prohibited payments beyond expenses. Nonetheless, in 2009, the state-funded stem cell program in New York decided to break with this near consensus. A liberal bioethicist on the agency's ethics board argued for payments by saying, "I think that we are an ethics committee, and I actually think that, if good science demands these oocytes, that we have the obligation to provide them." The sad lesson of this incident is that some liberals are allowing themselves to rubber stamp even dubious proposals from the scientific enterprise they are charged with monitoring.

TOWARD PROGRESSIVE BIOPOLITICS

We have come to expect technical innovations outside the life sciences—for example, nuclear energy, large hydroelectric dams, and synthetic pesticides—to elicit wide and sometimes vociferous debate. These controversies involve questions about safety, efficacy, appropriate regulation, and alternative approaches.... Human biotechnologies pose an additional challenge because of the moral dilemmas they raise. In the "culture wars" that have divided the American polity, these moral matters are too often reduced to a single concern—the status of human embryos. Liberals and progressives need to reject this limiting definition and cultivate broader and richer understandings about the meaning and morality of human biotechnologies.

In doing so, we can draw on a number of moral traditions: on progressive and liberal commitments to social justice and market regulation; on the environmentalist case for precaution in the face of powerful technologies that manipulate the material world; and on the important strands of moral philosophy and religious thought that counsel preferential treatment of the poor and forethought about future generations. Building on these rich legacies, we can craft a biopolitical imagination that is adequate to meet the unprecedented challenges we face, and we can develop a policy agenda that will ensure a healthy, just, and progressive human future.

NOTES

1. See Osagie K. Obasogie, "Playing the Gene Card?," *Center for Genetics and Society* (2009), 1–7, www.geneticsandsociety.org/downloads/complete_PTGC.pdf.

2. See E.J. Dionne Jr., *Souled Out: Reclaiming Faith and Politics after the Religious Right* (Princeton, NJ: Princeton University Press, 2008), 34–39.

3. Amy Sullivan, *The Party Faithful: How and Why Democrats Are Closing the God Gap* (New York: Scribner, 2008), vii.

4. Editor's Note, "Why Americans Hate Democrats—A Dialogue," *Slate*, November 4, 2004, www.slate.com/id/2109188/ (on file with the Harvard Law School library).

5. Robert Reich, "Gotta Have Faith," *Slate*, November 4, 2004, www.slate.com /id/2109 190 (on file with the Harvard Law School library).

6. Christopher Hitchens, *God Is Not Great: How Religion Poisons Everything* (New York: Hachette, 2007).

7. Barack Obama, "Keynote Address at the Call to Renewal Conference: Building a Covenant for New America" (June 28, 2006), www.barackobama.com/ 2006/06/28 /call_to_renewal_keynote_address.php.

8. See Exec. Order No. 13,505 § 1, 74 Fed. Reg. 10667 (March 11, 2009).

9. President Barack Obama, "Remarks at the Signing of Stem Cell Executive Order and Scientific Integrity Presidential Memorandum" (March 9, 2009), www .whitehouse.gov/the_press_office/remarks-of-the-president-as-prepared-for-deliverysigning-of-stem-cell-executive-order-and-scientific-integrity-presidential-memorandum/.

10. Id.

11. National Research Council of the National Academies of Sciences, *Guidelines for Human Embryonic Stem Cell Research* (2005), www.nap.edu/catalog .php?record_id = 11278 (hereinafter NAT'L ACADS.).

12. Id., 49–50.

13. Id., 50.

14. Id., 40.

15. Id., 50.

16. President George W. Bush, State of the Union Address (January 31, 2006), www.washingtonpost.com/wp-dyn/content/article/2006/01/31/AR2006013101 468.html.

17. NAT'L ACADS., supra note 11, 50.

18. Pharyngula, http://scienceblogs.com/pharyngula/2006/ 02/president_ panders_to_antimanim.php (February 1, 2006, 11:31 EST) (on file with the Harvard Law School library).

19. Paul Root Wolpe, review of *Playing God? Human Genetic Engineering and the Rationalization of Public Bioethical Debate*, by John H. Evans, *American Journal of Sociology* 109 (2003): 215, 216.

20. See Council Directive 98/44, art. 6, 1998 O.J. (L 213) 13, 18, http://europa .eu.int/eur-lex/pri/en/oj/dat/1998/1_213/1_21319980730en00130021.pdf; see also European Union Fact Sheet, *Biotechnology Industry* 4 (2005), www.europarl .europa.eu/ftu/pdf/en/FTU_4.8.9.pdf. . . .

21. Council for Trade-Related Aspects of Intellectual Property Rights, Note by the Secretariat: Review of the Provisions of Article 27.3(b), Summary of Issues Raised and Points Made 10, IP/C/W/369/Rev.1 (March 9, 2006), www.wto.org/english /tratop_E/trips_ e/ipcw369r1.doc.

22. Council of Europe, "Convention for the Protection of Human Rights and Dignity of the Human Being with Regard to the Application of Biology and Medicine: Convention on Human Rights and Biomedicine," April 4, 1997, ETS No. 164, Preamble, http://conventions.coe.int/treaty/en/treaties/html/164.htm; see also Council of Europe, "Additional Protocol on the Prohibition of Cloning Human Beings," January 12, 1998, ETS No. 168, http://conventions.coe.int/treaty /en/treaties/html/168.htm.

23. Assisted Human Reproduction Act, 2004 S.C., ch. 2 (Can.), http://laws .justice.gc.ca/en/A-13.4/.

24. Id., 1.

25. Id., 2.

26. International Summit on Transplant Tourism and Organ Trafficking, "The Declaration of Istanbul on Organ Trafficking and Transplant Tourism," *Clinical Journal of American Society of Nephrology* 3 (2008): 1227, 1228.

27. See 42 U.S.C. § 274e (2006).

28. Universal Declaration of Human Rights, G.A. Res. 217A, ¶ 29, U.N. Doc. A/810 (December 10, 1948).

29. See John A. Robertson, *Children of Choice* (Princeton, NJ: Princeton University Press, 1996), 22–42. . . .

30. See Marcy Darnovsky, "Political Science," *Democracy*, Summer (2009): 36, 43–44.

31. "American Gays Looking to India Surrogate Industry to Have Children," *Hindustan Times*, October 12, 2009, www.dnaindia.com/health/report_american-gays-looking-to-indian-surrogate-industry-to-have-children_1297969. . . .

32. Abigail Haworth, "Womb for Rent," *Marie Claire*, August 2007, 124.

33. See Stephanie Saul, "Building a Baby with Few Ground Rules," *New York Times*, December 13, 2009, A1.

34. Haworth, supra note 32.

35. "Wombs for Rent," *The Oprah Winfrey Show*, aired October 9, 2007 (Harpo Productions, Inc.).

36. See posting of Marcy Darnovsky to *Biopolitical Times* (May 14, 2009), www .biopoliticaltimes. org/article.php?id = 4745 (on file with the Harvard Law School library).

37. Leon Jaroff et al., "Science: The Gene Hunt," *Time*, March 20, 1989, 62, 67.

38. Steve Connor, "Nobel Scientist Happy to 'Play God' with DNA," *Independent* (London), May 17, 2000, 7.

39. James D. Watson, "The Road Ahead: A Panel Discussion," in *Engineering the Human Germline*, ed. Gregory Stock and John Campbell (New York: Oxford University Press, 2000), 73, 85.

40. Gina Kolata, "Scientists Brace for Changes in Path of Human Evolution," *New York Times*, March 21, 1998, A1; Rick Weiss, "Engineering the Unborn: Genetic Cures That Cross Generations," *Washington Post*, March 22, 1998, A1.

41. See Lee M. Silver, *Remaking Eden: Cloning and Beyond in a Brave New World* (New York: Harper Perennial, 1997), 4–7.

42. Lee M. Silver, "Reprogenics," in *Engineering the Human Germline*, supra note 39, 57, 58.

43. See Jeffrey J. Fox, "Germline Gene Therapy Contemplated," *Nature Biotechnology* (1998): 407.

44. See David Malakoff, "White House Denies Playing Politics with Science," *Science* 303 (2004): 1446.

45. Chris Mooney, *Republican War on Science* (New York: Basic Books, 2005).

46. Darnovsky, supra note 30, 37–38.

47. Cloning-based stem cell research is also known as "therapeutic cloning," "research cloning," and "somatic cell nuclear transfer" (SCNT).

48. See Diane Beeson and Abby Lippman, "Egg Harvesting for Stem Cell Research: Medical Risks and Ethical Problems," *Reproductive Biomedicine Online* 13 (2006): 573, 573.

49. Ron Reagan Jr., "Speech to the Democratic National Convention" (July 27, 2004), www.nytimes.com/2004/07/27/politics/campaign/27TEXT-REAGAN .html).

50. See Roger Highfield, "Embryo-Free Stem Cell Research Gets Boost," *Telegraph*, June 4, 2008, www.telegraph.co.uk/science/science-news/3343533 /Embryo-free-stemcell-research-gets-boost.html (on file with the Harvard Law School library).

51. Gina Kolata, "A Cloning Scandal Rocks a Pillar of Science Publishing," *New York Times*, December 18, 2005, A28.

52. Roger Highfield, "Dolly Creator Prof Ian Wilmut Shuns Cloning," *Telegraph*, November 16, 2007, www.telegraph.co.uk/science/science-news/3314696/Dolly-creator-ProfIan-Wilmut-shuns-cloning.html (on file with the Harvard Law School Library); Sally Lehrman, "No More Cloning Around," *Scientific American*, July 31, 2008, 100.

53. NAT'L ACADS., supra note 36, 85–87; California Institute of Reproductive Medicine, "Guidance for CIRM Medical and Ethical Standards Regulations Governing Donation of Oocytes for CIRM-Funded Research 2" (2008), www.cirm.ca.gov/workgroups/pdf/ Guidance_Donation_Oocytes.pdf; see also Deborah Spar, "The Egg Trade—Making Sense of the Market for Human Oocytes," *New England Journal of Medicine* 356 (2007): 1289, 1290.

Afterword

Patricia J. Williams

ACT I—IMAGINATION

The genetic revolution presents us humans with a sea of sudden revelations in microbiology, chemistry, and physics. Discoveries are made; prizes will be had. But what makes the moment so genuinely exciting is the suspense we all feel about how this enormous body of knowledge will be applied; how it will be translated and distributed into new forms of social power; and how it will reform and transform hierarchies of relation and systems of legitimacy. After all is said and done, it is this social application that causes us to hover somewhere between exhilaration and deepest apprehension. The science will let loose its cascading interactions with utter impassivity; yet how we inhabit that knowledge will be a contest of the imagination, a sedimentation of political futures, a constructed infinity of worlds.

We *imagine* that we have never seen anything like this moment, even as we also tell ourselves that there is nothing new under the sun. We *imagine* that whatever happens will merge nature and nurture, body and mind, free will and determinism—indeed merge all binarism into a utopia of total rapture rather than a totalistic hell. And, all evidence to the contrary, we tend *not* to imagine ourselves as superstitious, faith-addicted magical thinkers.

I would like to offer three short stories about the limits of imagination in this context. The first story dates back to the beginning of my career, in 1980, when I was a consumer advocate specializing in health law, particularly the buying and selling of purported medicines whose promised healing was based on false claims. My favorite case involved a supposed cancer cure. It was a salve made from olive oil and the ash of Mount St. Helens, a volcano in Washington State that had recently erupted. It had been marketed so as to capitalize on the magical properties of Mount St. Helens specifically—something akin to a dip in the Ganges or a bottle of water from the River Jordan, perhaps. The quasi-religious nature of the advertising made it particularly interesting to prosecute, because the purveyors, who were New Age entrepreneurs selling New Age remedies alongside astrology books, mood rings, and rose crystals, asserted that they were selling hope for a cure, not a cure itself, and thus the product was no different from their stock of handwoven prayer cloths. But, alas, the label said otherwise. It asserted very specific medicinal properties; it promised to cure cancer. So with a roll of our eyes and a laugh, we snatched a product off the market that was deliciously redolent of rosemary and eucalyptus, and actually very soothing, if in a medieval kind of way.

This brings me to the Precision Medicine Initiative, which also promises to cure cancer, in exchange for the simple price of data collected from as many people as possible. President Obama's January 2016 State of the Union message underscored the enormity of the project: alongside corporate behemoths like Google and 23andMe, the pharmaceutical industry, and university research labs, the government desires to collect millions of human genomes—as well as data about lifestyle drawn from health records, chat rooms, and wearable activity trackers. Once aggregated and analyzed, these data will be worth trillions of dollars in patents and profits, but there is no plan to distribute that wealth among the citizenry. Rather, the purchase of all that information about ourselves will be compensated merely by the promise of a "moonshot" to "cure cancer."

I believe that there is a degree of misappropriation in that hand-off, no less concerning than a cancer cure whose active ingredient turns out to be volcanic ash. Of course the promise of medicinal enhancements through genetic manipulation has more potential than the Mount St. Helens example. Still, delivery of any such cure will be sold on the market; for all the trappings of prayerful altruism right now, patients will pay. My point is that, at this point in time, the Precision Medicine Initiative's promise of cures for everything from cancer to schizophrenia is wrapped in a fair amount of confabulation and hyperbole. As a lawyer, I am concerned about the legitimizing impact of myths of inevitability and infallibility surrounding genetic technologies, myths that are already shaping policy and law. They are often quite distanced from underlying science.

That disconnect is what we lawyers are trained to taxonomize as false advertising, which can include not only outright lies but also mislabeling and exaggeration. False advertising applies to the misuse of statistical models as well; direct-to-consumer companies like 23andMe have been taken to task by the FDA for selling predictions of disease risk based on faulty data sets and inconsistent stochastic methods. This concern was the subject of a Government Accountability Office report that came out in 2010.[1] Similarly, there is little coordination or oversight of the handling and analysis of forensic DNA—whether through cold hits or familial tracking—at the state level, in federal databases, or globally.

Careless media assertions of associations among genes and disease, physical traits, temperament, IQ, musical ability, or athleticism further complicate this picture. Even quite flawed representations about "genes for" this or that nevertheless have power—as does anything dressed up as "genetic." Some courts use associative work, like the Caspi study linking MAOA deficiencies to potential susceptibility for aggressive acts,[2] to both shorten *and* lengthen sentences. This seeming inconsistency is not attributable to different interpretations of the science, but to different *ideologies* underlying its application. For example, in Italy, a sentence was mitigated because of the understanding that one couldn't help oneself if one had a gene for aggression, whereas in the United States courts are thinking about lengthening sentences, because those who possess the gene ought to have known, or they ought to have put themselves in a position of being able to prevent a crime before it happens.[3]

Finally, much genetic misrepresentation is not straightforwardly intentional, but rather inspired by the corruption from rank and pervasive conflict of interest. Researchers, university administrators, and venture capitalists want to believe in their hopes, because they are praying to the great gods of profit, funding, equity, and fame—or maybe as the *Car Talk* guys put it, just trying to figure out how to make those boat payments. It's not that they are lying; they really want to believe.

My second story is about techno-political possession by eugenic demons. More than a decade ago, Rebecca Mead wrote in the *New Yorker* about a young woman who had broken a record for the price she received from the harvest of her eggs.[4] That young woman, whom I will call Jane, had answered an ad for "Ivy League eggs" that was searching for ova-producers who were enrolled in an Ivy League institution, stood at least 5'10" tall, were fair-haired and blue- or green-eyed, had SAT scores of at least 1400, and had a good health history. Jane had been paid $50,000, a price that quickly thereafter marked the low end for Ivy League eggs.

I had the interesting experience of actually having a "Jane" in my class on law and social justice at the time this article came out. (I write this with her permission; and although she was not private about it, I will not use her real name.)

Here, I would like to interrupt the mellifluous flow of my own narrative and ask you to draw a picture of how you see Jane—i.e., what or who you are imagining her to be. What would she look like? Would you be able to easily pick her out of a crowd of Columbia Law students?

On paper, Jane's profile didn't differ much from most students; but Jane did indeed stand out among my students, as her promised naturally blonde hair had been dyed purple and was twisted in some very long dreadlocks. Her skin was tattooed everywhere with engaging nests of writhing snakes. She had an unusual number of piercings on every visible part of her body, which visible parts included her navel, because she wore very short, glittery little bustiers to class. Jane was a stand out in other ways as well. She had been an abused child, an abused woman, and had started a halfway home and prostitutes' union. And she was one of the smartest and most eloquent advocates I have ever had the pleasure of knowing.

Jane also considered herself a pragmatic agent, a controller of her own body, and so she sold her eggs much more frequently than was medically recommended (but not regulated as a market matter). As a result, she was laid low by several serious infections. In the end, she got what she wanted, which was the ability to pay her way through law school and then put aside a little to buy real estate. She acquired her degree, and was able to leverage that knowledge into all sorts of benefits for the cause and the constituents for whom she still works.

Yet I still have problems with the individualized rational-actor, economic-agent storyline by which she lived. First, selling one's body in exchange for an education that will help you rescue other women who are selling their bodies and whose bodily integrity has been violated is a very complicated phenomenon.

Second, I am intrigued by the wish list of ingredients that the egg purchasers or putative parents had consigned. If you go back to your drawing of Jane imagined from her fulfillment of said egg batches and compare it to the embodied Jane, well, there is probably some distance. In other words the actual gulf between identified traits and their expression (genetic, environmental, or otherwise) is particularly vivid and vividly ironic.

Thirdly, Jane's best friend was a young black law student who had all the attributes Jane had, but for the blonde hair and blue eyes. She too would have liked to sell her eggs to put herself through law school, but there were no takers, so she was busy filling out the forms for scholarships and loans.

Nearly two decades later, the industry of egg harvesting has been more or less overshadowed by other forms of assisted reproductive technology, particularly preimplantation genetic diagnosis and selection. Today, a few of my current students are also the "products" of such early fertility technologies; that, plus the steep rise of libertarianism among young people, means that classroom conversations about faith-based Ivy League bio-determinism have shifted with time: most embrace the eugenics of human perfection as uncorrupted, and even uncorruptable. They see the project of designing a baby as responsible behavior, an act of good parenting.

Nevertheless, I offer this as a cautionary tale. The joys of self-commodification are too often premised on the continued invisibility of certain actors or groups who are not valued in the marketplace. And that exclusion, whether *de jure* or *de facto,* describes a wall around the reproductive festivity, so to speak—an invisible sieve whose porousness is transgressed only by those who possess the valued traits of a given temporal era or given culture or given physical specification. Some players will insist that it's all about personal preference and improved options and freedom of market choice. Some players will insist that it's all about donation and altruism and generosity and helping those who desperately want children, family, and love, love, love. But I worry about the cultural underpinnings that shape such market desire—particularly when it comes to the aesthetics of racial identity. As a collective or normative matter, we don't just want children; we want socially acceptable children. We want perfect children. And we don't just want intelligent children, or my black law student would have had egg seekers lining up outside the door. We want racialized children. And we don't just want healthy children, or we wouldn't be insisting on tall or 5'10" eggs. We want the willowy, toweringly, supra-height-normative children of our dreams. This is an invisible component of what so many purchasers seek in the market for a child—a child otherwise described as a child we can love.

New reproductive technology, in other words, is no longer only about reproducing ourselves. It's about crafting something—someone—bigger and taller and more foolproof. We chase "better"—always that comparative to an invisible referent. We chase "better" as cipher for the driving but unspoken mission of sailing into the best preschools, through the best colleges, and onto the ultimate golf course of life.

None of this is new, of course; but like the well-monitored breeding of royal lineages throughout the eons, it is to very great degree about retaining material power through notions of inherent superiority, often with progeny as the crucible of that sense of superiority. In stating it this way I wish to underscore the overtones of entitlement. This is an old story that structures the root of our most insidious "isms": nativism, narcissism, classism, racism, big-brotherism, and colonialism.

This is not only a worry about class power or bio-colonialism; it's also a problem for family life. When one has paid fifty thousand or a hundred thousand dollars for a perfect egg that is Ivy League-guaranteed and supposedly has, or is sold as having, the gene for violin-playing, and when that child comes out petulant and dyspeptic, takes up the bongos, and is just not terribly grateful for the price tag, one is actually investing in the conceptual groundwork for lawsuits premised on breach of promise, product liability, and wrongful birth. In short, one no longer expects one's child to be a fountain of surprises or to hold one hostage to fate. Rather genetic code is read as literally and as narrowly as a fundamentalist reads the bible. The child is a product, whose warrantable features may be standardized with a great and terrible faith.

My third and final story is about translation among the various disciplines of genetics—or the Tower of Babel problem. I currently teach a class at Columbia that is offered not just to law students but to graduate students from medicine, biology, journalism, and elsewhere throughout the university. The seminar is called "Human Identity, DNA, and the Scientific Revolution," and it's a lot of fun. For some years I cotaught it with a member of the biology department. In various iterations, my co-instructors have been professors from other disciplines as well—usually one from psychiatry and one from philosophy.

One of the things that we are constantly working around is our very different senses of reference. Take the word *nature.* It is a cellular reference to the biologist; it is a chemical/pharmacological reference to the psychiatrist; to our philosopher, it invokes a moral ordering based in religious codes of natural law, like canon law. As a lawyer, I hear it as a normative social descriptor.

Or take the word *human.* To the biologist it is a species boundary. To the psychiatrist, it is a fluid set of behaviors and intelligences that are not necessarily limited by the ability to mate as with a species, and can include parrots, apes, dolphins, elephants, and octopi. To the philosopher, the human describes the bounds of certain vulnerabilities. To me, as a lawyer, it is a complicated status that invokes a set of principles and sometimes rights that are derived largely from international conventions premised on the notion of dignity as a kind of penumbra around the body, a social status that invokes respect and a particular kind of demand for interpersonal dynamic. In American law, however, the *human* is actually less important than the *person.* A *person* is an entity, biologically alive or imaginary, human or fictive or dead, to whom certain kinds of legal protections are owed. Hence, corporations, municipalities, and universities can be persons in that they can sue and be sued. They have standing and protection and recognition in our judicial and political system. So when one of our law students wanted to write a paper about how dolphins should be extended personhood for purposes of bringing suits to protect their habitat and right to exist, the biologist was utterly confounded, and his response was that dolphins aren't human. But in US law, humans are not the only persons.

Now, if there is this much subtlety and confusion among just this very small group of like-minded academics about what a human being is—even before the question of genetic modification or transhumanist possibilities, well. . . . This does raise questions about how scientists, academics, researchers, and doctors imagine the power of genetics. Here too, it's a kind of cognitive dissonance, a problem of internal translation. Sometimes I begin a semester by asking my students to draw what their DNA looks like. And no matter how sophisticated their scientific backgrounds, what they draw is almost relentlessly premodern: It's a set of drones circulating just beneath the skin. It's a little womb in the stomach, with a fully formed self, curled in a fetal position. It's a tiny scroll in a gold box just behind the thorax. It's a minibrain with a little steam engine that's churning all the time. It is a biological Torah in the Ark of the body. It's a homunculus in the alchemists' oven.

In other words there's a clash between what science tells us and the unfettered cultural associations at play in the untamed gardens of magical thinking.

ACT II—VALUES

As I mentioned in act I, when I ask my students to visualize the DNA in their bodies, they have offered endearing images of little scurrying workers, Leonardo da Vinci's Vitruvian man, pictures of the Book of Life, or a Harvard beanie drawn in glitter pen, perched atop a yearning heart. It is not only the clarity and persuasiveness of these framing metaphors and tropes but their relentless reassertion, year after year and class after class, in the minds of even the most secular and scientific souls, that I find intriguing. These are images of religious faith and karma and lead-into-gold and holy text and the resurrection of the body—as well as of color, race, and class entitlement and endowment. Hence, I want to excavate the values embedded in the words with which we wrap new technologies and their figurations of human destiny.

The etymology of words like *genes, genius, genus* are rooted in very deep and unconscious ancient myths. The word "eugenic" captures a huge sense not only of personal entitlement, but of *human entitlement:* of destiny, faith, unstoppable inevitability, and what allows us to render invisible all sense of boundary. Exorcised and exercised, hubris trammels dignity, and ultimately mutilation occurs in the name of improvement. Do no harm ends up having no meaning where doing harm in the name of progress or self-improvement means never having to say you are sorry.

Let me offer an example of this concern about value. Many readers will probably remember the Ashley X case, the "Pillow Angel," who came to public attention in 2006. She suffered a form of encephalopathy that caused her brain to stop developing at about the age of three to eight months. She smiled and she seemed at times to recognize her family members and to enjoy music. But she couldn't move by herself, and would never learn to speak. When she was six, Ashley's parents subjected her body to a series of extensive interventions designed to keep her small, easy to lift and thus less prone to bedsores, and to render her permanently childlike.

To these ends her breast buds were removed, her appendix was removed, because there was fear she wouldn't be able to communicate her distress if she developed appendicitis. She was given sufficiently high quantities of estrogen to ensure that her growth plates would fuse, limiting her height. Her uterus was removed to spare her the pain of menstrual cramps and to relieve her parents' concerns about pregnancy in the event of rape. Since it is against the law to sterilize a minor without appointing a guardian or representative to protect the best interests of the child, one has to ask how Ashley's parents just teamed up with doctors to execute a design of what they wanted and then had it delivered up in the scarifying modification of their child's body, like so much Kentucky Fried Chicken with a side of fries. It was surprising as a matter of procedure and ethics that experimentation with hormones and breast bud removal and unnecessary appendectomy sailed on through the oversight committee and the hospital's review board.

My instinct is that philosophers like Nikolas Rose are quite right to point out that we have increasingly displaced the do-no-harm care ethic of "doctor-patient" with the choice-driven ethic of buyer-seller, i.e., service-provider/consumer preference.[5] Consider the case of parents who "gift" their teenage daughters with nose jobs or breast enhancement; or the family who adopted a baby girl from China, and then subjected her to plastic surgery in order to "Westernize" her eyes.[6] The pervasive availability of elective plastic surgery is just one example of how issues of social stigma have been minimized by treating them as matters of contract.

What happened to Ashley is, I think, a growing tendency to see everything from a consumerist perspective. Ashley's parents effectively positioned themselves as *purchasers,* and the doctors therefore became *providers.* And the outcome for them, the purchasers, was a less burdensome *package.* If, however, Ashley had been a patient, if she had been wrapped in the vocabulary of patient rather than the object of a contract, there might have been some recollection that her body, in fact, was healthy. The inconvenient weight of it was not *per se* a disease, but an inefficiency for her caretakers. I don't question how much Ashley's parents love their daughter or how overwhelming their responsibilities must be. I do, however, believe that the hospital allowed ethical questions about Ashley's long-term care and comfort to be privatized by deferring so unquestioningly to her parents' positive love. The hospital created an extreme presumption in favor of often cash-strapped caretakers that was heedless of medical necessity. Indeed, much of the debate about Ashley's treatment since 2006 has tended to leave the decision to parental determination of risk-benefit analysis, and to dismiss regulatory oversight as some kind of invasion of privacy.

More recently, the debate about privatizing such decisions has reemerged in the context of the Zika virus's link to sharply increased incidence of encephalopathy among broad populations. In Brazil, El Salvador, Colombia, and now in some southern parts of the United States where mosquito infestation is a threat, public health officials have recommended that women postpone pregnancy until the danger is controlled. It has been interesting to see how these discussions intersect with others about the legality of abortion, and access to contraception and health insurance; it remains to be seen whether the "best interests" of encephalopic children born during an epidemic will be left to the private decision making of families, or if we might expect the intervention of some broader public program of support. Says Tarah Demant, senior director of Amnesty International's Identity and Discrimination Unit, "It's putting women in an impossible place, by asking them to put the sole responsibility of public health on their shoulders by not getting pregnant, when over half [in Latin America] don't have that choice."[7]

As medical ethicist Harriet Washington points out in her book, *Medical Apartheid,* the very notion of privacy is inflected by the aesthetics of gender, race, and class.[8] Ashley is white, middle-class, and a now-perpetually-little girl. That embodiment evokes a very particularly-imagined social response. But one wonders if the debate about how we treat such children will shift if the numbers of such children rise from rarity to a more diverse demographic. Will doctors so compliantly agree to suppress testosterone in boys—or effectively chemically castrate them—in order to keep them less disruptively aggressive, or to prevent them from developing secondary sex characteristics? Similarly, I wonder if poor or black children will be so easily romanticized as "pillow angels." Alternatively, what will it mean if the privatized decisions of parents of "burdensome" children like Ashley become public policy?

The glib libertarianism of "Who are you to judge?" masks not only inequalities of social response but also our failure to grapple with the woeful state of a health care system that leaves all Americans, even middle-class families like Ashley's, so burdened. The United States is still the wealthiest nation on earth, yet we cannot find the resources to provide the common medical devices that would have better enabled Ashley's family to care for her, unaltered, in their home: a simple hoist, mattresses that prevent bedsores, the assistance of home health care workers. Ashley's parents apparently felt driven to such lengths because they did not wish to institutionalize her as she grew older, bigger, more cumbersome. They feared her institutionalization with good reason, that fear reflecting but a fraction of the anxiety generated by our public health crisis.

If we reimagined Ashley's humanity as something larger than a private burden to be borne by a single family, we might align her debilitation with that of Alzheimer's patients, or the severely mentally ill, or veterans whose bodies or minds have been shattered by war, or whole populations who have been cognitively disabled by lead poisoning enabled by privatized cost-cutting metrics, as is alleged to have happened in Flint, Michigan. Unlike Ashley, their bodies cannot be always surgically miniaturized or pixied up with heavenly pet-name metaphors. They are full-grown, complex, their bodies heavy with sorrow, with need. Perhaps it is they who will provoke a collective reexamination—a call to judgment—of our polity's obligations to broader notions of human dignity. Perhaps then the public health issues would be a bit more obvious. Perhaps then we might not turn so quickly to carving up the body as a response to the scandalous deficiencies of our public hospital system, and the scandalous costs of our private one.

This habit of thought has shifted our attention in quiet but powerful ways away from the hard political work of maintaining our right to exist in the world without having to disguise, apologize, or suffer for our raced, gendered, or non-normative bodies.

ACT III—PERSONHOOD

In one of the Tarrytown Meeting sessions, Stuart Newman spoke of Craig Venter's institute's attempt to model the entire life cycle of a cell. Dr.

Newman described how the actual paper makes it very clear that the shape of the simulated cell is amorphous. But to render it graphically for a magazine article, it had been re-presented as a cylinder with two big bowls on either end. To me it sounded distinctly like a barbell—talk about governing metaphors in a brave new world of branding! Framing emergent genetic and cellular technology is a matter of depiction, of taking what is amorphous and putting it into concrete shapes that direct our thought one way or the other. It's visual and not just verbal.

I recently attended a conference at which Anne Wojcicki, founder and CEO of the personal genomics company 23andMe, was a featured speaker. Even before she opened her mouth, the poster for the conference was powerfully suggestive of the values embedded in her talk. Her talk was entitled "Deleterious Me: Whole-Genome Sequencing, 23andMe, and the Crowd-Sourced Health Care Revolution." The poster depicted the double helix as a spiral staircase with little Lego-like people climbing upward, ever upward, toward a darkly glooming heaven.

These images provide a very effective metaphor for our concerns about this entire subject, this industry, this bio-prospective enterprise. DNA is figured in the popular imagination as an inevitably uplifting stairway to heaven, an infallible path to higher truth doing the heavy barbell lifting toward Utopia. This, I worry, leads to a credulous suspension of both ethics and caution.

Consider again the peculiar locution of the conference's very title, "*Deleterious Me.*" It posited the intimacy of "Me," the individual, as inherently self-destructive. It's an odd but very effective and increasingly ubiquitous recasting of mortality as autoimmunity. One's essence, the Me, is framed as noxious, diseased, and decaying. Health care and health, by contrast, are positioned on the other side of that colon, "*Whole-Genome Sequencing . . . and Crowd-Sourced Health Care Revolution.*" They are located squarely in the geography of crowd as source. If the individual is framed as dangerous, lonely, and self-annihilating, its rescue lies in the comfort of crowds, safety in numbers, and collective shelter from the harmful Me. There is power in this conjoint set of idealized genetic references, a poignant longing for embodied self-perfection, yet fear and loathing of the assured self-betrayal.

There is something very nearly Shakespearian about the tension—tremulously human, mythically themed, with just a hint of hovering tragedy. Indeed the urgencies of our technological revolution beg for philosophizing our negotiation on some theatrical public stage, some Faustian oratorio where narrative and necromancy meet for a solemn duet. That's why I love performances in artwork. We need more of that as a unifying and creative expression for what is amorphous and needs to be shaped humanely.

In the twenty-first century, however, our greatest passion plays are placed in the realm of private contract rather than in public good or participative democracy. So it is that privately held companies like 23andMe can own, store, and resell to anyone the most elemental biological markers of individual identity while marketing themselves as direct-to-consumer purveyors of "personal self-knowledge."

Richard Hayes and the team from the Center for Genetics and Society have written a statement entitled "Values for a New Biopolitics." Those values are generous values, values of social justice, human rights, the common good, a precautionary approach, and democratic governance. In contrast, contract jurisprudence is styled as the realm of personal choice, needing little regulation or oversight because self-interest will save the day. It is a jurisprudence that also imagines that the objects of contract exchange are inanimate things, industrial, fungible, without feeling or value other than what the parties to the contract choose to embody in the price for that thing.

But at the other end, constitutional jurisprudence looks at the duties of care we owe each other and to the common geography we share—with the culture that constitutes us, the constitutive values that cohere us as a society. Violations of constitutional rights are injuries to identity as a legal person, as a free subject whose citizenship is beyond price and sale, whose autonomy is inalienable rather than alienable. The remedy to these violations requires regulating or restoring political order, civic enfranchisement, and civil rights, including by revoking or amending laws.

In response to a question after her talk, Anne Wojcicki said that there is no difference between a customer, consumer, purchaser, and citizen. But there is. And if we try to fight any of our battles on behalf of consumers, customers, and purchasers, we are instantly limited by a contract frame. We will already have acceded to a market model with the least moral stake in any kind of state regulation because consumers are contractors and contractors act alone. They are rugged individualists; they choose their fate with little discernible detriment or interaction with others.

Legal and political actions styled as human rights, or constitutional law, on the other hand, will tend to have a greater societal and therefore structural impact than those focused on contract. In order to recognize the human interest, the personhood of people, we need to root claims in language that invokes the values of citizenship and the values listed in the

Center for Genetics and Society's position paper. I offer this as something to bear in mind as we frame either legal claims or political action.

We live in a world where actual enfleshed human beings, citizen beings, have become more and more dispossessed in the allocations of research and medical dollars, exiled by exclusive obsession with and exploitation of microbiomes. In the brave new world of CRISPR-Cas9's immense transformative powers, we are failing to take into account the degree to which the privileges and immunities of citizenship are depleted by propitiatory sacrifice not merely to *homo economicus,* nor even *homo faber* (or man the creator), but to man the manufactured—including some potentially very unhappy cyborgs and automatons.

In H. G. Wells' *Island of Dr. Moreau,* a mad doctor attempts to construct a human being by vivisecting and stitching together bits and pieces of random animals. One such creature, called Sayer of the Law, lives in a cave and recites rules randomly in seeming mockery of human hypocrisy: "I sit in the darkness and say the Law." Just so, as we explore our magic islands of genomic divination, we must not do so in ethical darkness, insensate to the generous ideals that bind us, "we the people" who are.

NOTES

With thanks to the Institute for Critical Social Inquiry at The New School; *Columbia Journal of Gender and Law*; and *The Nation* magazine, for which portions of this paper have been presented in public discussions.

1. Gregory Kutz, *Direct-to-Consumer Genetic Tests: Misleading Test Results Are Further Complicated by Deceptive Marketing and Other Questionable Practices,* testimony before the Subcommittee on Oversight and Investigations, Committee on Energy and Commerce, House of Representatives, GAO-10–847T (Washington, DC: US Government Accountability Office, July 22, 2010), www.gao .gov/products/GAO-10–847T.

2. Avshalom Caspi, Joseph McClay, Terrie E. Moffitt, Jonathan Mill, Judy Martin, Ian W. Craig, Alan Taylor, and Richie Poulton, "Role of Genotype in the Cycle of Violence in Maltreated Children," *Science* 297, issue 5582 (August 2, 2002): 851–54, accessed November 18, 2016, www.sciencemag.org/content/297 /5582/851.full.

3. Emiliano Feresin, "Lighter Sentence for Murderer with 'Bad Genes,'" *Nature,* October 30, 2009, accessed November 18, 2016, www.nature.com/news /2009/091030/full/news.2009.1050.html; University of Texas at Dallas, "Genes Influence Criminal Behavior, Research Suggests," *ScienceDaily,* January 26, 2012, accessed November 18, 2016, www.sciencedaily.com/releases/2012/01/120125151841 .htm.

4. Rebecca Mead, "Eggs for Sale," *New Yorker,* August 9, 1999, accessed November 18, 2016, www.newyorker.com/magazine/1999/08/09/eggs-for-sale.

5. Nikolas Rose, *The Politics of Life Itself: Biomedicine, Power and Subjectivity in the Twenty-First Century* (Princeton, NJ: Princeton University Press, 2007).

6. Alicia Ouellette, "Eyes Wide Open: Surgery to Westernize the Eyes of an Asian Child," *Hastings Center Report* 39, no. 1 (2009): 15–18, www.thehastingscenter.org/Publications/HCR/Detail.aspx?id = 3116.

7. Charlotte Alter, "Why Latin American Women Can't Follow the Zika Advice to Avoid Pregnancy," *Time,* January 28, 2016, accessed November 19, 2016, http://time.com/4197318/zika-virus-latin-america-avoid-pregnancy/.

8. Harriet Washington, *Medical Apartheid: The Dark History of Medical Experimentation on Black Americans from Colonial Times to the Present* (New York: Doubleday, 2007).

Contributors

Marcia Angell is a member of the faculty of Global Health and Social Medicine at Harvard Medical School and former editor-in-chief of the *New England Journal of Medicine.*

Adrienne Asch (1946–2013) was director at the Center for Ethics and Edward and Robin Milstein Professor of Bioethics at Yeshiva University.

Tom Athanasiou is executive director at EcoEquity and codirector of the Climate Equity Reference Project.

Derek Ayeh has a master of science in bioethics from Columbia University.

Ronald Bayer is professor of sociomedical science at Columbia University.

Edwin Black is the award-winning *New York Times* bestselling author of *War against the Weak—Eugenics and America's Campaign to Create a Master Race* (Dialog 2012).

Deborah A. Bolnick is associate professor of anthropology at the University of Texas at Austin.

Naomi Cahn is Harold H. Greene Professor at George Washington University Law School.

Nathaniel Comfort is professor of the history of medicine at Johns Hopkins University.

Richard S. Cooper is Anthony B. Traub Chairman of Public Health Sciences at Loyola University.

Jessica Cussins is a master in public policy candidate at Harvard Kennedy School.

Beth Daley is senior investigative reporter at the New England Center for Investigative Reporting *The Eye.*

MARCY DARNOVSKY is executive director at the Center for Genetics and Society.

LENNARD J. DAVIS is Distinguished Professor of Arts and Sciences in the departments of English, Disability and Human Development, and Medical Education at the University of Illinois at Chicago.

DONNA DICKENSON is emeritus professor of medical ethics and humanities at the University of London.

DAVID DOBBS is an award-winning author who writes regularly for *National Geographic*, the *New York Times*, *Slate*, and other publications.

TROY DUSTER is Emeritus Chancellor's Professor of Sociology at the University of California, Berkeley.

CARL ELLIOTT is professor in the Center for Bioethics at the University of Minnesota.

JOHN H. EVANS is professor of sociology at the University of California, San Diego.

JOAN H. FUJIMURA is professor of sociology at University of Wisconsin-Madison.

DUANA FULLWILEY is associate professor of anthropology at Stanford University.

SANDRO GALEA is Robert A. Knox Professor and dean at the Boston University School of Public Health.

EVELYNN M. HAMMONDS is Barbara Gutmann Rosenkrantz Professor of the History of Science and professor of African American studies at Harvard University.

ALLEN M. HORNBLUM is a historian and the author of books on topics ranging from organized crime and Soviet espionage to medical ethics.

RUTH HUBBARD (1924–2016) was professor of biology at Harvard University and an activist on feminist and social issues.

J. BENJAMIN HURLBUT is assistant professor in the School of Life Sciences at Arizona State University.

MARA HVISTENDAHL is a contributing correspondent for *Science* magazine and an Eric & Wendy Schmidt Fellow at *New America*.

LISA CHIYEMI IKEMOTO is Martin Luther King Jr. Professor of Law at the University of California, Davis, School of Law.

KARUNA JAGGAR is executive director at Breast Cancer Action.

SHEILA JASANOFF is Pforzheimer Professor of Science and Technology Studies at the Harvard Kennedy School.

JONATHAN KAHN is James E. Kelley Professor of Law at the Mitchell Hamline School of Law.

MICHAEL B. KATZ (1939–2014) was professor of history at the University of Pennsylvania and an influential voice challenging the roots of poverty.

JAY S. KAUFMAN is professor of epidemiology, biostatistics, and occupational health at McGill University.

WENDY KRAMER is cofounder and director of the Donor Sibling Registry.

ERIC S. LANDER is president and founding director of the Broad Institute of MIT and Harvard Universities.

MARGARET OLIVIA LITTLE is professor of philosophy and director of Georgetown University's Kennedy Institute of Ethics.

LAURA MAMO is professor and associate director of San Francisco State University's Health Equity Institute.

JONATHAN MARKS is professor of anthropology at the University of North Carolina at Charlotte.

ANN MORNING is associate professor of sociology at New York University.

CATHERINE MYSER is director of global health and ethics at Rosalind Franklin University.

GAUTAM NAIK is a freelance journalist and former reporter at the *Wall Street Journal*.

ALONDRA NELSON is dean of social science and professor of sociology at Columbia University.

STUART NEWMAN is professor of cell biology and anatomy at New York Medical College.

OSAGIE K. OBASOGIE is Haas Distinguished Chair and professor of bioethics in the Joint Medical Program and School of Public Health at the University of California, Berkeley.

PILAR OSSORIO is professor of law and bioethics at the University of Wisconsin-Madison and the Ethics Scholar-in-Residence at the Morgridge Institute for Research.

EFTHIMIOS PARASIDIS is associate professor of law and public health at Ohio State University and a faculty affiliate of Ohio State's Center for Bioethics and Medical Humanities.

COREY PEIN is a writer, with a forthcoming book on Silicon Valley. His work can be found at coreypein.net.

DOUGLAS PET is a former Center for Genetics and Society staff member and medical student at Vanderbilt University School of Medicine.

JENNY REARDON is professor of sociology and director of the Science and Justice Research Center at the University of California, Santa Cruz.

SUSAN M. REVERBY is professor of women's and gender studies at Wellesley College.

DOROTHY ROBERTS is George A. Weiss University Professor of Law and Sociology at the University of Pennsylvania.

KRISHANU SAHA is assistant professor of biomedical engineering at the Wisconsin Institute for Discovery at the University of Wisconsin-Madison.

REBECCA SKLOOT is an award-winning writer and bestselling author of *The Immortal Life of Henrietta Lacks.*

DEBORA SPAR is president of Barnard College.

ALEXANDRA MINNA STERN is professor of American culture, history, women's studies, and obstetrics and gynecology at the University of Michigan. She is the author of *Eugenic Nation: Faults and Frontiers of Better Breeding in Modern America* (University of California Press, 2nd ed., 2015).

KIMBERLY TALLBEAR is Canada Research Chair in Indigenous Peoples, Technoscience, and Environment at the University of Alberta.

HARRIET A. WASHINGTON is a writer and medical ethicist.

JAMES Q. WHITMAN is the Ford Foundation Professor of Comparative and Foreign Law at Yale Law School.

PATRICIA J. WILLIAMS is James L. Dohr Professor of Law at Columbia University.

GREGOR WOLBRING is associate professor of community rehabilitation and disability studies in Community Health Sciences at the University of Calgary.

MIRIAM ZOLL is an award-winning writer, a public speaker, and an international health and human rights advocate and educator.

Index